# Mining Technologies Innovative Development II

# Mining Technologies Innovative Development II

Editors

**Sergey Zhironkin**
**Dawid Szurgacz**

MDPI

Basel • Beijing • Wuhan • Barcelona • Belgrade • Novi Sad • Cluj • Manchester

*Editors*
Sergey Zhironkin
Open Pit Mining Department
T.F. Gorbachev Kuzbass State
Technical University
Kemerovo
Russia

Dawid Szurgacz
DOH Hydraulik Center
Rybnik
Poland

*Editorial Office*
MDPI
St. Alban-Anlage 66
4052 Basel, Switzerland

This is a reprint of articles from the Special Issue published online in the open access journal *Energies* (ISSN 1996-1073) (available at: www.mdpi.com/journal/energies/special_issues/mining_technologies_innovative_development_II).

For citation purposes, cite each article independently as indicated on the article page online and as indicated below:

Lastname, A.A.; Lastname, B.B. Article Title. *Journal Name* **Year**, *Volume Number*, Page Range.

**ISBN 978-3-7258-1026-0 (Hbk)**
**ISBN 978-3-7258-1025-3 (PDF)**
**doi.org/10.3390/books978-3-7258-1025-3**

# Contents

# About the Editors

**Sergey Zhironkin**

Sergey Zhironkin is a Professor at National Research Tomsk Polytechnic University, Siberian Federal University, and the T.F. State Technical University of Gorbachev Kuzbasa, all leading universities in Siberia, where he focuses on Russian mining clusters. Zhironkin specializes in geotechnology, the neoindustrial development of mining clusters, and sustainable development. He serves on the Editorial Board of many internationally renowned peer reviewed scientific journals

**Dawid Szurgacz**

Dawid Szurgacz is a manager of research and development projects at the Center of Hydraulics DOH Ltd., a leading company globally. On a daily basis, he works in Polska Grupa Górnicza, the largest mining company in Europe. He specializes in machines and equipment in an underground hard coal mining complex. In this area, he also gives lectures to students from the Faculty of Geoengineering, Mining and Geology, Wroclaw University of Science and Technology.

*energies*

MDPI

*Editorial*

# Mining Technologies Innovative Development II: The Overview

**Sergey Zhironkin [1,2,*] and Dawid Szurgacz [3]**

1 Institute of Trade and Services, Siberian Federal University, 79 Svobodny Av., 660041 Krasnoyarsk, Russia
2 Open Pit Mining Department, T.F. Gorbachev Kuzbass State Technical University, 28 Vesennya St., 650000 Kemerovo, Russia
3 DOH Hydraulics Center, Polska Grupa Górnicza S.A., ul. Powstańców 30, 40-039 Katowice, Poland; dawidszurgacz@vp.pl
* Correspondence: zhironkinsa@kuzstu.ru

Dear colleagues—readers and authors of Energies journal, we present the Special Issue "Mining Technologies Innovative Development II", which continues the previous issue, developing the ideas of sustainable mining of fossil energy sources, concentrating around the innovative modernization of the mineral resource sector in the context of achieving sustainable development goals [1].

Today the innovative development of fossil energy sources mining coincides in its imperative with the expansion of renewable energy production [2], bringing modern society closer to a low-carbon economy, thanks to post-mining and full extraction of minerals and their recycling. The research development in the field of convergence of mining, digital, nature-saving and managerial technologies is currently taking place within the framework of Tripple and Quadruple Helix—new forms of implementation of innovative cooperation between firms and universities, local communities and governments [3]. In this regard, the Special Issue "Mining Technologies Innovative Development II" aims at creating a discussion platform for the global dissemination of advanced scientific ideas in the field of sustainable and innovative development of the mining sector.

The success of mining innovative development as a part of the global transition to sustainable development is determined by interdisciplinary research that combines innovations in the extraction of traditional fossil fuels, the production of energy from alternative sources, in geophysics and geochemistry, and information technology. Therefore, collected papers have contributed to the transition of mining to the circle of sectoral leaders in innovative development.

Like the previous one, the current Special Issue "Mining Technologies Innovative Development II" aims to progress the pluralism in the discussion of the problems concerning increasing the contribution of the mining sector to sustainable development by bringing together representatives of the research community from different countries.

Below there is a summary of each article in the Special Issue "Mining Technologies Innovative Development II", which underwent a thorough peer-review and was selected by the Editors from a number of publications devoted to solving the most urgent organizational, technical, technological and environmental problems of mineral resource sector development.

A. Wróblewski, A. Macek, A. Banasiewicz, S. Gola, M. Zawiślak, A. Janicka in their article consider a model of mining equipment operator's cabin using CFD analysis of the forced airflow and temperature distribution in the air [4]. The authors rightly assert that the primary temperature of rocks is the main source of thermal airflow, but diesel engines of mining equipment are an equally important source of heat. Therefore, the cabin geometric model proposed in the article takes into account heat transfer, according to three conditions: the current actual temperature, the predicted temperature of the conditioned air, and the



**Citation:** Zhironkin, S.; Szurgacz, D. Mining Technologies Innovative Development II: The Overview. *Energies* **2023**, *16*, 5668. https://doi.org/10.3390/en16155668

Received: 3 June 2023
Accepted: 17 July 2023
Published: 28 July 2023

simulated flow rate to ensure the thermal comfort of the operator. The achievement of the authors is the calculation and analysis of the personal mean vote (PMV) index. With its help, a conclusion was made about the optimal parameters of conditioned air (flow rate $2.4 \times 10^{-2}$ $m^3/s$, temperature 10.00 °C) for the temperature in the mine working of 35.00 °C; the air temperature in mining machine cabin will be 20.40 °C. In turn, if the air temperature in the mine working rises to 38.00 °C due to increased heat release from the rock array or more powerful machine, it is required to lower the temperature of the conditioned air to 8.00 °C or increase the flow rate to $3.14 \times 10^{-2}$ $m^3/s$. In this case, the thermal comfort of the operator in the cabin will correspond to a constant value of the PMV index.

The article by M. Pikuła, K. Chudy, M. Worsa-Kozak, M. Czop presents the results of the analysis of the deep-lying rocks hydraulic parameters, in relation to the deepening of mine shafts [5]. As the main method, the authors used an assessment of the drainage potential of Triassic sandstones during the analysis of core parameters. The authors proved that laboratory tests underestimate the hydraulic parameters of porous-fractured rocks, while the part of all core samples with the maximum value of hydraulic conductivity can be representative of the aquifer. Along with this, the authors showed that the currently used methods for calculating the water inflow to the mineshaft are characterized by low accuracy and reliability due to unrepresentative values of hydraulic conductivity. Therefore, the clarification of hydrogeological conditions when predicting the inflow to the mineshaft largely affects the pace of its construction, labor safety and safety of capital investments. Based on the results of laboratory studies and calculations, the authors recommend using the highest values of hydrogeological parameters of rocks based on a pessimistic scenario for predicting water inflow to conducted mineshafts. As a limitation of the study, the authors highlighted the applicability of the proposed method exclusively for porous and porous-fractured rocks.

A. Banasiewicz, P. Śliwiński, P. Krot, J. Wodecki, R. Zimroz presented a 4th order statistical polynomial model for 11 and 10 input variables with modeling accuracy of 8% and 13% respectively (measurements were carried out using SYN-APSA systems) in their article devoted to forecasting emissions of nitrogen oxides (NOx) by diesel transport vehicles in mines [6]. For comparison, the authors consider sensor accuracy of 10% and 20% for stable and transient operation modes. The importance of the study is confirmed by the fact that as the mine workings deepens and the mining front moves away from the ventilation shafts, the gas content increases, and the risk of reaching critical values increases. According to the results of the study presented in the article, flexible planning of the ventilation system is possible to optimize the consumption of electricity by transport and other mining equipment in deep underground workings. The authors rightly position the developed model as a "soft sensor", promising for monitoring and predicting NOx emissions from diesel vehicles, including dump trucks that are not equipped with harmful gas sensors, as well as for providing faster ventilation of underground mine workings along which diesel locomotives move.

The original method for detecting blockages of the belt conveyor transfer point was proposed by P. Bortnowski, H. Gondek, R. Król, D. Marasova, M. Ozdoba [7]. This method is based on the use of RGB camera for data acquisition and CNN Autoencoder for data analysis and interpretation. The authors proceed from the need to solve the widespread problem of failure of the transfer point of the conveyor belt due to clogging of the transfer chute and disruption of the flow of material by oversized pieces of rock or other objects, as a result of which the entire transport route can be disabled. In this regard, the use of a neural network allows restoring images of the operation of conveyor equipment for comparison with standard operating conditions. The authors concluded that the best results are possible when using Gaussian filters for image selection, thresholding and transformation using morphological operators. In particular, the neural network quite accurately detected transfer point blocking when oversized pieces of rock and anchor bolts hit (the most common reasons for a decrease in conveyor throughput). It will be useful for readers to know that the future research of the authors promises to be focused on real-time processing of RGB images using a neural network.

D. Szurgacz, B. Borska, R. Diederichs, A.J.S. Spearing, S. Zhironkin in their article consider new possibilities of internal leaks minimizing in hydraulic prop of a powered roof support [8], starting from its three functions: equipment control, overload protection—the result of rock pressure, minimization of leaks in props and in entire hydraulic system. To do this, the authors propose replacing the existing support block with a double block with charging, which allows increasing the load capacity of the prop by 10–50% by maintaining pressure in the under-piston space of at least 250 bar. The leakage also can be minimized. The authors found that the block prototype maintains the required load capacity of the props, despite leaks, which makes it possible to avoid costly and time-consuming replacement of the props. The article presents the results of bench tests of the prototype block, which confirmed the correctness of the working hypothesis, as well as tests in real conditions, which approved the authors in the assumption of the operability of the power support prop in case of an internal leak. Consequently, the results of the study of double block with charging, presented in the article, allow replacing existing blocks in the real conditions of mines already today.

A. Smirnova, K. Varnavskiy, F. Nepsha, R. Kostomarov, S. Chen in their article consider the use of methane from coal mine as a primary energy source for power supply of mine consumers and, above all, data processing centers [9]. The authors rightly point out that the development of coal deposits by the underground mining releases up to 8% of methane into the atmosphere (the contribution to greenhouse gas emissions is up to 17%). In this regard, the authors proposed the integration of methane extraction from coal seams with the development of mine data centers (concept "Coal-Energy-Information") in three variants. They are cogeneration—burning the methane at gas power generating utilities; burning for use in absorption systems for data center equipment cooling; trigeneration—the use of power generating unit and absorption refrigerator for power supply to mines' data centers. As a result of a comparative analysis of capital and operating costs, profitability and savings for each option, the authors determined the payback period for investments: five years for the first, seven years for the second and six years for the third variant. This made it possible to single out the cogeneration of electricity from coalmine methane as the most promising way to supply power to data centers at mining enterprises. There is no doubt that the implementation of the author's idea will allow coalmines to enter the information technology market and diversify their activities.

M. Siami, T. Barszcz, J. Wodecki, R. Zimroz consider design and prototyping of the infrared image processing pipeline for robotic inspection, designed for monitoring conveyors at surface mines [10]. The authors rightly argue that traditional methods of transport equipment inspecting are laborious and dangerous, which fully applies to the thermography of intensively operating units. On this basis, the article considers a robot inspector moving along a conveyor, receiving, processing, analyzing and interpreting infrared images for automatic detection and analysis of overheated rollers. The authors identified significant potential for timely recognition of temperature anomalies analyzed using a histogram of infrared images obtained by the robot. The authors declare a limitation of the proposed method associated with false thermography results obtained from a mirror object in the RGB shooting field, which can lead to false alarms. The authors see overcoming the limitation in the use of pre-processing algorithms, including CLAHE, adaptive gamma correction to reduce unwanted gray level variations. The authors associate future research in this area with the widespread introduction of robotic devices and infrared image scanners in conveyor systems equipped with artificial intelligence for the transition to new synthesis methods.

The problem of underground structures geometry evaluation was considered by A. Wróblewski, J. Wodecki, P. Trybała, R. Zimroz in the context of using point cloud data for multivariate parameterization and multidimensional analysis [11]. Without a doubt, the implementation of underground mine workings and large-scale structures (gallery, bunkers, chambers) in strong rocks by drilling and blasting or mechanically leads to the formation of surfaces that are uneven in terms of geometry. In addition, the resulting vibrations and seismic phenomena violate the stability of previously created structures. At the same time, it is difficult to use stationary underground monitoring systems due to the constant blasting, operation of mining and transport mining equipment. In this regard, the authors proposed an original method for monitoring the geometric parameters of underground workings, based on measurements using LiDAR/Terrestrial Laser Scanner. The resulting digital point cloud is used to create a 3D model of the tunnel geometry, processed repeatedly to obtain a homogeneous structure, followed by segmentation to separate cross sections with the required resolution. An important achievement by the author is the standardization of the operational structure of 3D data, which simplifies data analysis and reduces their required volume, and significantly increases the efficiency of post-processing.

The article by J. Janus, P. Ostrogórski is devoted to the modeling of underground mine workings using data obtained from laser scanning, in comparison with manual measurements [12]. The paper discusses the advantages and disadvantages of several methods for measuring the cross-sectional area of a mine working: the empirical method; use of CAD; approximation by a semi-ellipse, including those with attached straight sections. It has been established that the use of CAD software gives the best approximation than other methods. The results show the maximum differences between the semi-ellipse and laser scanning methods, and the approximation method at the level of 5.9%. The authors argue that in all methods the shape of the lower part of the working is modeled as a straight line, which often leads to an unacceptable simplification, therefore, for the deformed floor, it is recommended to take into account its height marks. This allows adjusting the length of the straight parts of the arc for better modeling of the shape of an underground mine working. Finally, the article convincingly proves that the use of the entire family of CAD methods makes it possible to speed up the modeling process, in particular, calculations of the cross-sectional area of mine workings. At the same time, for modeling each individual mine working, it is necessary to make a decision on the choice of an appropriate method, which is dictated by the inevitable deformations of the tunnels under the action of rock pressure and seismic effects.

A. Wróblewski, P. Krot, R. Zimroz, T. Mayer, J. Peltola in their article provide a detailed review of the energy-saving and environment-friendly solution for linear electric motor hammers [13], which are widely used for crushing oversized pieces of blasted rocks, as well as for disintegration of concrete structures. Today, along with electric hammers, hydraulic hammers are widely used, which have their own advantages and disadvantages. In this regard, the article provides an overview of existing hammers with a linear electric motor in order to select more energy-saving and environmentally friendly equipment. The authors argue that the shorter payback period of electric hammers (1–2 years) makes them especially profitable in enterprises with a large fleet of machines. However, their operation requires additional costs for the purchase of cables and power points, which, due to their versatility, can be used in enterprises for other purposes. The authors set promising scientific tasks for future research on electric hammers, related to the improvement of their design for use at elevated temperatures and humidity in underground mines, where full maintenance of machines is difficult.

Y. Yang, X. Xu, C. Wang in their article, devoted to forming an approach to the study of host rocks deformation in the development of coal seams [14], note that the mechanism of rocks deformation during the close mining of coal seams in coal mines is not sufficiently reflected in the studies. Therefore, the authors proposed a model for studying the distribution of stresses at different distances between deformation zones. The article presents the results of a significant number of measurements of rock pressure, which indicate that the degree of damage to a single column in the workings of the upper and lower coal seams is quite low (no more than 5% and 1%, respectively). With a roof settlement in the upper layer of the stratum in the range of 0.74–1.33 m, the authors revealed the occurrence of sliding instability. Accordingly, if the size of the subsidence exceeds the values of 1.33 m, the massif within the underground mine working will be deformed, causing an increase in rock pressure. The staggered distance of 40 m between the upper and lower faces allows reducing the pressure on the face when mining the lower coal seam. Without a doubt, the results of the study presented by the author are the contribution to improving the method of reducing damage to underground mine workings and stabilizing coal production.

Feeling proud for the opportunity to participate in the editing of the Special Issue "Mining Technologies Innovative Development II" and select the best articles, it is necessary to thank all the Reviewers, whose work has allowed strengthening the expression of scientific thought in the field of sustainable development and mining–important modern energy carriers. One can be sure that further progress in the field of energy supply in modern economy and society will be stimulated by the popularization of innovative ideas and know-how in mining.

**Author Contributions:** Conceptualization, S.Z. and D.S.; methodology, S.Z. and D.S.; validation, S.Z. and D.S.; writing—original draft preparation, D.S.; writing—review and editing, S.Z.; supervision, S.Z. and D.S. All authors have read and agreed to the published version of the manuscript.

**Funding:** This research received no external funding.

**Conflicts of Interest:** The authors declare no conflict of interest.

## References


1. United Nations. Sustainable Development Goals. 17 Goals to Transform Our World. Available online: https://www.un.org/sustainabledevelopment/ (accessed on 30 May 2023).
2. Cehlar, M. The structural role of convergent technologies in the modern economy. *Econ. Innov. Manag.* **2019**, *3*, 24–31. [CrossRef]
3. Khloptsov, D.M.; Gasanov, M.A.; Potyagail, S.V. Evolution of the technological platform of structural shifts in the economy. *Econ. Innov. Manag.* **2022**, *1*, 20–32. [CrossRef]
4. Wróblewski, A.; Macek, A.; Banasiewicz, A.; Gola, S.; Zawiślak, M.; Janicka, A. CFD Analysis of the Forced Airflow and Temperature Distribution in the Air-Conditioned Operator's Cabin of the Stationary Rock Breaker in Underground Mine under Increasing Heat Flux. *Energies* **2023**, *16*, 3814. [CrossRef]
5. Pikuła, M.; Chudy, K.; Worsa-Kozak, M.; Czop, M. Scale Effect on Hydraulic Properties of Pore-Fissure Deep Rock Formations and Its Importance for the Mining Shaft-Sinking Process. *Energies* **2023**, *16*, 2263. [CrossRef]
6. Banasiewicz, A.; Śliwiński, P.; Krot, P.; Wodecki, J.; Zimroz, R. Prediction of NOx Emission Based on Data of LHD On-Board Monitoring System in a Deep Underground Mine. *Energies* **2023**, *16*, 2149. [CrossRef]
7. Bortnowski, P.; Gondek, H.; Król, R.; Marasova, D.; Ozdoba, M. Detection of Blockages of the Belt Conveyor Transfer Point Using an RGB Camera and CNN Autoencoder. *Energies* **2023**, *16*, 1666. [CrossRef]
8. Szurgacz, D.; Borska, B.; Diederichs, R.; Spearing, A.J.S.; Zhironkin, S. Minimizing Internal Leaks of a Powered Roof Support's Hydraulic Prop Based on Double Block with Charging. *Energies* **2023**, *16*, 1341. [CrossRef]
9. Smirnova, A.; Varnavskiy, K.; Nepsha, F.; Kostomarov, R.; Chen, S. The Development of Coal Mine Methane Utilization Infrastructure within the Framework of the Concept "Coal-Energy-Information". *Energies* **2022**, *15*, 8948. [CrossRef]
10. Siami, M.; Barszcz, T.; Wodecki, J.; Zimroz, R. Design of an Infrared Image Processing Pipeline for Robotic Inspection of Conveyor Systems in Opencast Mining Sites. *Energies* **2022**, *15*, 6771. [CrossRef]
11. Wróblewski, A.; Wodecki, J.; Trybała, P.; Zimroz, R. A Method for Large Underground Structures Geometry Evaluation Based on Multivariate Parameterization and Multidimensional Analysis of Point Cloud Data. *Energies* **2022**, *15*, 6302. [CrossRef]
12. Janus, J.; Ostrogórski, P. Underground Mine Tunnel Modelling Using Laser Scan Data in Relation to Manual Geometry Measurements. *Energies* **2022**, *15*, 2537. [CrossRef]
13. Wróblewski, A.; Krot, P.; Zimroz, R.; Mayer, T.; Peltola, J. Review of Linear Electric Motor Hammers—An Energy-Saving and Eco-Friendly Solution in Industry. *Energies* **2023**, *16*, 959. [CrossRef]
14. Yang, Y.; Xu, X.; Wang, C. Study on the Mechanism of Surrounding Rock Deformation and Its Control for Roof Cutting Retained Gob-Side Entry in Close-Distance Coal Seams Co-Mining. *Energies* **2023**, *16*, 4379. [CrossRef]

energies

MDPI

Article

# Study on the Mechanism of Surrounding Rock Deformation and Its Control for Roof Cutting Retained Gob-Side Entry in Close-Distance Coal Seams Co-Mining

Yongkang Yang [1,2,*], Xuecong Xu [1] and Chenlong Wang [1]

[1] Key Laboratory of In-Situ Property-Improving Mining of Ministry of Education, Taiyuan University of Technology, Taiyuan 030024, China
[2] State Key Laboratory of Coal Resources and Safe Mining, China University of Mining & Technology (Beijing), Beijing 100083, China
* Correspondence: yangyongkang@tyut.edu.cn; Tel.: +86-132-034-13056

**Abstract:** Sustainable development in coal mining requires a continuous and efficient method of coal extraction. Research shows that gob-side entries retained through roof cutting retained gob-side (RCGE) are vital for improving mining efficiency, enhancing coal recovery rates, and enabling continuous production. However, the mechanism of surrounding rock deformation during close-distance co-mining of coal seams with this technique is not yet clear. For the Jiaokou coal mine in China, due to an unreasonable stagger distance between upper and lower working faces, the gob-side entries retained at the 9102 tailgate and 10102 headgate experience severe rock pressure, leading to significant prop damage and a sharp reduction in the cross-section of the entry. This greatly hampers the reuse of these entries. To investigate this issue, we established a model to study the stress distribution of surrounding rocks at different stagger distances (20 m, 40 m, 60 m, 80 m, and 120 m) through numerical simulation and optimized the support parameters for the retained entries. Our research found that when the subsidence of the roof in the upper coal seam exceeds 0.74 m but is less than 1.33 m, there is sliding instability in the mining body. When the subsidence exceeds 1.33 m, the mining body will rotate and deform, causing significant mining pressure within the retained entry. A stagger distance of 40 m between the upper and lower working faces can reduce pressure on the face during the mining of the lower coal seam. Extensive field measurements of rock pressure revealed that the damage rate of the single column in the gob-side entries of the upper and lower coal seams does not exceed 5% and 1%, respectively. In summary, this study provides a practical method to reduce damage to entries during the mining process, thereby increasing the continuous production capability of the coal mine. This is critical for the sustainable development of coal mining.

**Keywords:** close-distance coal seams co-mining; roof pre-fracturing; gob-side entry retained by roof cutting; surrounding rock control



**Citation:** Yang, Y.; Xu, X.; Wang, C. Study on the Mechanism of Surrounding Rock Deformation and Its Control for Roof Cutting Retained Gob-Side Entry in Close-Distance Coal Seams Co-Mining. *Energies* **2023**, *16*, 4379. https://doi.org/10.3390/en16114379

Academic Editors: Sergey Zhironkin and Dawid Szurgacz

Received: 7 May 2023
Revised: 20 May 2023
Accepted: 24 May 2023
Published: 28 May 2023



## 1. Introduction

With the growth of global energy demand, coal, as an important source of energy, has seen an increase in both consumption and production [1,2]. However, due to the increasing scarcity of coal resources, coal mines around the world are facing mining difficulties, and mining has caused varying degrees of environmental damage [3–6].

In some areas, such as the United States and India, room and pillar mining is often used in underground mines [7,8]. The room and pillar mining method is a kind of pillar system mining method. Coal pillars of different shapes are left in the coal room during coal mining, and the coal pillars temporarily support the roof in the coal mining room [9–11]. In shallow coal seams, the thickness of the covering layer should not exceed 500 m, the inclination angle should be small, single coal seams or non-short-distance coal seams are used, the roof is more than moderately stable, and the floor is hard. Under these conditions, the coal quality is medium hard or above medium hard. There is no hard gangue; the gas emission is low, and the coal seam will easily spontaneously ignite, etc. This approach is more suitable. Many scholars have conducted research on mine pressure appearance and corresponding support measures within the mining process [12–14].

Compared to thick and easily mined coal seams, the mining conditions and geological situations of close-distance coal seams are more complex [15,16], yet they account for a significant portion of coal reserves. Co-mining of close-distance coal seams is an effective mining method to address these challenges [17,18]. However, the traditional co-mining of close-distance coal seams presents significant risks and low recovery rates, making the development of safer and more efficient mining technologies a current priority [19,20].

Gob-side entry retaining technology involves preserving some goaf in the entry, where the collapsed goaf naturally forms one side of the entry. The successful implementation of this technology not only increases coal recovery rates but also greatly enhances safety. Specifically, the roof of the gob-side entry can be effectively preserved, improving the safety of mining operations [21]. Many scholars have studied the gob-side entry retaining technology under different roof and burial conditions [22,23]. Especially in deep areas, the successful use of gob-side entry retaining has solved the serious threat brought by enormous geo-stress to the entry.

However, in areas with shallow burial, due to the thin and multi-layered nature of coal seams, traditional co-mining of close-distance coal seams presents many problems. The close proximity of the roadways leads to multiple disturbances, resulting in difficult-to-support roadways. Therefore, we combined gob-side entry retaining technology with co-mining of upper and lower coal seams and conducted research on roof cutting retained gob-side entry in close-distance coal seams co-mining. Considering that in recent years, with the development of technology, the technical condition systems for monitoring reservoirs and mines have been significantly improved [24–26]. These systems aim to monitor various parameters of mines in real-time, including geo-stress, gas pressure, coal seam temperature, etc., to ensure the safety of miners and improve the operating efficiency of mines. This provides us with more detailed geological production data and safety guarantees as we attempt to solve these issues.

In close-distance coal seams combined mining, a key issue is determining the reasonable offset between upper and lower coal seam entries [27,28]. This is because the floor may undergo plastic failure after the upper coal seam is mined, which makes the roof of the lower coal seam entry prone to collapse. The interaction between upper and lower coal seams during mining can affect the surrounding rock structure and stress distribution. Too small of an offset can make it difficult to control the surrounding rock, while too large of an offset can affect production continuity and output stability. Therefore, the layout of the lower coal seam entry is divided into stable pressure zone layout and depressurized zone layout, both of which aim to reduce the impact of disturbance caused by the upper coal seam mining on the stability of the lower coal seam entry [29,30]. However, in actual engineering, a reasonable entry layout is difficult to achieve due to geological conditions and mining plans. Entries that are too close can lead to problems such as surrounding rock deformation, mining pressure, increased support difficulty, and dangerous gas concentration [31–35].

However, there is currently no clear conclusion on the mechanism of cutting the top and unloading pressure along the gob-side entry in close range coal seams, so it is necessary to conduct in-depth research on the characteristics of overall roof movement of upper and lower coal seams, the deformation mechanism of surrounding rock, and the mining pressure law of the interaction between the roadway and surrounding rock.

Taking the 9102 tailgate and the 10102 headgate in Jiaokou Coal Mine as the background, analyze the spatiotemporal evolution of the overlying strata structure and the deformation of surrounding rock. By constructing a mechanics model, analyze the theoretical mechanism of the roof, floor, and surrounding rock in the cutting of the top and unloading pressure along the gob-side entry at different stages. Using FLAC$^{3D}$ simulation software, study the stress field variation during the mining process of the No. 9 and No. 10 coal seams, and explore the influence of the upper and lower coal seam working faces on the stress distribution of the surrounding rock in the gob-side entry. Finally, optimize the support scheme and apply it to engineering.

## 2. Field Practice of Gob-Side Entry Retained

In Jiaokou Coal Mine, the method of co-mining and roof cutting gob-side entry retaining is adopted for close-distance coal seams, where the horizontal distance between the 9102 tailgate and the 10102 headgate is 6 m, the vertical distance is 10 m, and the cross sections of the 9102 tailgate and the 10102 headgate are 4700 mm × 3000 mm and 4700 mm × 3500 mm, respectively. First, the 9102 working face starts mining, and the 9102 tailgate begins gob-side entry retaining. When the working face advances 120 m, the 10102 working face starts mining, and the 10102 tailgate begins gob-side entry retaining, with a staggered distance of 120 m between the working faces, as shown in Figure 1.

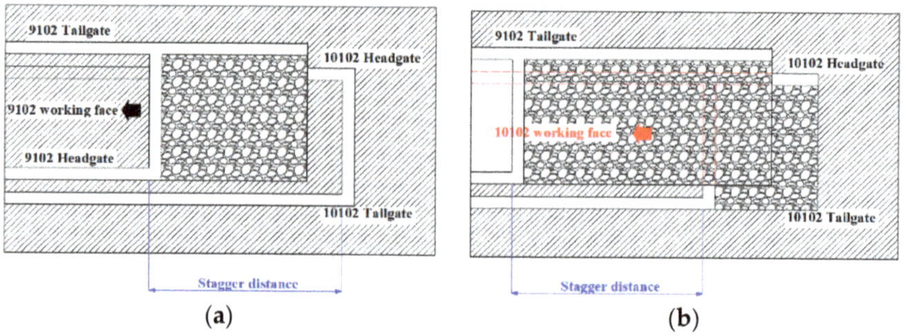


**Figure 1.** Relationship between goaf of upper and lower coal mining faces and gob-side entry. (**a**) 9102 working face location (**b**) 10102 working face location.

As shown in Figure 2, the slit holes are located 200 mm above the roof, with a spacing of 600 mm. The advanced support method adopts 4.2 m $\pi$-shaped beams + single columns (one beam, three columns) with single column spacings of 2800 mm and 1200 mm, while the lagging support method adopts 4.2m $\pi$-shaped beams + single columns (one beam, eight columns) with single column spacings of 1000 mm, 1000 mm, 400 mm, 400 mm, 400 mm, 400 mm, and 400 mm. The advanced support distance for the 9102 tailgate is 20 m, and the lagging support distance is 240 m. As the working face advances, the single columns lagging by 240 m are partially recovered for pedestrian and transportation purposes. The advanced support distance for the 10102 headgate is 80 m, and the lagging support distance is 200 m. As the working face advances, the single columns lagging by 200 m are partially recovered and reused for pedestrian and transportation purposes.

Due to unscientific historical mining in the No. 9 coal seam of Jiaokou Coal Mine, the setup entry of the 9103 working face cannot be excavated, as shown in Figure 3a. Therefore, ventilation was improved by excavating a roadway at the setup entry to connect the 9102 tailgate and 10102 headgate. This allows the 10102 headgate, 10102 tailgate, and 9102 headgate to be ventilated simultaneously while the 9102 tailgate returns the airflow. That is, the No. 9 and No. 10 coal seams share a common return airway, as shown in Figure 3b.

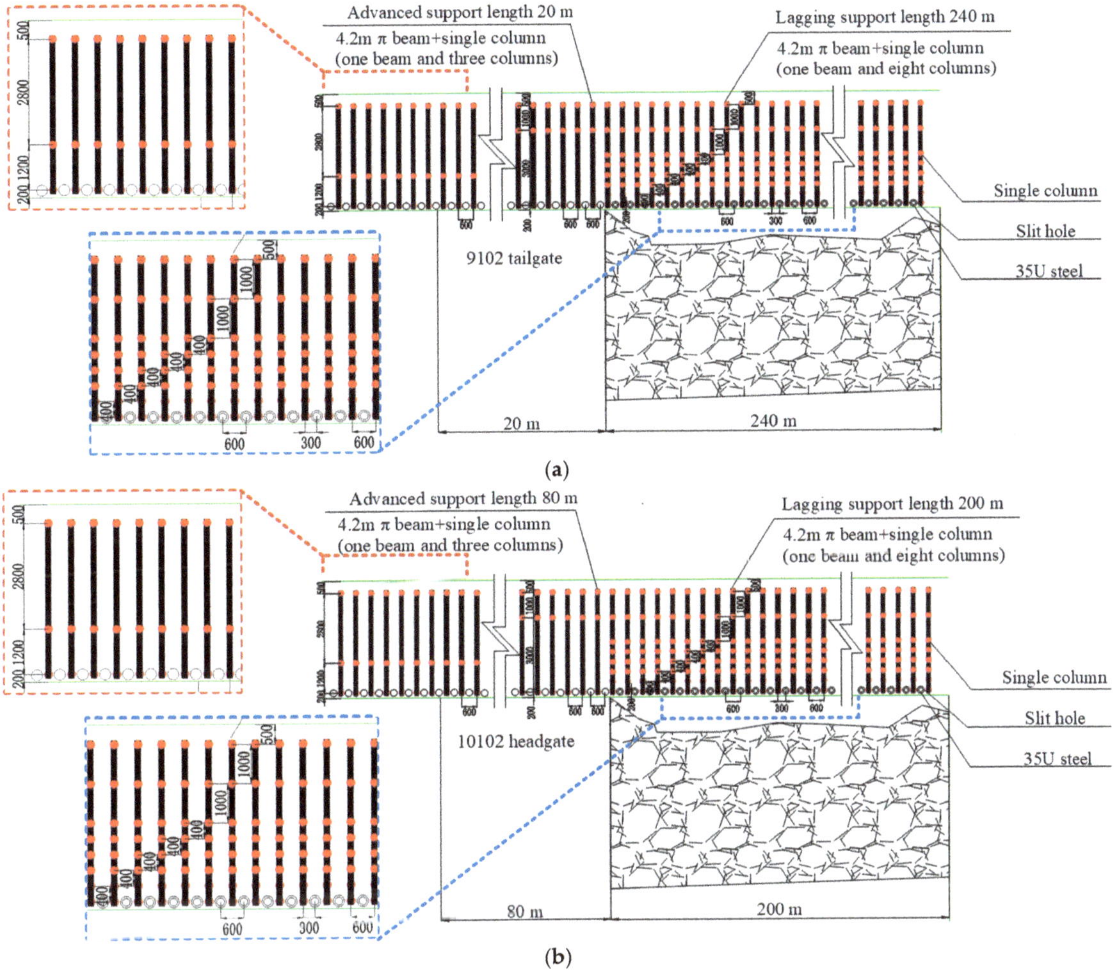


**Figure 2.** Advance support and lagging support gob-side entry. (**a**) 9102 tailgate advance support and lagging support gob-side entry (**b**) 10102 headgate advance support and lagging support gob-side entry.

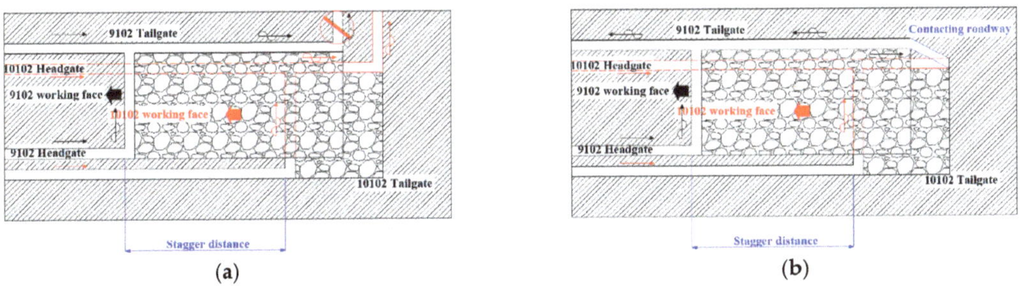


**Figure 3.** Ventilation method in the mining area. (**a**) Traditional ventilation gob-side entry (**b**) Actual ventilation gob-side entry.

The lithology of the roof and floor of 9102 and 10102 working faces is shown in Figure 4. The thickness of the No. 9 coal seam is 0.30 m to 1.75 m, averaging 1.04 m, with a simple structure, generally containing 0 to 1 layer of gangue, and is a stable and mostly mineable coal seam in the whole area. The roof is L1 limestone, and the floor is mudstone and sandy mudstone. The thickness of the No. 10 coal seam is 1.74 m~3.93 m, with an average of 3.04 m, and a simple structure containing 1~3 layers of gangue, stable thickness, and is mineable across the whole area.

| Lithology | Rock descriptions | Depth/m | Thickness/m | Column |
|---|---|---|---|---|
| Mudstone | Dark gray, thinly stratified, with plant fossils, top with coal line | 268.50 | 3.04 | |
| Limestone | Gray, medium-thick stratified, fine-crystalline structure, with animal fossils, massive | 274.90 | 6.40 | |
| Mudstone | Dark gray, thinly laminated, containing a few fossilized plant fragments | 275.40 | 0.50 | |
| Limestone | Gray, medium-thick stratified, fine-crystalline structure, with animal fossils | 279.59 | 4.19 | |
| 9# Coal | Black, semi-bright type, simple structure | 280.47 | 0.88 | |
| Sandy mudstone | Gray-black, massive, with fossilized plants and fine grains of pyrite, interspersed with thin layers of dark gray siltstone | 287.49 | 7.02 | |
| 10# Coal | Black, semi-bright type, simple structure, interspersed with a layer of black mudstone interlayer | 290.46 | 2.97 | |
| Sandy mudstone | Gray-black, massive, with pyrite nodules | 294.27 | 3.81 | |
| Fine Grain Sandstone | Light gray, medium-thick stratified, composition dominated by quartz | 296.42 | 2.15 | |
| Sandy mudstone | Dark gray, thinly layered, with fossilized plant fragments and small amounts of pyrite | 297.60 | 1.18 | |
| Fine Grain Sandstone | Light gray, medium-thick stratified, composition mainly quartz, followed by feldspar | 301.32 | 3.72 | |

**Figure 4.** Borehole diagram.

## 3. Overlay Transport Characteristics and Surrounding Rock Deformation Mechanism

### *3.1. Deformation and Damage Characteristics of the Roof*

Figure 5 demonstrates the roof damage and key block changes during the mining process of the upper and lower group coal seam working face. As the coal seam mining advances, the overlying rock subsidence leads to changes in key blocks A, B, and C. Among them, the position of A is stable, C moves down, and B is affected by A and C and rotates and sinks. When the lower group of the coal seam is mined, the roof sinking increases and the structural stability of key block E is weakened, which in turn affects the structural stability of key block B, and the mine pressure shows obviously. The damage to the bottom plate has a significant impact on the stability of the key block, and the degree and extent of the damage are related to the difficulty of safe mining of the coal seam of the lower group. In order to clarify the deformation mechanism of the surrounding rock, it is necessary to analyze the sliding instability and rotary deformation mechanism of key block B in the mining process of the upper and lower group coal seams.

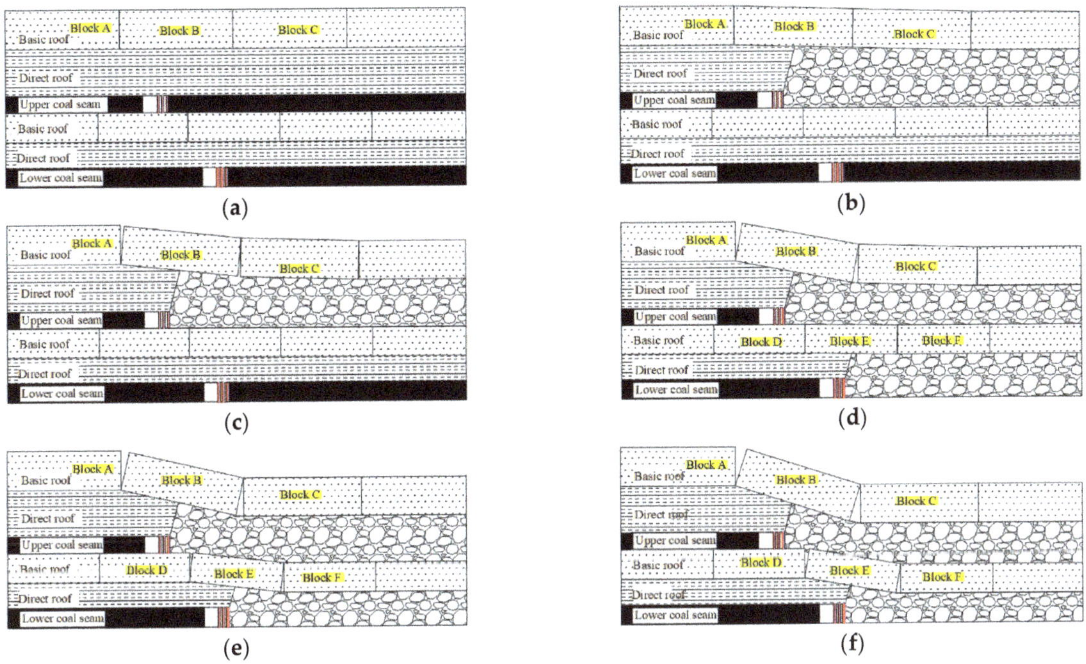


**Figure 5.** The process of roof failure during mining at the working face. (**a**) End of upper and lower roadway excavation (**b**) Start of upper group coal seam mining (**c**) End of upper and lower roadway excavation (**d**) Start of upper group coal seam mining (**e**) End of upper and lower roadway excavation (**f**) Start of upper group coal seam mining.

### 3.2. Research on the Damage Mechanism of the Surrounding Rock Structure

In the process of the workover of the upper and lower groups of coal seams in the near coal seam with cutting top unloading pressure along the air stay, the force analysis of the key blocks A, B, and C is carried out as shown in Figure 6. The force of the key blocks (assumed to be A, B, and C) of the overlying roof rock of the upper group coal seam is analyzed. Assuming that these key blocks are articulated, it is necessary to analyze the slip instability and rotary deformation of key block B. Here, it is simplified to a two-dimensional problem, and some basic parameters are set for the blocks. The force analysis parameters and the key parameters of the rock masses are shown in Table 1.

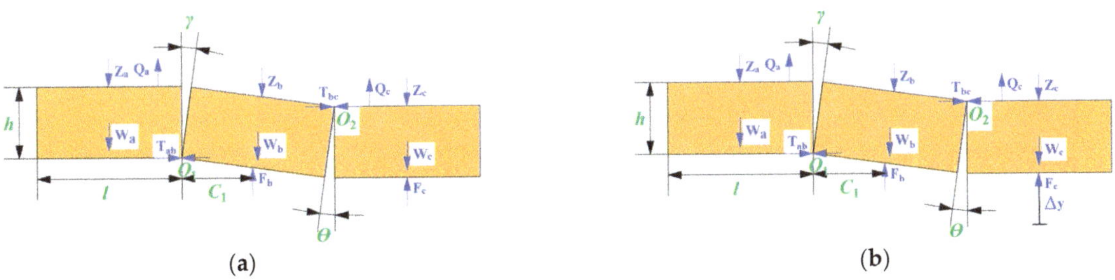


**Figure 6.** Stress analysis of the roof of the upper and lower coal seams. (**a**) End of upper and lower roadway excavation (**b**) Start of upper group coal seam mining.

**Table 1.** Stress analysis and key parameters of rock mass.

| Parameter Symbols | Physical Meaning | Unit |
|---|---|---|
| $K$ | Key block C mining gangue stiffness | N/m |
| $S$ | Key block C mining area gangue displacement | m |
| $W_a$, $W_b$, $W_c$ | Self-weight of key blocks A, B, and C | kN |
| $Q_a$, $Q_b$, $Q_c$ | Key block A, B, C structure friction shear reaction force | kN |
| $Z_a$, $Z_b$, $Z_c$ | Key blocks A, B, and C are loaded | kN |
| $F_b$ | Support force transmitted by the bracket underneath the critical block B | kN |
| $C_1$ | Distance of the support transfer key block B | m |
| $F_c$ | Key block C mining area gangue support force | kN |
| $T_{ab}$, $T_{bc}$ | Horizontal binding force of the structure on A and B, B and C | kN |
| $l$ | Length of key blocks A, B, C | m |
| $h$ | Height of key blocks A, B, C | m |
| $\gamma$ | Turning angle of key block B with respect to A | ° |
| $\theta$ | Turning angle of key block B with respect to C | ° |
| $k_{p1}$ | Crushing and swelling coefficient | |
| $k_{p2}$ | Residual swelling factor | |
| $h_i$ | Thickness of the *i*th collapsed rock layer | m |

According to the hydrostatic equilibrium equation for the point $O_1$ we have Equation (1).

$$\begin{cases} -Q_a + T_{ab}\sin\gamma = 0 \\ Z_b + T_{ab}\cos\gamma - Z_a = 0 \\ -Q_a h + Z_b C_1 - T_{ab}\cos\gamma\, l = 0 \end{cases} \tag{1}$$

For the $O_2$ point there is Equation (2).

$$\begin{cases} -Q_c + T_{bc}\sin\theta = 0 \\ Z_a + T_{bc}\cos\theta - Z_c = 0 \\ -Q_c h + Z_a C_1 - T_{bc}\cos\theta l = 0 \end{cases} \tag{2}$$

Slip destabilization analysis of key block B during the mining of the upper group of coal seams, according to the balance of lateral forces on the key block B, Equation (3) can be listed.

$$T_{ab} + T_{bc} + F_b = Z_b \tag{3}$$

where, $T_{ab}$ and $T_{bc}$ are the horizontal restraint force of the articulated structure on the key block B, respectively, $F_b$ is the support force transmitted by the bracket below the key block B, $Z_b$ is the load borne by the key block B. Consider that the maximum shear force on key block B occurs when it is in the sliding instability state, which satisfies so that Equation (3) can be simplified to Equation (4).

$$2T_{bc} + F_b = Q_b \tag{4}$$

Based on:

$$T_{bc} = S_{bc}K_{bc}L_{bc} \tag{5}$$

where $S_{bc}$ is the shear area between the key blocks C and B, $K_{bc}$ is the shear stiffness, and $L_{bc}$ is the hinge distance. Substituting into Equation (3) yields Equation (6).

$$2S_{bc}K_{bc}L_{bc} + F_b = Q_b \tag{6}$$

In order to ensure that the key block B does not slip and destabilize, it is necessary to ensure that the horizontal binding force $T_{bc}$ of the articulated structure on B is greater than or equal to the friction force $Q_b$ on the key block B. To this end, Equation (7) is established.

$$T_{bc}\mu \geq Q_b \tag{7}$$

Then, according to the moment balance equation, we can obtain Equation (8).

$$T_{bc}C_1 + F_b h\sin\gamma - Q_b l - Z_b\frac{h}{2} = 0 \tag{8}$$

The condition for determining whether the key block B slips and destabilizes is Equation (9).

$$(F_b + Z_b\frac{h}{2})\frac{1}{\mu} + F_b h\sin\gamma - Z_b\frac{h}{2}C_1 \geq 0 \tag{9}$$

Simplify and organize to obtain Equation (10).

$$(\mu_{ab} - 1)T_{ab} + (\mu_{bc} - 1)T_{bc} \geq \mu_{ab}g\sin\gamma + \mu_{bc}g\sin\theta - Q_a \tag{10}$$

The condition for the key block B not to undergo rotary deformation during mining of the upper group coal seam is Equation (11).

$$\frac{Z_a + W_a + T_{ab} - T_{bc}}{Z_b + W_b} \geq \tan\gamma + \tan\theta \tag{11}$$

Equation (11) simplified and organized as Equation (12).

$$\frac{S\cos(\gamma+\theta)(h - \frac{2}{3}h_i)}{\tan\gamma} + \frac{T_{bc}C_1}{h}l - \frac{1}{2}k_{p1}\rho g S h\sin\theta \geq 0 \tag{12}$$

The slip instability and rotary deformation conditions of key block E in the lower group coal seam are almost the same as the derivation of key block B in the upper group coal seam mining, so no derivation is made.

Since it is assumed that the block C sinks as a whole $\Delta y$, as in Figure 4b, then $Z_a$ and $Z_c$ decrease and increase $\rho gh\Delta y$, respectively, where $\rho$ is the rock density, kg/m$^3$; $g$ is the acceleration of gravity, kg/m$^2$, at which point there are Equations (13) and (14).

$$\begin{cases} -Q_a + T_{ab}\sin\gamma = 0 \\ Z_b + T_{ab}\cos\gamma - (Z_a - \rho gh\Delta y) = 0 \\ -Q_a h + Z_b C_1 - T_{ab}\cos\gamma\, l = 0 \end{cases} \tag{13}$$

$$\begin{cases} -Q_c + T_{bc}\sin\theta = 0 \\ Z_a + \rho gh\Delta y + T_{bc}\cos\theta - Z_c = 0 \\ -Q_c h + Z_a C_1 - T_{bc}\cos\theta\, l = 0 \end{cases} \tag{14}$$

Slip destabilization analysis of key block B during mining of the lower group coal seam, according to the balance of lateral forces on key block B and previous derivation, the condition for lower group coal seam mining to satisfy key block B without slip destabilization is Equation (15).

$$T_{ab} \geq Q_a\mu_a + Q_c\mu_c + F_b \tag{15}$$

where, $T_{ab}$ is the horizontal binding force of the articulated structure on the key block A and B, $Q_a$ and $Q_c$ are the frictional shear reactions of the articulated structure of the key block A and C, respectively, $\mu_a$ and $\mu_c$ are the friction coefficients between the rocks, and $F_b$ is the support force transmitted by the brace below the key block B.

Considering the overall sinking of block C by $\Delta y$, the horizontal and articulated forces on the key block B will also change. According to the principle of mechanical equilibrium, the condition that the key block B does not slip and destabilize is satisfied by Equation (16).

$$T_{ab} \geq Q_a\mu_a + Q_c\mu_c + F_b + Z_b\tan\theta - Z_c\tan\theta - Z_b\tan\gamma + Z_a\tan\gamma - K\Delta y \tag{16}$$

Equation (16) is expanded and organized to obtain Equation (17).

$$T_{bc} - \frac{\rho lh}{2}(1 - k_{p1})\Delta y + T_{ab}\cos\gamma - \frac{\rho glh}{2}(1 - kp_1)\Delta y\sin\gamma \geq \frac{\rho glh}{2}(1 - k_{p2})\mu(Q_A + Q_B)\cos\theta \quad (17)$$

In the lower group coal seam mining, when block B undergoes rotary deformation, Equation (18) needs to be satisfied.

$$T_{ab}\tan\gamma + T_{bc}\tan\theta \geq Z_b + G_b + F_b - F_a - F_c - k_{p1}S \quad (18)$$

Equation (18) is simplified and organized to obtain Equation (19).

$$\frac{2\left(k_{p1} - k_{p2}\right)\sum_{i=1}^{n} h_i\sin\gamma}{l} \geq \frac{(1-k_{p1})(\frac{Z_a}{l} + \frac{Z_c}{l} + g\rho h)(\frac{h}{3} - \frac{\Delta y}{3}) + \frac{T_{ab}\tan\mu}{l}(\frac{h}{2} - \frac{\Delta y}{3})}{\frac{2}{3}\rho g(\frac{h}{2} - \frac{\Delta y}{3}) + \frac{2}{3}\tan\mu(\frac{h}{2} - \frac{\Delta y}{3})} \quad (19)$$

According to the production geology and filed practice, take $h$ = 4.17 m, $l$ = 6.2 m, $k_{p1}$ = 1.4, $k_{p2}$ = 0.2, $E$ = 22 GPa, $\gamma = \theta = 15^\circ$, $\rho$ = 2450 kg·m$^{-3}$, $g$ = 9.8 kg·m$^{-2}$, $\mu$ = 0.35, and $\sum_{i=1}^{n} h_i$ = 275 m, $Z_a$ = 4.18 MPa, $Z_c$ = 3.9 MPa, $T_{ab} = T_{bc}$ = 3.16 MPa. By substituting them into Equations (17) and (19), it can be obtained that $\Delta y \geq 0.74$ m, $\Delta y \geq 1.33$ m.

Finally, when the subsidence $\Delta y$ of the goaf roof in the upper group of coal seams falls within the range of 0.74 m $\leq \Delta y \leq$ 1.33 m, block B experiences sliding instability. When $\Delta y \geq 1.33$ m, block B will rotate and deform, and both situations will cause severe mining pressure in the roadway, leading to large deformation of the surrounding rock and posing a threat to the stability of the next group of coal seam roadways and the safety of mining work. Therefore, in the mining of the lower group of coal seams, it is necessary to further consider the reasonable stagger distance between the upper and lower working faces to reduce the stress concentration caused by the mining impact of the upper group of coal seams.

## 4. Numerical Simulation of Rock Pressure in Gob-Side Entry

In order to study the effect of different working face stagger distances on gob-side entry stability under the same mining situation of upper and lower coal seams, we designed a numerical simulation plan for coal seam excavation under different working face stagger distances. The advancing distance of the upper coal seam working face is larger than that of the lower coal seam, and the stagger distances of the upper and lower coal seam working faces are 20 m, 40 m, 60 m, 80 m, and 120 m.

Combined with the lithology, spatial distribution, and the mining scheme of the 9102 and 10102 chutes of the Jiaokou coal mine, a numerical model of 335 m × 320 m × 50 m was established (see Figure 7), taking into account parameters such as the size of the coal roadway, the width of the working face, and the horizontal and vertical distances. The rock mechanics parameters (see Table 2) and contact parameters (see Table 3) were defined, and the boundary conditions were set after pre-splitting. Pre-splitting and roof cutting are achieved by using the interface unit in FLAC$^{3D}$ to define the cracking behavior. The model contains two support schemes, scheme I and scheme II, which adopt different anchor spacing, row spacing, and number of anchors, respectively. For scheme I, anchor spacing is 1100 mm, row spacing is 1600 mm, it has one anchor cable in the center of the roof, and 3 and 4 anchor cables in the 9102 return wind chute and 10102 transport chute of the roadway gang, respectively. For scheme II, two anchor cables are added on the basis of scheme I, the distance between its anchors is 700 mm, and the distance between its rows is 800 mm. After adding support, the coal seams 9 and 10 were excavated in steps, and the changes in the surrounding rock stresses in the two roadways during the retrieval process were studied.

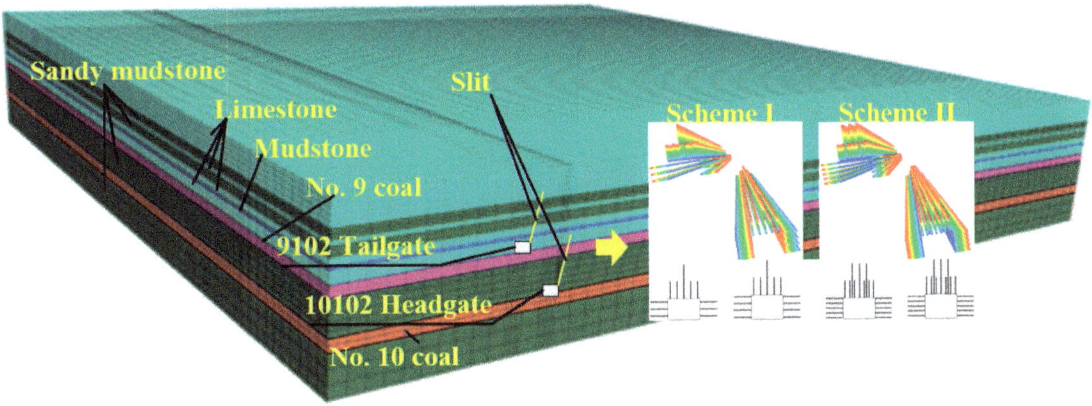


**Figure 7.** Numerical model.

**Table 2.** Rock mechanics parameters.

| Lithology | Density/kg·m$^{-3}$ | Bulk Modulus/GPa | Shear Modulus/GPa | Cohesion/MPa | Friction Angle/° | Tensile Strength/MPa |
|---|---|---|---|---|---|---|
| Sandy mudstone | 2450 | 9.8 | 5.6 | 1.8 | 27 | 10.5 |
| Coal | 1600 | 1.4 | 0.6 | 0.7 | 20 | 5.9 |
| Limestone | 2750 | 17 | 9.2 | 2.6 | 29 | 23.6 |
| Mudstone | 1900 | 7.2 | 3.4 | 1.4 | 25 | 8.9 |

**Table 3.** Contact mechanics parameters of slit.

| Friction Angle/° | Cohesion/MPa | Modulus of Elasticity/GPa | Tensile Strength/MPa | Shear Strength/MPa | Cracking Toughness/MPa·m$^{1/2}$ |
|---|---|---|---|---|---|
| 27 | 1.5 | 22 | 4.9 | 7.3 | 0.3 |

### *4.1. The Influence of the Upper and Lower Group of Working Face Retrieval on the Stress Distribution of the Gob-Side Entry Surrounding Rock*

In Figure 8, contour lines are drawn for the stress distribution of the upper and lower group coal seam along the gob-side entry at the different working face stagger distances. According to Figure 8a–e, it is known that the stress value of the lower group of coal roadways is smaller than that of the upper group of coal roadways. It can be seen that the upper group coal seam roof cut destroys the roof stress transfer, and with the stagger distance increase in the working face's advance, the stress on the solid coal side and the roof increases, and the roadway gang is always in a high-stress state. However, when the working face stagger distance is 40 m, see Figure 8b, the stress of the upper coal seam gob-side entry solid coal side is 9 MPa, the stress of the lower coal seam gob-side entry solid coal side is 8 MPa, the difference between the two is not significant. This indicates that the mining activity of the upper coal seam face has the greatest impact on the lower part at this time. When the working face is staggered by 20 m, the roof of the upper coal seam working face has not yet fully collapsed. For the safety and efficiency of mining the lower coal seam, we believe that the suitable working face offset is 40 m.

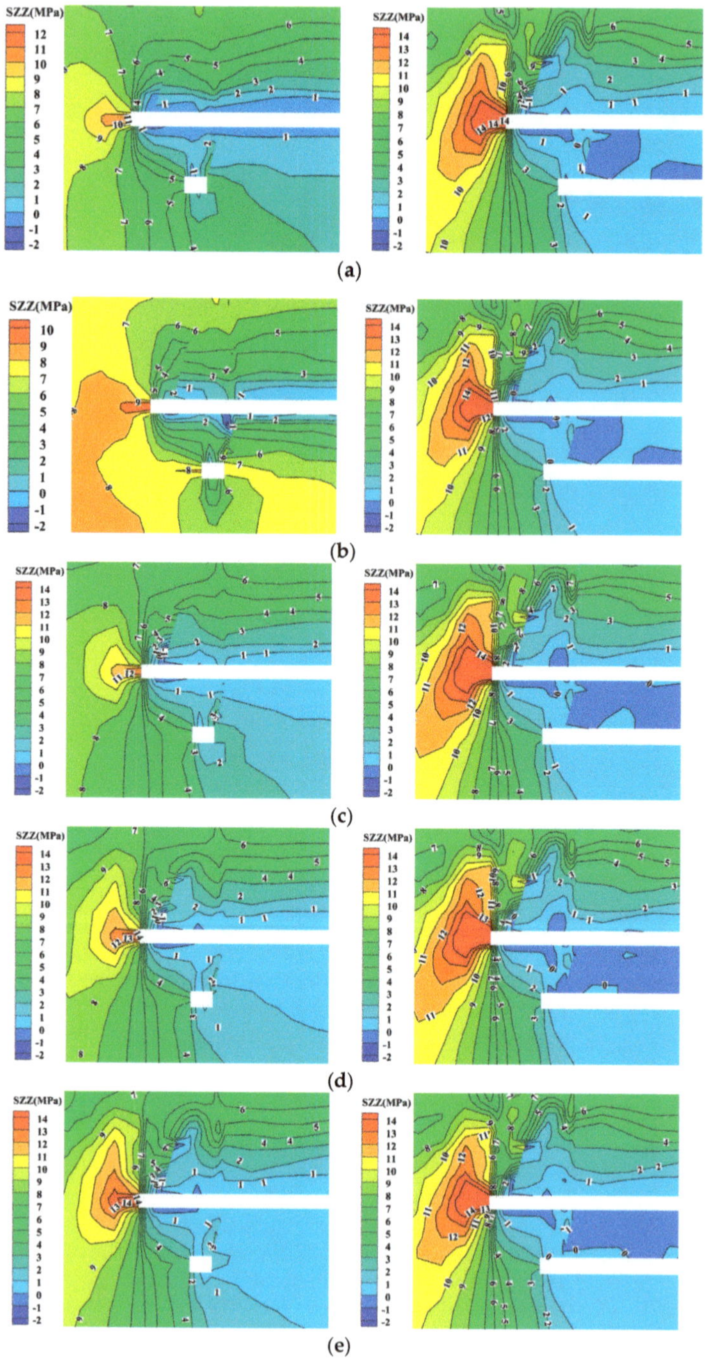


**Figure 8.** Stress distribution of gob-side entry under different working face stagger distances. (**a**) working face stagger distance is 20 m (**b**) working face stagger distance is 40 m (**c**) working face stagger distance is 60 m (**d**) working face stagger distance is 80 m (**e**) working face stagger distance is 120 m.

When the lower coal seam is retrieved, the horizontal to vertical stress distribution of the top plate gob-side entry and mining of the lower group coal seam will affect the stress distribution of the upper group coal seam gob-side entry but will not change the stress peak. With the increase in the working face of the lower group coal seam, the high-stress range of the surrounding rock of the upper group coal gob-side entry increases. The effect of the slit along the empty roadway of the lower group coal seam is different from that of the upper group coal seam, and the slit gob-side entry of the lower group coal seam maintains the original stress state of the roof plate under the activity of overburdened rock in the mining area and the change in surrounding rock of upper group coal seam roadways so that no stress concentration is generated.

### *4.2. Quarry Stress Distribution Characteristics*

In Figure 9, the stress magnitude along the slit side of the gob-side entry is compared and analyzed with the back of the mining area. In Figure 9a, the stresses on the solid coal side with a slit are larger compared with the stress values on the uncut side. This indicates that the bending and sinking of the upper mining area during the lower group of coal seam retrieval makes the solid coal side with cut seams suffer greater support stresses. The reason for this phenomenon is that the overburden activity in the lower mining area leads to an increase in the overburden activity space in the mining area of the upper coal seam. In this case, the overburden in the upper mining area will continue to sink, destroying the stability of the original articulated structure. As the overburden in the upper mining area rotates and slips, the stress will be redistributed and reach a new equilibrium state, resulting in differences in stress values. The overburden activity weaken the regulating effect of upper group coal seam cuttings on stress distribution, which increases the supporting stress on the solid coal side.

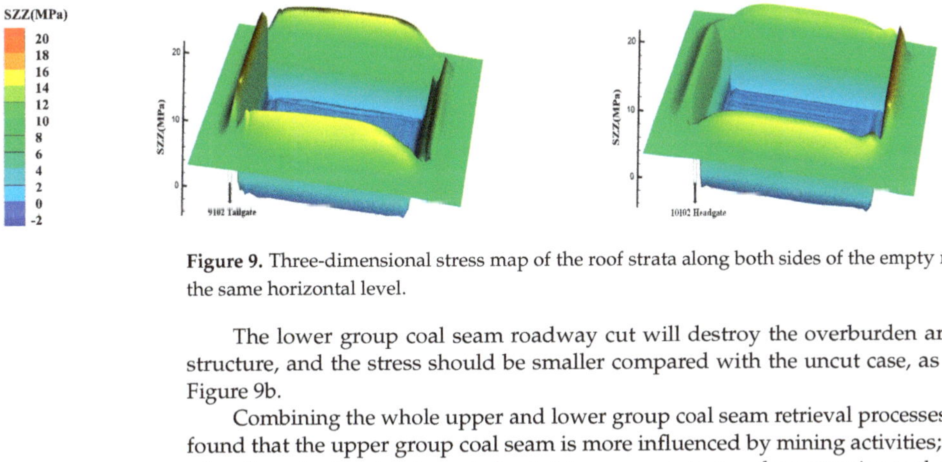


**Figure 9.** Three-dimensional stress map of the roof strata along both sides of the empty roadway at the same horizontal level.

The lower group coal seam roadway cut will destroy the overburden articulation structure, and the stress should be smaller compared with the uncut case, as shown in Figure 9b.

Combining the whole upper and lower group coal seam retrieval processes, it can be found that the upper group coal seam is more influenced by mining activities; therefore, when the actual project is carried out, it is necessary to pay close attention to the potential influence of the mining activities of the lower group coal seam on the stability of the upper gob-side entry and take appropriate support measures to ensure the safety and stability of the roadway.

### *4.3. Comparative Analysis of Different Support Schemes*

According to Figure 10a,b, in support scheme I, the maximum settlement of the roof slab of the 9102 return wind chute is 17.9 cm, which is located at model 95.7 m; while under scheme II, the maximum settlement of the roof slab is 15.3 cm. The short-term support effect of scheme II is 14.5% higher than that of scheme I. In the analysis of the settlement of the roof slab of the 10102 transport chute in the No. 10 coal seam, the maximum settlement of scheme I is 1.4 cm, and that of scheme II is 1.28 cm, with a difference of 0.16 cm, and the short-term support effect of scheme I is only improved by 7.8%. The support effect of both options meets the engineering requirements.

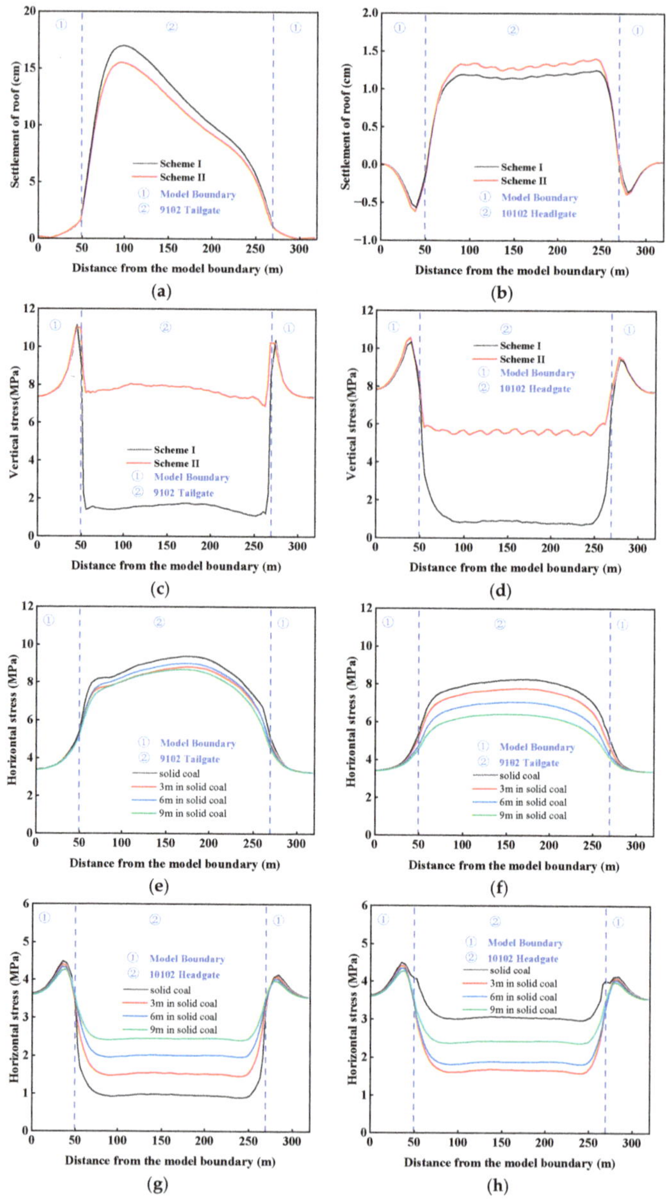


**Figure 10.** Comparison of support schemes. (**a**) 9102 tailgate roof settlement (**b**) 10102 headgate roof settlement (**c**) 9102 tailgate roof vertical stress (**d**) 10102 headgate roof vertical stress (**e**) Stress on solid coal side of 9102 tailgate in scheme I (**f**) Stress on solid coal side of 9102 tailgate in scheme II (**g**) Solid coal side stress of 10102 headgate in scheme I (**h**) Solid coal side stress of 10102 headgate in scheme II.

As shown in Figure 10c,d, the vertical stress distribution of the roof plate of the 9102 tailgate and the 10102 headgate shows that the minimum vertical stress of the 9102 roof plate is 1.05 MPa under scheme I and 6.9 MPa under scheme II. The support effect of scheme II is reduced by 37.3% and still has a better effect under the influence of subsequent overburden transport and mining movement; the support effect of scheme I is reduced by 88.2%, and it can hardly play a supporting role under the influence of subsequent overburden transport. In the 10102 headgate, the support effect of the two support schemes is reduced by 91.1% and 35.7%, respectively.

Although the change in the horizontal stress of the roadway gang directly affects the stability of the roadway, and the presence or absence of anchor rods and anchor cables does not affect this conclusion, in Figure 10e–h, after comparing the horizontal stress of the roadway gang in the upper and lower along-air roadway, it is found that: in the close-distance coal seam co-mining with pre-split gob-side retaining project, the horizontal high-stress distribution on the solid coal side of the upper group coal seam can be made under the condition of appropriately increasing the support strength. The range is reduced. Because the range of horizontal mining-induced stress changes reduced, thereby reducing the probability of rock bursts and other hazards, which is conducive to the long-term stability of the gob-side entry.

## 5. Industrial Trials

By analyzing the numerical simulation research results of the overburden transport characteristics and the mineral pressure performance law gob-side entry, the engineering application of the technology of close distance coal seams roof cutting gob-side entry retaining in the same mining in the close coal seam is carried out.

The 9102 tailgate and the 10102 headgate are both rectangular in the cross section, with dimensions of 4700 mm × 3000 mm and 4700 mm × 3500 mm, and the 9102 tailgate is cut into the bottom of the coal seam, while the 10102 headgate is cut along the top of the bottom. The cross section and plan of the gob-side entry roof reinforcement support are shown in Figure 11.

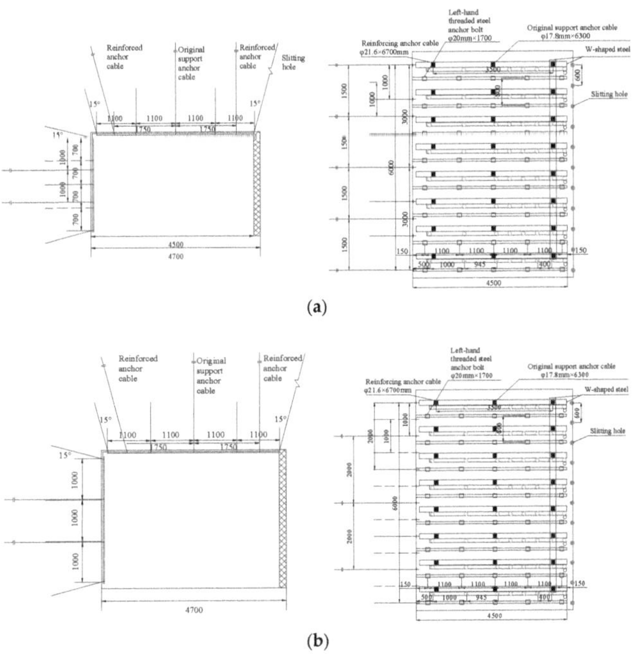


**Figure 11.** Design of roof reinforcement support for gob-side entry (**a**) 9102 tailgate roof reinforcement design (**b**) 10102 headgate roof reinforcement design.

After using this design scheme on site, the single column breakage of upper group coal does not exceed 5%, and the lower group coal basically does not exceed 1%, which indicates that the engineering application of near coal seam co-mining top cutting and pressure removal along the empty stay is successful.

## 6. Discussion

This study presents an in-depth examination of deformation mechanisms and rock pressure behavior during gob-side entry retaining in close distance co-mining of upper and lower coal seams. Our results highlight the significant impact of staggered distances between the working faces on the stress distribution of the surrounding rock, with a distance of 40 m, demonstrating optimal performance.

Despite these findings, there are limitations in our study that warrant further research First, our model is reliant on accurate geological data and complex numerical simulations, which may not always be feasible. Second, while our model primarily considers staggered distances, other factors that might influence deformation and rock pressure behavior are not extensively addressed. Future studies should explore these influences for a more comprehensive understanding of gob-side entry stability.

## 7. Conclusions

In conclusion, this research focused on understanding the deformation mechanisms and rock pressure behavior in gob-side entry retaining during close distance co-mining of upper and lower coal seams. A structural model was developed, and numerical simulations were performed to assess the impact of varying working face staggered distances on the stress distribution of surrounding rock.

The impact of roof subsidence ($\Delta y$) on the stability of gob-side entry retaining during the co-mining of upper and lower coal seams was investigated. Additionally, it was found that when the subsidence of the goaf roof in the upper group of coal seams falls within the range of 0.74 m $\leq \Delta y \leq$ 1.33 m, block B experiences sliding instability. In cases where $\Delta y \geq$ 1.33 m, block B undergoes rotation and deformation. Both of these scenarios result in severe mining pressure in the roadway, leading to significant deformation of the surrounding rock.

A working face staggered distance of 40 m was identified as the most effective choice, leading to significant improvements in the performance of support schemes. This optimal distance resulted in single column damage rates of no more than 5% for the upper coal seam group and 1% for the lower coal seam group.

These insights can inform the development of strategies to enhance mining safety and efficiency in gob-side entry retaining for close distance co-mining projects.

**Author Contributions:** Writing—original draft, X.X.; Writing—review & editing, C.W.; Project administration, Y.Y. and C.W. All authors have read and agreed to the published version of the manuscript.

**Funding:** This work was supported by the National Natural Science Foundation of China (Grant No. 51404167); Natural Science Foundation For Young Scientists of Shanxi Province (Grant No. 201901D211066); Scientific Research Grant Project for Returned Overseas Students in Shanxi Province (HGKY2019038); Teaching Reform Innovation Project for Higher Education Institutions in Shanxi Province (J2019055); China postdoctoral science foundation funding project (Grant No. 2016M590151).

**Data Availability Statement:** The article data used to support the findings of this study are included within the article.

**Conflicts of Interest:** The authors declare no conflict of interest.

## Nomenclature

| | |
|---|---|
| $K$ | Key block C mining gangue stiffness, N/m |
| $S$ | Key block C mining area gangue displacement, m |
| $W_a$, $W_b$, $W_c$ | Self-weight of key blocks A, B and C, kN |
| $Q_a$, $Q_b$, $Q_c$ | Key block A, B, C structure friction shear reaction force, kN |
| $Z_a$, $Z_b$, $Z_c$ | Key blocks A, B and C are loaded, kN |
| $F_b$ | Support force transmitted by the bracket underneath the critical block B, kN |
| $C_1$ | Distance of the support transfer key block B, m |
| $F_c$ | Key block C mining area gangue support force, kN |
| $T_{ab}$, $T_{bc}$ | Horizontal binding force of the structure on A and B, B and C, kN |
| $l$ | Length of key blocks A, B, C, m |
| $h$ | Height of key blocks A, B, C, m |
| ﬂ | Turning angle of key block B with respect to A, ° |
| ` | Turning angle of key block B with respect to C, ° |
| $k_{p1}$ | Crushing and swelling coefficient |
| $k_{p2}$ | Residual swelling factor |
| $h_i$ | Thickness of the ith collapsed rock layer, m |
| $\Delta y$ | Distance of key block C sinks as a whole, m |
| $S_{bc}$ | Area of the shear between the key blocks C and B, $m^2$ |
| $K_{bc}$ | Shear stiffness of key block C and B |
| $L_{bc}$ | Hinge structure length of key block C and B, m |

## References


1. IEA. *World Energy Outlook 2022*; IEA: Paris, France, 2022.
2. Hudeček, V.; Šancer, J.; Zubíček, V.; Golasowski, J. Experience in the Adoption of Room & Pillar Mining Method in the Company OKD, a.s., Czech Republic. *J. Min. Sci.* **2017**, *53*, 99–108. [CrossRef]
3. Tajduś, K.; Misa, R.; Sroka, A. Analysis of the Surface Horizontal Displacement Changes Due to Longwall Panel Advance. *Int. J. Rock Mech. Min. Sci.* **2018**, *104*, 119–125. [CrossRef]
4. Jiránková, E.; Mučková, J.; Jadviščok, P.; Vochta, R.; Molčák, V.; Havlicová, M. Geodetic Monitoring of the Surface within the Trial Operation of the Room and Pillar Mining Method in the Ostrava-Karvina Coal District. *Int. J. Clean Coal Energy* **2016**, *5*, 37–44. [CrossRef]
5. Jongpradist, P.; Tunsakul, J.; Kongkitkul, W.; Fadsiri, N.; Arangelovski, G.; Youwai, S. High Internal Pressure Induced Fracture Patterns in Rock Masses Surrounding Caverns: Experimental Study Using Physical Model Tests. *Eng. Geol.* **2015**, *197*, 158–171. [CrossRef]
6. Majdi, A.; Hassani, F.P.; Nasiri, M.Y. Prediction of the Height of Destressed Zone above the Mined Panel Roof in Longwall Coal Mining. *Int. J. Coal Geol.* **2012**, *98*, 62–72. [CrossRef]
7. Ghasemi, E.; Ataei, M.; Shahriar, K. An Intelligent Approach to Predict Pillar Sizing in Designing Room and Pillar Coal Mines. *Int. J. Rock Mech. Min. Sci.* **2014**, *65*, 86–95. [CrossRef]
8. Shekhar, N.; Pal, S.; Jaiswal, A.; Hazara, P. Numerical Modelling for Prediction of Ground Subsidence Over Room and Pillar Mining in an Underground Coal Seam. In *Soil Dynamics, Earthquake and Computational Geotechnical Engineering*; Muthukkumaran, K., Ayothiraman, R., Kolathayar, S., Eds.; Springer Nature: Singapore, 2023; pp. 155–165.
9. Zhou, N.; Du, E.; Li, M.; Zhang, J.; Dong, C. Determination of the Stability of Residual Pillars in a Room-and-Pillar Mining Goaf under Eccentric Load. *Energy Rep.* **2021**, *7*, 9122–9132. [CrossRef]
10. Waclawik, P.; Ptacek, J.; Konicek, P.; Kukutsch, R.; Nemcik, J. Stress-State Monitoring of Coal Pillars during Room and Pillar Extraction. *J. Sustain. Min.* **2016**, *15*, 49–56. [CrossRef]
11. Skrzypkowski, K. The Influence of Room and Pillar Method Geometry on the Deposit Utilization Rate and Rock Bolt Load. *Energies* **2019**, *12*, 4770. [CrossRef]
12. Mallı, T.; Yetkin, M.E.; Özfırat, M.K.; Kahraman, B. Numerical Analysis of Underground Space and Pillar Design in Metalliferous Mine. *J. Afr. Earth Sci.* **2017**, *134*, 365–372. [CrossRef]
13. Alejano, L.R.; Arzúa, J.; Castro-Filgueira, U.; Malan, F. Strapping of Pillars with Cables to Enhance Pillar Stability. *J. S. Afr. Inst. Min. Metall.* **2017**, *117*, 527–540. [CrossRef]
14. Skrzypkowski, K.; Korzeniowski, W.; Zagórski, K.; Zagórska, A. Flexibility and Load-Bearing Capacity of Roof Bolting as Functions of Mounting Depth and Hole Diameter. *Energies* **2019**, *12*, 3754. [CrossRef]
15. Tan, Y.L.; Zhao, T.B.; Xiao, Y.X. In Situ Investigations of Failure Zone of Floor Strata in Mining Close Distance Coal Seams. *Int. J. Rock Mech. Min. Sci.* **2010**, *47*, 865–870. [CrossRef]
16. Qin, Y.; Xu, N.; Zhang, Z.; Zhang, B. Failure Process of Rock Strata Due to Multi-Seam Coal Mining: Insights from Physical Modelling. *Rock Mech. Rock Eng.* **2021**, *54*, 2219–2232. [CrossRef]

17. Zhang, Y.; Zhang, C.L.; Wei, C.C.; Liu, Y.D.; Zhang, S.Q.; Zhao, J.J. The Study on Roadway Layout in Coordination of Mining Coal Seams Base on Failure of Floor Strata. *Adv. Mater. Res.* **2014**, *889–890*, 1362–1374. [CrossRef]
18. Cheng, Z.; Ma, H.; Sang, C.; Liu, Y. Experimental Research on Dynamic Evolution Characteristics of Roof Movement and Mining-Induced Stress of Superimposed Mining in a Close Distance Coal Seam Group. *Geotech. Geol. Eng.* **2021**, *39*, 13–24. [CrossRef]
19. Lian, Z.; Wang, J.; Hao, C. Numerical Simulation and Experimental Research of Surrounding Rock Deformation of Floor Roadway under Short-Distance Coal Seam Group Combined Mining. *J. Coal. Sci. Eng. China* **2010**, *16*, 230–234. [CrossRef]
20. Wu, G.; Fang, X.; Bai, H.; Liang, M.; Hu, X. Optimization of Roadway Layout in Ultra-Close Coal Seams: A Case Study. *PLoS ONE* **2018**, *13*, e0207447. [CrossRef]
21. Liu, H.; Zhang, B.; Li, X.; Liu, C.; Wang, C.; Wang, F.; Chen, D. Research on Roof Damage Mechanism and Control Technology of Gob-Side Entry Retaining under Close Distance Gob. *Eng. Fail. Anal.* **2022**, *138*, 106331. [CrossRef]
22. Tan, Y.L.; Yu, F.H.; Ning, J.G.; Zhao, T.B. Design and Construction of Entry Retaining Wall along a Gob Side under Hard Roof Stratum. *Int. J. Rock Mech. Min. Sci.* **2015**, *77*, 115–121. [CrossRef]
23. Fan, D.; Liu, X.; Tan, Y.; Yan, L.; Song, S.; Ning, J. An Innovative Approach for Gob-Side Entry Retaining in Deep Coal Mines: A Case Study. *Energy Sci. Eng.* **2019**, *7*, 2321–2335. [CrossRef]
24. Martirosyan, A.V.; Ilyushin, Y.V. Modeling of the Natural Objects' Temperature Field Distribution Using a Supercomputer. *Informatics* **2022**, *9*, 62. [CrossRef]
25. Martirosyan, A.V.; Ilyushin, Y.V. The Development of the Toxic and Flammable Gases Concentration Monitoring System for Coalmines. *Energies* **2022**, *15*, 8917. [CrossRef]
26. Skrzypkowski, K. Case Studies of Rock Bolt Support Loads and Rock Mass Monitoring for the Room and Pillar Method in the Legnica-Głogów Copper District in Poland. *Energies* **2020**, *13*, 2998. [CrossRef]
27. Zhang, Z.; Deng, M.; Bai, J.; Yan, S.; Yu, X. Stability Control of Gob-Side Entry Retained under the Gob with Close Distance Coal Seams. *Int. J. Min. Sci. Technol.* **2021**, *31*, 321–332. [CrossRef]
28. Li, Q.; Wu, G.; Kong, D. Study on Stability of Stope Surrounding Rock under Repeated Mining in Close-Distance Coal Seams. *Geofluids* **2022**, *2022*, e9630942. [CrossRef]
29. Shen, F.; Song, Y.; Zheng, J.; Zhao, W.; Yang, M.; Ma, H.; Shao, Z. Study on Stress Evolution Law and Control Technology of Roadway Formed by Roof Cutting during Crossing Overlying Remained Coal Pillar. *Energy Explor. Exploit.* **2023**, *41*, 745–765. [CrossRef]
30. Zhang, Z.; Wang, W.; Li, S.; Bai, J.; Hao, S.; Wu, H.; Yu, X. An Innovative Approach for Gob-Side Entry Retaining With Thick and Hard Roof: A Case Study. *Teh. Vjesn.* **2018**, *25*, 1028–1036. [CrossRef]
31. He, F.; Xu, X.; Qin, B.; Li, L.; Lv, K.; Li, X. Study on Deformation Mechanism and Control Technology of Surrounding Rock during Reuse of Gob Side Entry Retaining by Roof Pre-Splitting. *Eng. Fail. Anal.* **2022**, *137*, 106271. [CrossRef]
32. Fan, D.; Liu, X.; Tan, Y.; Song, S.; Gu, Q.; Yan, L.; Xu, Q. Roof Cutting Parameters Design for Gob-Side Entry in Deep Coal Mine: A Case Study. *Energies* **2019**, *12*, 2032. [CrossRef]
33. Zhang, Y.; Wang, X.; Zhang, F.; Li, M.; Wang, G.; Chen, D.; Li, G.; Zhao, X. Retracement Ground Pressure Appearance and Control of the Working Face under the Overlying Residual Pillar: A Case Study. *Energies* **2023**, *16*, 1701. [CrossRef]
34. Wang, K.; Huang, Y.; Gao, H.; Zhai, W.; Qiao, Y.; Li, J.; Ouyang, S.; Li, W. Recovery Technology of Bottom Coal in the Gob-Side Entry of Thick Coal Seam Based on Floor Heave Induced by Narrow Coal Pillar. *Energies* **2020**, *13*, 3368. [CrossRef]
35. Xinjie, L.; Xiaomeng, L.; Weidong, P. Analysis on the Floor Stress Distribution and Roadway Position in the Close Distance Coal Seams. *Arab. J. Geosci.* **2016**, *9*, 83. [CrossRef]

energies

MDPI

Article

# CFD Analysis of the Forced Airflow and Temperature Distribution in the Air-Conditioned Operator's Cabin of the Stationary Rock Breaker in Underground Mine under Increasing Heat Flux

Adam Wróblewski [1,*], Arkadiusz Macek [2], Aleksandra Banasiewicz [1], Sebastian Gola [1,3], Maciej Zawiślak [2] and Anna Janicka [2]

1 Faculty of Geoengineering, Mining and Geology, Wroclaw University of Science and Technology, Na Grobli 15, 50-421 Wroclaw, Poland
2 Faculty of Mechanical Engineering, Wroclaw University of Science and Technology, Łukasiewicza 5/7, 50-370 Wroclaw, Poland
3 KGHM Polska Miedź S.A., O/ZG Polkowice-Sieroszowice, 59-101 Kazimierzów, Poland
* Correspondence: adam.wroblewski@pwr.edu.pl

**Abstract:** The exploitation of natural resources is associated with many natural hazards. Currently, the copper ore deposits exploited in Polish mines are located at a depth of about 1200 m below the surface. The primary temperature of the rocks in the exploited areas reaches 48 °C, which constitutes a major source of heat flux to the mine air. However, another important source of heat is the machine plant, which mainly consists of machines powered by diesel engines. Following the results of in situ measurements, boundary conditions for a simulation were determined and a geometric model of the cabin was created. Furthermore, an average human model was created, whose radiative heat transfer was included in the analysis. Three cases were studied: the first covering the current state of thermal conditions, based on the measurement results, and two cases of forecast conditions. In the second case, the temperature of the conditioned air was determined, and in the third, the flow velocity required to ensure thermal comfort was found. The results of the simulation indicated that for the microclimatic conditions established based on the measurements (ambient air temperature in the excavation 35.0 °C, air-conditioned airflow $2.4 \times 10^{-2}$ $m^3/s$, and temperature 10.0 °C), the temperature of the air inside the air-conditioned operator's cabin would be 20.4 °C. Based on the personal mean vote (PMV) index, it was concluded that the thermal sensation would range from neutral to slightly cool, which confirmed the legitimacy of the actions taken to reduce the adverse impact of the microclimatic conditions on workers in the workplace. However, for the case of predicted conditions of enhanced heat flux from strata and machinery, resulting in an average ambient temperature increased to 38.0 °C, it would be necessary to lower the temperature of air from the air conditioner to 8.00 °C or increase the flow rate to $3.14 \times 10^{-2}$ $m^3/s$ to maintain thermal comfort at the same level of PMV index.

**Keywords:** CFD; cabin interior; cabin air conditioning analysis; heat transfer; thermal hazard



**Citation:** Wróblewski, A.; Macek, A.; Banasiewicz, A.; Gola, S.; Zawiślak, M.; Janicka, A. CFD Analysis of the Forced Airflow and Temperature Distribution in the Air-Conditioned Operator's Cabin of the Stationary Rock Breaker in Underground Mine under Increasing Heat Flux. *Energies* **2023**, *16*, 3814. https://doi.org/10.3390/en16093814

Academic Editors: Sergey Zhironkin and Dawid Szurgacz

Received: 20 March 2023
Revised: 20 April 2023
Accepted: 23 April 2023
Published: 28 April 2023



## 1. Introduction

Based on many years of experience in production activities in the area of the Polish copper basin, various procedures have been developed, which in general can be described as deposit mining technology. The individual elements of this process have been adapted based on various factors, such as the deposit mining conditions and natural hazards [1]. The method of mining the deposit in the mines belonging to KGHM Polska Miedź S.A. is exploitation with the use of explosives. The excavated material, initially detached from the rock mass, is loaded and transported using LHD's (load, haul, dump) and HT's (haul trucks) to transfer points, located at a certain distance from the mining front, called the unit discharge screen. In these places, the excavated material is cyclically unloaded from the machines [2], and the excavated material is reloaded onto a belt conveyor [3].

The multiplicity and complexity of the factors that influence the rock mining process with the use of explosives make production very diverse, in terms of the size of rock chunks In the hauled material there are many chunks of a very large size. Their appearance not only causes problems from the point of view of ore processing, due to the additional energy expenditure incurred in the first stage of crushing the material, but can also cause failures of the technical infrastructure during reloading, in particular the belt conveyors [4–6] at unit transfer points [7].

To avoid damage to the components of the conveyor belt at the unit transfer points, preliminary crushing of oversized chunks is carried out with the use of a stationary rock breaker. Copper ore is unloaded by LHDs and HTs onto the so-called screen, which is designed to separate smaller pieces from those that are too large [8]. The operation principle is similar to a sieving process. When the size of the screen mesh is adjusted, the maximum size of fragments that can be loaded onto the conveyor belt without damaging it is defined. The remaining chunks on the screen are crushed into smaller pieces using hammers placed on manipulators that form the stationary rock breaker [9]. Preliminary crushing of oversized chunks with this method is also found in the exploitation of other minerals in Poland, such as gypsum and anhydrite or salt.

In many cases, the stationary rock breaker is controlled by an operator directly from a cab located at the unit transfer point. As a result of their proximity to the mining area, transfer points are often affected by many natural hazards [10,11] due to their closeness to the mining fronts and due to the current depth of exploitation of Polish copper ore deposits, especially thermal hazards [12]. Measurements carried out near a selected operator cabin showed that the average dry bulb temperature reached 35 °C. The primary temperature of the rock mass, which locally exceeded 45 °C, was responsible for the largest increase in air temperature in the operator's workplace. It is worth mentioning that in underground mines there are also other heat sources, which have a significant impact on the increase in the temperature of the mine air. In the case under consideration, the machines and devices that periodically appear at the unit discharge screen, where the ore is being discharged, emit significant amounts of heat into the atmosphere. To protect the stationary rock breaker operator from significant exposure to high temperatures and humidity, and to improve the thermal conditions in the workplace, the cabin in which the operator works is air-conditioned.

This article presents an experimental study to determine the forced air flow and temperature distribution inside an air-conditioned operator's cabin of a stationary rock breaker during a standard working shift and in conditions characteristic of unit transfer points in an underground copper ore mine. Based on technical design, a CAD model of the cabin was created, in which a human model based on the dimensions of the 50th percentile of the male population was placed. In situ measurements determined the microclimate parameters, both outside (dry and wet bulb temperatures) and inside (temperature and airflow velocity at the outlets of the air conditioning system), which were used as boundary conditions. The simulation was carried out using ANSYS FLUENT software, based on the results of which a thermal comfort assessment was made, utilizing a predicted mean vote (PMV) methodology. Due to a planned expansion of exploitation, the depth of exploitation and, thus, the climate threat are expected to intensify. Therefore, a second case was simulated, in which a higher air temperature was assumed for the excavation, and the values of the inlet air parameters (temperature and velocity) were determined, which are necessary for maintaining the thermal comfort of the operator following the PMV methodology. This article presents an innovative approach to determining environmental and microclimatic parameters in confined spaces, such as machinery cabins, particularly the difficult conditions encountered in underground mining, but also in other industries; the provision of which will allow creating acceptable working conditions.

## 2. State of the Art

Due to the often difficult environmental conditions in mining, work parameters such as temperature, humidity, gas concentrations, and dust levels are controlled in the workplace [13–16]. Mining machine operators generally work in conditions partially isolated from the direct impact of the environmental conditions. However, closed cabins equipped with filters and an air conditioning system only partially protect miners from these hazards, especially in extreme cases. For this reason, inspections include dust measurements outside and inside cabins to estimate the filtration efficiency in machines with different filters and intake configurations [17]. The contribution of diesel particle matter (DPM) from individual engine emissions to total emissions, cabin integrity, positive pressure, and air filtration systems is validated, to assess exposure to this factor [18], while quantitative levels of exposure to diesel exhaust expressed by elemental carbon (EC) are estimated, together with a description of the excess risk of lung cancer that can result from these levels [19]. Similarly, thermal comfort and the threshold range of airflow supply parameters are assessed for different types of work, including shearer drivers [20].

The exploitation of deposits located deep underground, which is increasing its share in the global mining industry, due to the exploitation of shallower deposits, is inextricably linked to the occurrence of climatic threats. This problem affects the mining of various mineral resources at different depths globally; for example, metal ores or coal mines in China [21–23], India [24], South Africa [25], and Poland [26,27], driving scientists to look for new solutions to combat this threat. A method for achieving this goal may be thermal insulation, the application of which was studied in [28,29].

The subbranch of fluid mechanics in which numerical methods are used to solve fluid description problems is called computational fluid dynamics (CFD). Through discretization and numerical determination of partial differential equations, it is possible to determine approximate distributions of physical quantities, such as velocity, pressure, temperature, and other characteristic flow parameters [30]. In the literature, many examples of successful CFD applications to solve flow-related problems in the mining industry [31] and in tunneling can be found. Ren and Balus [32] conducted a review of the literature in the field of solving health and safety problems in mining with the use of CFD. Solving ventilation problems with CFD at working faces and in mine tunnels was presented by Yi et al. [33], while an analysis of ventilation safety using CFD in hard coal mining was summarized in [34]. Roadway fire accidents, considering smoke diffusion, temperature distribution, and CO concentration and distribution in four working conditions, were numerically analyzed by Peng et al. [35].

The reduction of the dead zone with the definition of an appropriate ventilation system and the optimization of the relative airflow velocity inside a tunnel boring machine (TBM) utilizing CFD was presented in [36]. Based on numerical simulations, Liu et al. [37] studied the effect of dust control in TBM construction tunnels with different dust extraction configurations and flow rates. The influence of ventilation parameters on dust pollution in TBM tunnels under different air suction volumes was also investigated in [38], where different distances from the dust removal device ports from the mining face were considered, and where CFD analysis allowed the establishment of the optimal suction rate and distance.

Researchers have devoted a great deal of attention to the analysis of the diesel particle matter (DPM) distribution in mines. Xu et al. [39] conducted a numerical study of the DPM distribution in two different operational scenarios: when the machine worked in isolated zones upstream and downstream. The results allowed the upgrading of the auxiliary ventilation designs to achieve smaller concentrations. As proven in [40,41], optimizing auxiliary ventilation through the utilization of computational fluid dynamics allows minimizing human carcinogenic DPM intakes. One of the most interesting applications of CFD in mining developed recently is the analysis of mine airflow and its components, using data from laser scanning to create high-accuracy geometric models of excavations [42] and thanks to which the accuracy of the results obtained increases. An example from hard coal mining can be found in [43], where airflow analysis was conducted at the longwall and at the goafs, as well as in a section of a mine gallery in [44]. The results showed the great impact of the infrastructure elements present in the excavation on the turbulent airflow behavior, which should not be neglected.

Taking into account the application of CFD in the literature, one can find many examples of its use to predict and analyze the air distribution inside closed spaces, such as aircraft cabins. In the last century, Aboosaidi [45] proved the validity and role of this tool in application, by performing experimental and analytical tests using the example of two three-dimensional cabin airflow configurations. Zhang et al. [46] analyzed the air velocity, air temperature, and $CO_2$ concentration distributions in a Boeing 767 aircraft cabin using a validated CFD tool, selecting the optimal air distribution system to counteract draft risk. The global transport process of contaminated air within a cabin was investigated in [47] using experiments and CFD simulations to understand the mechanism of the spread of diseases. Li [48] studied airflow and contaminant concentration fields in a cabin section, applying two turbulence models and three types of air supply model. In their review of the literature, Liu et al. [49] indicated that due to developments in the performance and affordability of computing, numerical simulations for studying airflow and contaminant distributions in aircraft cabins have become very practical.

A closer relation to the topic discussed in this paper can be found in the automotive industry, where CFD is very often used to model the distribution of air and its components in vehicle interiors. The size and distribution of volatile carcinogens inside a vehicle were analyzed in [50], while the concentrations of volatile organic compounds were analyzed in [51,52]. Exposure of the driver to the spread of toxic compounds, such as the hydrocarbons emitted by the engine was investigated in [53]. Taking heat transfer as a factor, Hadi et al. [54] analyzed variations in temperature and airflow in the interior of a parked car, considering different solar intensities at various angles, while Mao et al. did the same for a parked and an electric car driven in summer and winter conditions [55]. Analysis using CFD tools often covers the temperature distribution and energy efficiency analysis of air conditioning systems in electric vehicles, in which, especially in winter, energy consumption has a key impact on range, as shown by Basciotti et al. [56].

Limiting the internal space of a cabin to the case where only one person is inside, one can mention examples of the use of CFD to model airflow in the cabins of agricultural and industrial machinery. Singh and Abbassi [57] created a thermal model of a combine-harvester air-conditioning system, to investigate the instantaneous thermal state of cabin air, combining a 3D CFD model with a neural network and updating the thermal state of the air based on data on the performance of the refrigeration cycle. A similar work presented by Oh et al. [58] studied and analyzed the dependence of thermal comfort on airflow and the location of the air conditioner in a tractor. The temperature and airflow distribution in a tractor cabin was also modeled using CFD by Akdemir et al., considering summer [59] and winter [60] conditions. The issue of thermal comfort in the cabin of construction equipment was raised in [61], where the authors proposed a thermal comfort performance index for the position of the excavator operator, evaluating the location of the air conditioning vents using CFD modeling. Ghorpade et al. [62] dealt with the design, selection, and fabrication of the components of air-conditioning systems by calculating the heat load of operation under hot conditions and using CFD analysis of the heat and airflow distribution in manipulator cabins.

## 3. Materials and Methods

The following section presents the methodology applied for the simulation of forced air and temperature distribution in the cabin. Section 3.1 presents the governing equations, while Section 3.2 presents the applied turbulence model and its description. In Section 3.3, the description and dimensions of the created geometry are presented. The type and quality of the mesh developed is described in Section 3.4, and the boundary conditions applied for the model and their determination method are given in Section 3.5.

### 3.1. Governing Equations

For the steady compressible laminar flow of a Newtonian fluid, the equations of motion in Cartesian coordinates can be written as follows [63]:

$$\frac{\partial(\rho u)}{\partial x}+\frac{\partial(\rho v)}{\partial y}+\frac{\partial(\rho z)}{\partial z}=0 \qquad P=\rho RT \tag{1}$$

$$\rho\left(u\frac{\partial u}{\partial x}+v\frac{\partial u}{\partial y}+w\frac{\partial u}{\partial y}\right)=\rho g_x-\frac{\partial P}{\partial x}+\frac{\partial}{\partial x}\left(2\mu\frac{\partial u}{\partial x}+\lambda\vec{\nabla}\cdot\vec{V}\right)+\frac{\partial}{\partial y}\left[\mu\left(\frac{\partial u}{\partial y}+\frac{\partial v}{\partial x}\right)\right]+\frac{\partial}{\partial z}\left[\mu\left(\frac{\partial w}{\partial z}+\frac{\partial u}{\partial z}\right)\right] \tag{2}$$

$$\rho\left(u\frac{\partial v}{\partial x}+v\frac{\partial v}{\partial y}+w\frac{\partial v}{\partial z}\right)=\rho g_y-\frac{\partial P}{\partial y}+\frac{\partial}{\partial x}\left[\mu\left(\frac{\partial v}{\partial x}+\frac{\partial u}{\partial y}\right)\right]+\frac{\partial}{\partial y}\left(2\mu\frac{\partial v}{\partial y}+\lambda\vec{\nabla}\cdot\vec{V}\right)+\frac{\partial}{\partial z}\left[\mu\left(\frac{\partial v}{\partial z}+\frac{\partial w}{\partial y}\right)\right] \tag{3}$$

$$\rho\left(u\frac{\partial w}{\partial x}+v\frac{\partial w}{\partial y}+w\frac{\partial w}{\partial z}\right)=\rho g_z-\frac{\partial P}{\partial z}+\frac{\partial}{\partial x}\left[\mu\left(\frac{\partial w}{\partial x}+\frac{\partial u}{\partial z}\right)\right]+\frac{\partial}{\partial y}\left[\mu\left(\frac{\partial v}{\partial z}+\frac{\partial w}{\partial y}\right)\right]+\frac{\partial}{\partial z}\left(2\mu\frac{\partial w}{\partial z}+\lambda\vec{\nabla}\cdot\vec{V}\right) \tag{4}$$

$$\rho c_p\left(u\frac{\partial T}{\partial x}+v\frac{\partial T}{\partial y}+w\frac{\partial T}{\partial z}\right)=\beta T\left(u\frac{\partial P}{\partial x}+v\frac{\partial P}{\partial y}+w\frac{\partial P}{\partial z}\right)+\vec{\nabla}\cdot(k\vec{\nabla}T)+\Phi \tag{5}$$

where $u$, $v$, and $w$—the velocity components in the $x$, $y$, $z$ directions; $P$—the pressure; $\rho$—the mass density; $R$—the specific ideal-gas constant; $T$—the temperature; $g$—the gravitational acceleration; $\mu$—the viscosity; $\lambda$—the second coefficient of viscosity; $\vec{\nabla}$—divergence; $\vec{V}$—velocity of the fluid; $c_p$—specific heat at constant pressure; $\beta$—the thermal expansion coefficient of air; $k$—thermal conductivity; and $\Phi$—dissipation function.

The given equations are partial differentials of conservation of mass Equation (5), momentum in three directions Equations (2)–(4), and continuity Equation (1) left, while the algebraic equation is the ideal-gas law Equation (1) right true for laminar flow. Considering high flow speeds, the flow would be treated as turbulent and the equations should be modified to include a turbulence model.

### 3.2. Turbulence Model Applied and Its Description

The turbulent flow inside the cabin was modeled using the realizable $k$-$\epsilon$model. As in the standard $k$-$\epsilon$ model, two partial differential (transport) equations were solved: the turbulent kinetic energy $k$ and the turbulence eddy dissipation $\epsilon$. The difference with the standard $k$-$\epsilon$ model is that it contains an improved formulation for turbulent viscosity, and the transport equation for the dissipation rate $\epsilon$ was derived from an exact equation for the transport of the mean-square vorticity fluctuation [64]. Realizable means satisfying certain mathematical constraints on Reynolds stresses following the physics of turbulent flows. The model used allows an increased accuracy in predicting the distribution of the dissipation rate of flat and round streams. Better prediction is also provided for the characterization of the boundary layer with large pressure gradients and separated and recirculating flow [65]. The transport Equations (6) and (7) in the realizable k-$\epsilon$ model may be written as follows:

$$\frac{\partial k}{\partial t}+\frac{\partial k u_i}{\partial x_i}=\frac{\partial}{\partial x_i}\left(Dk_{eff}\frac{\partial k}{\partial x_i}\right)+G_k-\epsilon \tag{6}$$

$$\frac{\partial \epsilon}{\partial t} + \frac{\partial \epsilon u_i}{\partial x_i} = \frac{\partial}{\partial x_i}\left(D\epsilon_{eff}\frac{\partial \epsilon}{\partial x_i}\right)\sqrt{2}C_{1\epsilon}Sij\epsilon - C_{2\epsilon}\frac{\epsilon^2}{k+\sqrt{v\epsilon}} \quad (7)$$

where $G_k$—generation of turbulent kinetic energy due to mean velocity gradients; $Dk_{eff}$ and $D\epsilon_{eff}$—effective diffusivity for $k$ and $\epsilon$ (8):

$$Dk_{eff} = v + \frac{v_t}{\sigma_k} \quad D\epsilon_{eff} = v + \frac{v_t}{\sigma_\epsilon} \quad (8)$$

where $\sigma_k$ and $\sigma_\epsilon$—turbulent Prandtl number for $k$ and $\epsilon$ equals 1.0 and 1.2 [57].

Turbulent viscosity (9):

$$v_t = C_\mu \frac{k^2}{\epsilon} \quad (9)$$

$$C_\mu = \frac{1}{A_0 + A_s\frac{kU^*}{\epsilon}} \quad (10)$$

$$U^* = \sqrt{S_{ij}S_{ij} + \tilde{\Omega}_{ij}\tilde{\Omega}_{ij}} \quad (11)$$

$$\tilde{\Omega}_{ij} = \bar{\Omega}_{ij} - \epsilon_{ijk}\omega_k - 2\epsilon_{ijk}\omega_k \quad (12)$$

where $\bar{\Omega}_{ij}$—mean rate of rotation tensor; $\omega_k$—angular velocity.

Constants $A_0$ and $A_s$ are determined as follows:

$$A_0 = 4, \quad A_s = \sqrt{6}\cos\varphi. \quad (13)$$

$$\varphi = \frac{1}{3}\arccos(min(max(\sqrt{6}W, -1), 1)) \quad (14)$$

$$W = \frac{S_ijS_jkS_ki}{\tilde{S}^2} \quad (15)$$

$C_1\epsilon$ is defined as:

$$C_1\epsilon = max\left(\frac{\eta}{5+\eta}, 0.43\right) \quad (16)$$

$$\eta = S\left(\frac{k}{\epsilon}\right) \quad (17)$$

Constant $C_2$ equals 1.9 [57].

### 3.3. Geometry

The air flow and temperature distribution in the operator's cabin, which is an integral part of the stationary rock breaker, were analyzed. A technical drawing of this device is shown in Figure 1.

The geometric model of the operator cabin is 1.8 m high, 1.4 m long (by the floor), and 1.3 m wide. The front wall of the cabin at a height of 0.940 m bends at an angle of approximately 162°, so that its length at the ceiling is approximately 1.676 m. Above the bend, there is a window through which the operator controls the hammer. In the symmetrical side walls, there are also windows secured with a lattice, as well as doors. In turn, there are a total of five openings in the rear wall: in the upper part, at a height of 1.580 m, at a distance of 0.250 m from the side edge and 0.400 m apart (center of the openings) three air intake holes with a diameter of 0.100 m each, while in the lower part, at a height of 0.255 m, at a distance of 0.400 m from the center of symmetry of the wall, two outlet holes with a diameter of 0.159 m. Inside the cabin, there is the necessary equipment in the form of a cabin frame made of steel profiles with dimensions of 0.120 × 0.120 m, diffusers that direct air through the inlet openings (preventing the inflow of cooled air directly to the operator), a control panel with dimensions of 0.860 × 0.305 × 0.175 m, as well as the necessary wiring and lighting. In the central part of the cabin, there is the operator's seat, on which the human model is placed. There is also a fire extinguisher in the lower corner of the cab, behind the operator's seat. A simplified CAD model created based on the technical sheet of the cabin dimensions is presented in Figure 2a (exterior) and Figure 2b (interior).

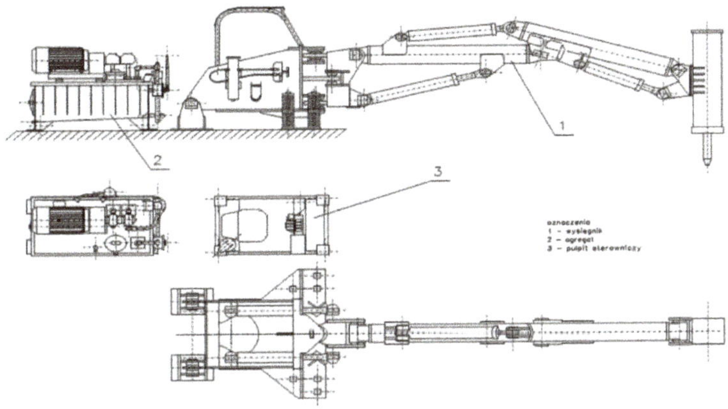


**Figure 1.** Stationary Rock Breaker Type URB/K (1—manipulator; 2—generator; 3—control panel in the operator's cabin [66]).

Following the guidelines described in [67], a model of the 50th percentile of the male population was made. The model was simplified as much as possible to shorten and maintain the stability of numerical calculations. The dimensions based on which the model used for numerical calculations was created are shown in Table 1.

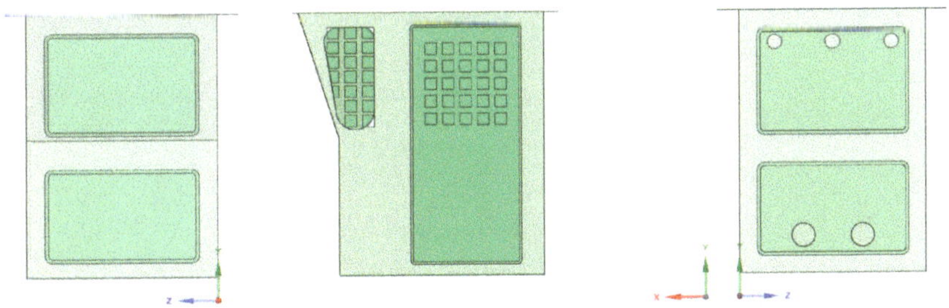

(**a**)

**Figure 2.** *Cont.*

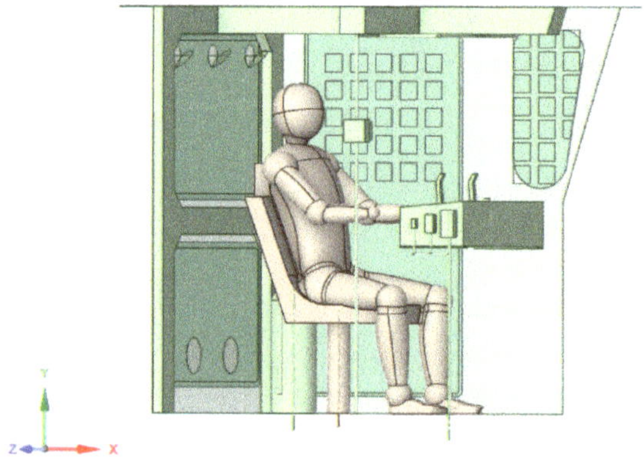


(b)

**Figure 2.** Simplified CAD model of the cabin with the human model. (**a**) Front, side, and rear view of the cabin exterior; (**b**) side view for the cabin interior.

**Table 1.** Dimensions for creation of the 50th percentile of the male population model (based on [67]).

| Description | Dimensions (mm) |
|---|---|
| Total Sitting Height | 884 |
| Shoulder Pivot Height | 513 |
| H-Point Height—ref. | 88.5 |
| H-Point from seat Back—ref. | 137 |
| Shoulder Pivot from Backline | 89 |
| Thigh Clearance | 147 |
| Elbow to Wrist Pivot | 297 |
| Shoulder to Elbow | 338 |
| Elbow Rest Height | 201 |
| Buttock to Knee | 592 |
| Popliteal Height | 442 |
| Knee Pivot to Floor | 493 |
| Buttock Popliteal Length | 465 |
| Chest Depth | 221 |
| Foot Length | 259 |
| Foot Width | 99 |
| Shoulder Width | 429 |
| Chest Circumference | 965.5 |
| Waist Circumference | 851 |

*3.4. Mesh*

The geometry of the model was simplified to improve the stability of the numerical calculations. For this purpose, the control panel was simplified and unnecessary equipment elements that do not much affect air flow in the operator cabin were removed. Due to the fact that the model is symmetrical in relation to the vertical plane passing through the center of the operator's cabin, only half of the model was meshed, to shorten the calculation time. A discrete model was made with a poly-hexcore mesh. This type of mesh provides a reduced calculation time with adequate precision [68]. The mesh consists of 2.1 million elements, of which 200 thousand and 1.9 million are hexa- and poly-elements, respectively (Figure 3). A skewness value of 0.80 was achieved and orthogonal quality—0.18.

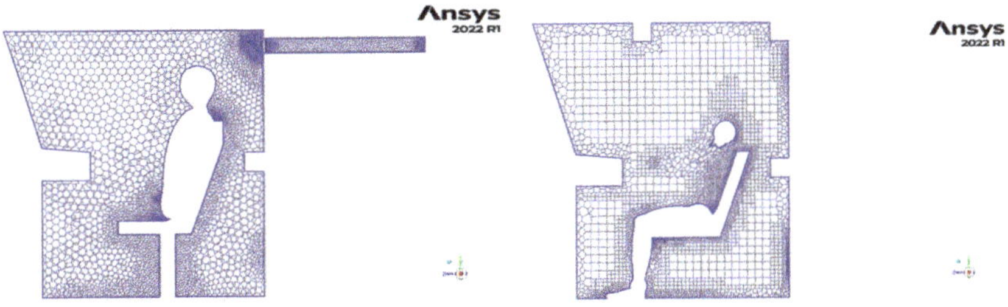


(a) (b)

**Figure 3.** Poly-hexcore mesh elements. (**a**) Isotropic poly-prisms in the boundary layer; (**b**) hexahedron elements in the bulk region.

*3.5. Boundary Conditions Applied and Their Determination*

Measurement of the airflow velocity and temperature in the air-conditioning vents (inlet air supply holes) was difficult, due to the presence of air-directing diffusers. For this reason, measurements were made using a special nozzle with a narrowed outlet (Ø10 cm) relative to the inlet (Ø30 cm) (Figure 4).

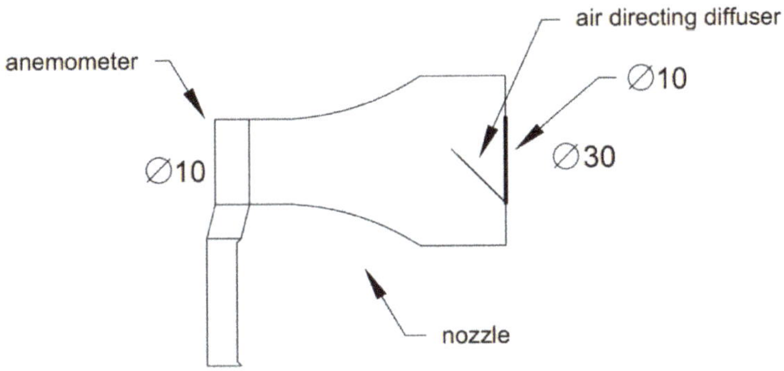


**Figure 4.** Airflow measurements with the nozzle.

Airflow measurements were conducted with a TSI Airflow Instruments LCA301 anemometer. Following the calibration certificate, the device was verified by the TSI manufacturer, certified to ISO 9001:2015 and accredited to 17025:2017. The measurement range of this device is 0.25–30 m/s in terms of velocity and 0.01–3000 $m^3/s$ in terms of volumetric, the accuracy of the measurements is +/−1% reading (+/−0.02 m/s) in terms of velocity. The test results showed that the airflow rates at the discharge ports were $2.30 \times$ t $10^{-2}$, $2.50 \times 10^{-2}$, and $2.40 \times 10^{-2}$ $m^3/s$, while the vents area was $7.85 \times 10^{-3}$ $m^2$. Thus, the air velocity at the vent was calculated as 2.93, 3.18, and 3.13 m/s, and the average values of 3.06 m/s and 3.00 m/s were applied as the boundary condition.

The ambient air temperature in the excavation was defined using an Assman psychrometer. The dry thermometer reads were 35.0 °C in the first measurement, 35.2 °C in the second measurement, and 34.8 °C in the third measurement, in 5 min periods; thus, the value of 35.0 °C was adopted as a radiative temperature boundary condition for all external boundary surfaces of the cabin. Furthermore, the calculations took into account the generation of heat from the human body, and the temperature of 36.6 °C was applied as the radiative temperature on its surfaces. The radiative heat transfer model used assumed an internal emissivity value of 1 at the boundaries. The applied boundary conditions are shown in Table 2.

**Table 2.** Boundary Conditions Applied.

| Name | Boundary Condition |
|---|---|
| **Inlets** | Velocity—3 m/s; Temperature—10 °C |
| **Outlets** | Pressure—0 Pa |
| **Human** | Temperature—36.6 °C, Radiation (internal emissivity—1) |
| **Cabin** | Temperature—35.0 °C, Radiation (internal emissivity—1) |

## 4. Results

Figure 5 shows the inner airflow streamlines following the velocity variable, while Figure 6a,b show the velocity magnitude field and vectors in the cross section. Figure 7 shows the temperature field in the same cross-section. In turn, Figure 8 shows the results of cabin cooling as the decrease in temperature as a function of time.

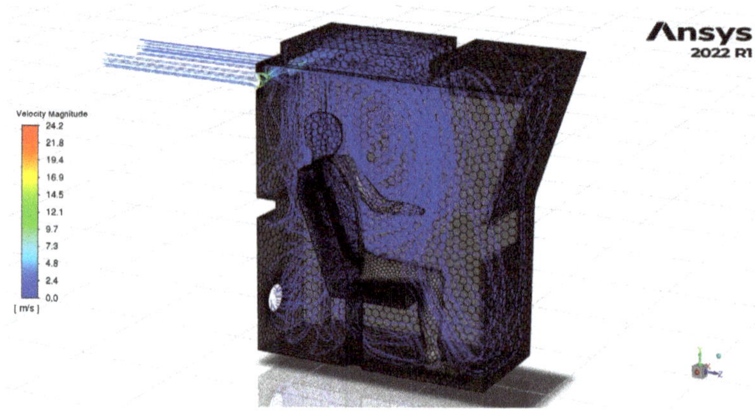


**Figure 5.** Airflow streamlines colored by the velocity variable.

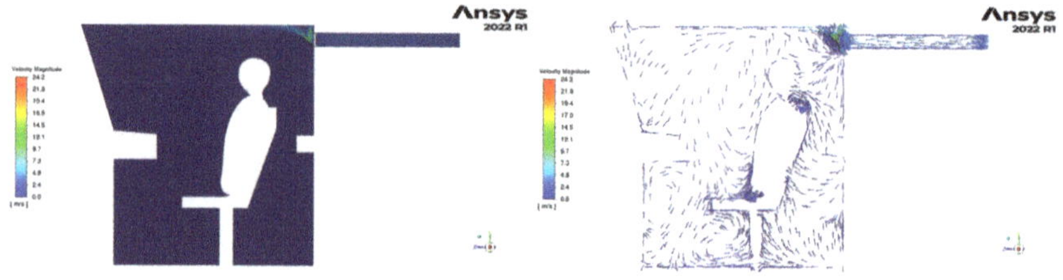


(**a**) (**b**)

**Figure 6.** Air velocity distribution inside the operator's cabin. (**a**) Air velocity magnitude field; (**b**) Air velocity magnitude vectors.

**Figure 7.** Temperature field inside the cabin.

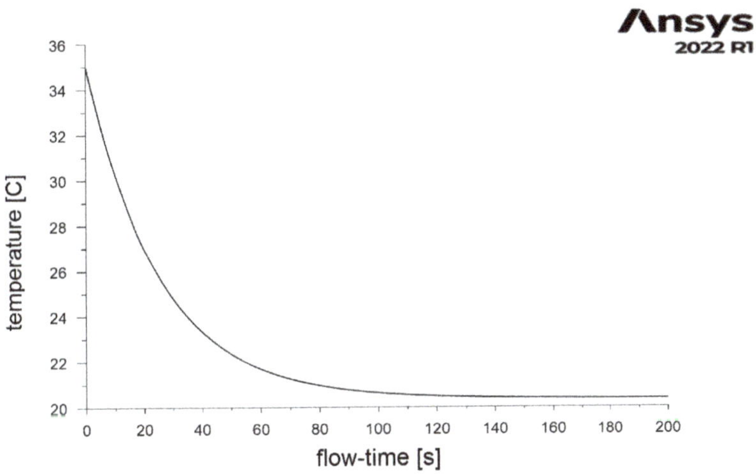


**Figure 8.** Decrease in average air temperature over time inside the cabin. $f(x) = 0.2711e^{-2.323x} + 293.5e^{1.297\times10^{-5}x}$, $R^2 = 1$, RMSE < 1%.

Following the trajectory of the conditioned air flowing out of the intake ports, based on airflow streamlines and air velocity magnitude vectors, we can see that there is a secondary vortex around the operator's head and chest. Another smaller vortex can be seen above and below the control panel, as well as behind the seat. Following the results presented above, the average air velocity in the operator cabin was 0.35 m/s, while the average air temperature was 20.40 °C. Considering the air velocity, it can be seen that its distribution is constant in the entire cabin, except for the area near the inlets and air diffusers. On the other hand, the air temperature distribution changes in the cross-section. Cooler temperatures of about 10 °C are observed around the diffusers and air-conditioned intakes of the cabin, as expected. In the immediate vicinity of the face and chest, the temperature is slightly warmer, about 18.3 °C, than in the greater part of the cabin toward the front window and the control panel (about 17.6 °C). This is a direct result of the path of the airflow that comes out of the inlets and circulates in the cabin when an obstacle, in the form of the seat and the operator, appears. In addition, a lower air temperature is noted in the further trajectory of the air-conditioned stream from the inlets under the operator's panel near his feet, where in the floor area the air stream is refracted and directed toward the outlets. In the remaining parts of the cabin, that is, under and behind the seat and above the control panel, the temperature was relatively constant and reached 21.9 °C. The results obtained were consistent with three verification measurements carried out after 15 min of starting the air conditioning and closing the door by the operator at three points in the cabin: above the control panel at the level of the inlets (average 22.7 °C), in front of the operator at head/chest height (average 19.5 °C), and below the control panel at leg height (average 17.4 °C).

Since the 1970s, the methodology for assessing thermal comfort based on the thermal comfort equation created by Fanger has been used globally [69]. In 1970, Fanger [70] defined PMV as an index to predict or represent the mean vote on a standard scale for thermal sensation for a large group of people for any given combination of thermal environmental variables, activities, and clothing levels. It allows predicting the optimum temperature that should be provided for a group in closed spaces. Based on PMV, the predicted percentage of dissatisfaction (PPD) can also be determined. The predicted mean vote (PMV) model has become the internationally accepted model for describing predicted mean thermal perceptions. Furthermore, based on this methodology, the International Organization for Standardization created a standard for the analytical determination of thermal comfort using the calculation of the PMV and PPD indices [71].

For a given parameters, the PMV value can be obtained in two ways. One way relies on calculating the value using the equation developed by Fanger and presented in [70]. However, following the guidelines that can be found there, the mathematical form of PMV is quite complicated and impractical; thus, its author suggests using a more practical method; that is, utilization of prepared tables. The authors used the suggested method to determine the PMV value, focusing primarily on the determination of the parameters needed.

Due to the climatic conditions that prevail in mines, employees are usually equipped with light clothing selected according to personal preferences: a flannel shirt or thermoactive shirt, trousers or thermoactive shorts, work shoes or rubber boots, underwear, and socks. Based on the values defined in [71], it was determined that the thermal insulation of the combination of workwear used by operators is 1 clo (0.115 $m^2K/W$), which corresponds to the average value for the set of underwear, overalls, socks, shoes, and for the set of underwear, shirt, overalls, socks, and shoes from the workwear group. Similarly, taking into account the metabolic rate, this value was determined to be 1.2 met (70 $W/m^2$), which corresponds to activity in a sitting position. Based on the results of this calculation, for further estimations, it was assumed that the average airflow velocity inside the cabin was 0.35 m/s, and the air temperature was 20 °C.

Considering the above, it was established that the value of the PMV index for the conditions defined in this way ranged from −0.67 to −0.76. This means that the operator's thermal sensation would range from neutral to slightly cool. The graph in Figure 8 suggests that the conditions referred to above would prevail inside the cab after about 180 s, assuming that the operator enters the cab and closes the door and that the air-conditioning system immediately starts working at full capacity.

Due to the planned development of the mine, the depth of exploitation will increase, and higher ambient air temperatures should be expected in the operator's workplace. By changing the boundary condition of the external boundary surface radiative temperature to 38 °C (which is the predicted dry bulb temperature of the mine air at the planned depths of exploitation of the mining areas currently being prepared) and the air-conditioned air temperature at the inlets, an attempt was made to obtain conditions similar to those that in the previous case gave the same result for the PMV index. This goal was achieved for the air temperature boundary condition at the inlets of 8 °C (Figure 9). These conditions, as in the first case and with the same assumptions, would be obtained by the operator after about 180 s (Figure 10). The temperature distribution obtained in the cabin in this case is very similar to the conditions simulated on the basis of the measurement data in the previous case. As the air velocity parameters did not change, the average air velocity remained at 0.35 m/s, while the average air temperature changed slightly to 20.35 °C. In the vicinity of the face and chest, the temperature decreased to 17.9 °C, while it increased to 17 °C in the further part of the cabin toward the front window and the controls. In the remaining parts of the cabin, the temperature remained constant, but increased to 22.2 °C. For these thermal conditions, the PMV index remained at the same level, and thus so did the operator's thermal sensation.

**Figure 9.** Temperature field inside the cabin—case 2.

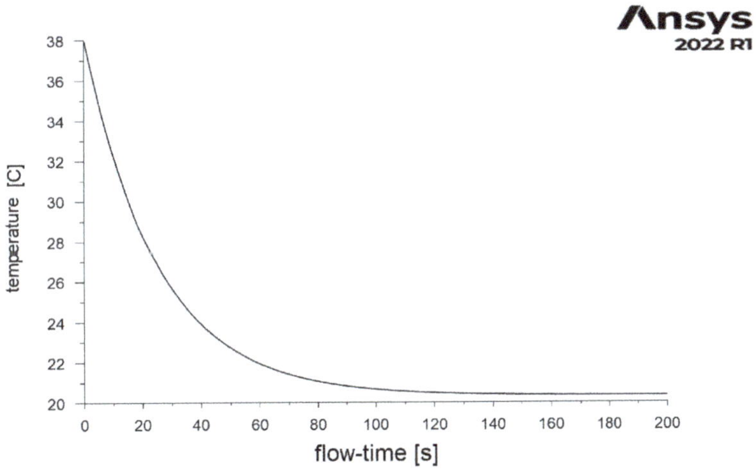


**Figure 10.** Decrease in average air temperature over time inside the cabin—case 2. $f(x) = 0.3226e^{-2.33x} + 293.5e^{1.378\times10^{-5}x}$, $R^2 = 1$, RMSE < 1%.

In the case of increased heat flux from the rock mass and intensified extraction, resulting in an increase in the average air temperature at the unit discharge point, it may turn out that the cooling power of the device will not allow for such a significant reduction in the temperature of the conditioned air at the inlets to the cabin. In such a case, an increase in the air temperature inside the cabin and a reduced thermal comfort of the operator should be expected. With limited possibilities for reducing the air temperature at the inlet, in order to maintain satisfactory thermal conditions of work at the station, its speed should be changed. In case 3, the climatic conditions within the cabin were simulated by changing the boundary conditions in the form of increasing the radiative temperature of the external boundary surface to 38 °C and the air-conditioned air velocity at the inlets to 4 m/s, while maintaining the temperature at the same level as in case 1 (10 °C). The temperature distribution in the cabin in this case is shown in Figure 11. The operator would obtain these conditions in a shorter period of about 150 s (Figure 12).

**Figure 11.** Temperature field inside the cabin—case 3.

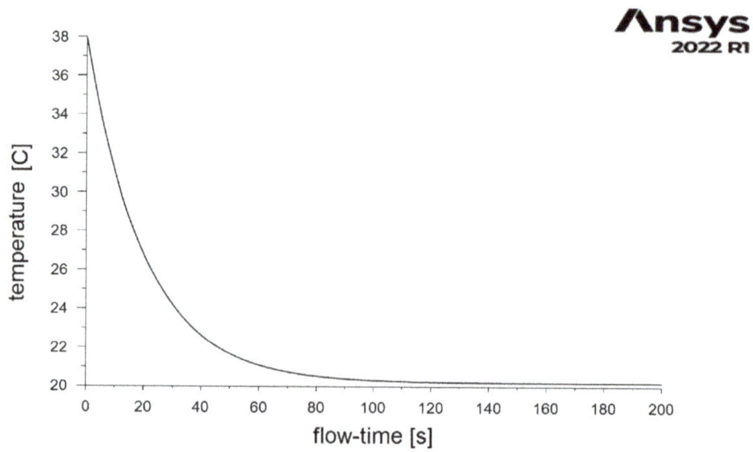


**Figure 12.** Decrease in average air temperature over time inside the cabin—case 3. $f(x) = 18.76e^{-0.04994x} + 293.4e^{-2.469\times10^{-7}x}$, $R^2 = 1$, RMSE < 1%.

Furthermore, in this case the temperature distribution obtained in the cabin was similar to that simulated in Case 1 for the conditions that currently prevail in the mine. The average air velocity increased to 0.47 m/s, while the average air temperature decreased to 20.20 °C. In the vicinity of the face and chest, the temperature decreased to 17.9 °C, while in further part of the cabin toward the front window and the control panel it increased to 17.3 °C. In the remaining parts of the cabin, the temperature remained constant and reached 21.5 °C. The ultimate goal of maintaining thermal comfort at a similar level, as assessed by the PMV index, was achieved.

## 5. Conclusions

In this study, in situ measurements of environmental parameters were performed to determine the boundary conditions, based on which CFD simulations were carried out to analyze the thermal conditions in the operator's workplace for a stationary rock breaker in an underground mine. Based on the simulation results, a thermal comfort assessment was performed using the PMV methodology.

However, taking into account a future planned expansion of exploitation, which would result in an increase in the depth of the works carried out and the intensification of the excavation, two additional cases were also analyzed. Assuming an increase in the average air temperature in the mine, an attempt was made to determine the temperature and velocity of the air-conditioned air flow at the outlet of the air-conditioning device required to maintain the climatic conditions at the same level of PMV.

The PMV value obtained suggests that the thermal sensation in the operator cabin for the microclimate conditions considered in case 1 (currently prevalent in the mine) would range from neutral (PMV = 0) to slightly cool (PMV = −1). On this basis, it can be concluded that the actions taken to reduce the negative impact of microclimatic conditions on the human body in the workplace of the stationary rock breaker operator had the intended effect. Thus, this PMV value was treated as a reference for further investigation of the inlet air parameters that should be ensured to maintain the thermal comfort of the operator.

The results of the CFD simulation showed that similar climatic conditions could be obtained within the operator's cabin, with a projected increase in the average air temperature to 38 °C. In the first of the cases considered, this effect was obtained by lowering the temperature of the air conditioning to 8 °C. However, taking into account the limitations in the cooling power of the devices, with a higher temperature in the second case, this effect was achieved by maintaining the current temperature value of 10 °C and by increasing the air flow velocity at the inlet to the cabins to 4 m/s.

Presentation of the simulation results in the form of the temperature distribution inside the cabin in the section passing through its center could be used to determine the location of the air temperature measurement points used in the mine to determine the operator's comfort using PMV, or another indicator determining the level of thermal comfort. Although the velocity distribution was constant in most of the cabin, except for the area near the inlets and air diffusers, the temperature distribution implied that several measurement points should be taken into account to obtain a reliable average cabin temperature. The extreme values found around the inlets and diffusers should not be considered.

Based on the analysis of the results, it was found that, with regard to the PMV index and URB operator cabin, it was reasonable to take measurements at six points. From the point of view of operator comfort, the temperature should be measured in the immediate vicinity of the face/chest. The second point should be located at the same height, in the path of the airflow that comes out of the inlets; and the third, also at the same height, should be located in the vicinity of the front glass. The fourth measurement point should be placed in the path of airflow directed from the inlets to the lower part of the cabin, at the level of the tibias. The last two points should be placed behind the seat, again at the height of the tibia and at the height of the center of the back. Similar temperature values were obtained in the area of the front window and seat (directly below and behind). Due to the difficulties in taking measurements around the seat, they were omitted; however, they should be included in the calculations, taking the value registered at the front window.

The created model of the operator's cabin allows the prospect of further analyses and simulations aimed at ensuring and maintaining optimal working conditions; especially taking into account other hazards, i.e., the adverse impact of harmful substances contained in mine air in gaseous or solid form, in particular harmful and dangerous gases, dust, and DPM. Future work by the authors will focus on determining the operating parameters of devices and the measurement characteristics needed to counteract these threats, to maintain comfort and comply with regulations in the workplace.

**Author Contributions:** Conceptualization, A.W., A.B., M.Z. and A.J.; methodology, A.W. and A.M.; validation, M.Z.; formal analysis, A.W.; investigation, A.W. and A.B.; resources, S.G.; data curation, A.M.; writing—original draft preparation, A.W.; writing—review and editing, A.W., A.B., M.Z. and S.G.; visualization, A.M.; supervision, M.Z., A.J. and S.G. All authors have read and agreed to the published version of the manuscript.

**Funding:** This activity received funding from the European Institute of Innovation and Technology (EIT), a body of the European Union, under the Horizon 2020, the EU Framework Programme for Research and Innovation. This work was supported by EIT RawMaterials GmbH under Framework Partnership Agreement No. 21016 (ECHO: Electrical Computerised Hammering Operator). The project is co-financed by the Polish Ministry of Education and Science within the framework of the program titled International Co-Financed Projects.

**Data Availability Statement:** Data available on request.

**Acknowledgments:** The authors would like to thank the company KGHM Polska Miedź S.A. and the authorities and employees of O/ZG Polkowice-Sieroszowice for enabling the performance of underground measurements, and their support and assistance in the implementation of this work.

**Conflicts of Interest:** The authors declare no conflict of interest.

## References

1. KGHM Polska Miedź S.A. Deposit Mining Technology. Available online: https://kghm.com/en/our-business/processes/ore-mining/ (accessed on 11 January 2023).
2. Krot, P.; Śliwiński, P.; Zimroz, R.; Gomolla, N. The identification of operational cycles in the monitoring systems of underground vehicles. *Measurement* **2020**, *151*, 107111. [CrossRef]
3. Bortnowski, P.; Gondek, H.; Król, R.; Marasova, D.; Ozdoba, M. Detection of Blockages of the Belt Conveyor Transfer Point Using an RGB Camera and CNN Autoencoder. *Energies* **2023**, *16*, 1666. [CrossRef]
4. Zimroz, R.; Król, R. Failure analysis of belt conveyor systems for condition monitoring purposes. *Min. Sci.* **2009**, *128*, 255.
5. Blazej, R.; Jurdziak, L.; Zimroz, R. Novel approaches for processing of multi-channels NDT signals for damage detection in conveyor belts with steel cords. In *Key Engineering Materials*; Trans Tech Publications Ltd.: Wollerau, Switzerland, 2013; Volume 569, pp. 978–985.
6. Trybała, P.; Blachowski, J.; Błażej, R.; Zimroz, R. Damage detection based on 3d point cloud data processing from laser scanning of conveyor belt surface. *Remote Sens.* **2020**, *13*, 55. [CrossRef]
7. Wróblewski, A.; Krot, P.; Zimroz, R.; Mayer, T.; Peltola, J. Review of Linear Electric Motor Hammers—An Energy-Saving and Eco-Friendly Solution in Industry. *Energies* **2023**, *16*, 959. [CrossRef]
8. Stefaniak, P.; Wodecki, J.; Jakubiak, J.; Zimroz, R. Development of test rig for robotization of mining technological processes—Oversized rock breaking process case. *IOP Conf. Ser. Earth Environ. Sci.* **2017**, *95*, 042028.
9. KGHM ZANAM. Stationary Rock Breaker Type URB/Klim. Available online: https://www.kghmzanam.com/wp-content/uploads/2020/06/URB_Klim_PL.pdf (accessed on 5 December 2022).
10. Tutak, M.; Brodny, J.; Szurgacz, D.; Sobik, L.; Zhironkin, S. The impact of the ventilation system on the methane release hazard and spontaneous combustion of coal in the area of exploitation—A case study. *Energies* **2020**, *13*, 4891. [CrossRef]
11. Szurgacz, D. Electrohydraulic control systems for powered roof supports in hazardous conditions of mining tremors. *J. Sustain. Min.* **2015**, *14*, 157–163. [CrossRef]
12. Wróblewski, A.; Banasiewicz, A.; Gola, S. Heat Balance Determination Methods for Mining Areas in Underground Mines—A Review. *IOP Conf. Ser. Earth Environ. Sci.* **2021**, *942*, 012011. [CrossRef]
13. Hebda-Sobkowicz, J.; Gola, S.; Zimroz, R.; Wyłomańska, A. Identification and statistical analysis of impulse-like patterns of carbon monoxide variation in deep underground mines associated with the blasting procedure. *Sensors* **2019**, *19*, 2757. [CrossRef]
14. Hebda-Sobkowicz, J.; Zimroz, R.; Wyłomańska, A.; Gola, S. Identification and statistical analysis of impulse-like patterns of carbon monoxide variation in deep underground mine. *AIP Conf. Proc.* **2020**, *2209*, 020005.
15. Hebda-Sobkowicz, J.; Gola, S.; Zimroz, R.; Wyłomańska, A. Pattern of H2S concentration in a deep copper mine and its correlation with ventilation schedule. *Measurement* **2019**, *140*, 373–381. [CrossRef]
16. Banasiewicz, A.; Śliwiński, P.; Krot, P.; Wodecki, J.; Zimroz, R. Prediction of NOx Emission Based on Data of LHD On-Board Monitoring System in a Deep Underground Mine. *Energies* **2023**, *16*, 2149. [CrossRef]
17. Garcia, J.J.; Gresh, R.E.; Gareis, M.B.; Haney, R.A. Effectiveness of cabs for dust and silica control on mobile mining equipment. In Proceedings of the 8th US Mine Ventilation Symposium, Rolla, MO, USA, 11–17 June 1999.
18. Haney, R.A.; Schultz, M.J.; Rude, R.L.; Tomko, D.M. Controls being used to reduce diesel particulate matter exposures in US underground metal and non-metal mines. In Proceedings of the Eighth International Mine Ventilation Congress, Brisbane, Australia, 6–8 July 2005.

19. Peters, S.; de Klerk, N.; Reid, A.; Fritschi, L.; Musk, A.B.; Vermeulen, R. Estimation of quantitative levels of diesel exhaust exposure and the health impact in the contemporary Australian mining industry. *Occup. Environ. Med.* **2017**, *74*, 282–289. [CrossRef]
20. Li, J.; Wu, L.; Chen, H. Analysis of thermal comfort and threshold range of airflow supply parameters for different types of work in humid-heat coal mines. *Case Stud. Therm. Eng.* **2023**, *44*, 102826. [CrossRef]
21. You, S.; Sun, J.; Ji, H.; Feng, Q. Analysis of Thermal Environment and Its Influencing Factors in Deep Stope of Metal Mine. *Geofluids* **2022**, *2022*, 6408714. [CrossRef]
22. Wei, D.; Du, C.; Lin, Y.; Chang, B.; Wang, Y. Thermal environment assessment of deep mine based on analytic hierarchy process and fuzzy comprehensive evaluation. *Case Stud. Therm. Eng.* **2020**, *19*, 100618. [CrossRef]
23. Wang, K.; Li, Q.; Wang, J.; Yang, S. Thermodynamic characteristics of deep space: Hot hazard control case study in 1010-m-deep mine. *Case Stud. Therm. Eng.* **2021**, *28*, 101656. [CrossRef]
24. Roy, S.; Mishra, D.P.; Bhattacharjee, R.M.; Agrawal, H. Genetic programming for prediction of heat stress hazard in underground coal mine environment. *Nat. Hazards* **2022**, *114*, 2527–2543. [CrossRef]
25. Bornman, W.; Dirker, J.; Arndt, D.C.; Meyer, J.P. Integrated energy simulation of a deep level mine cooling system through a combination of forward and first-principle models applied to system-side parameters. *Appl. Therm. Eng.* **2017**, *123*, 1166–1180. [CrossRef]
26. Slazak, N.; Obracaj, D.; Borowski, M. Methods for controlling temperature hazard in Polish coal mines. *Arch. Min. Sci.* **2008**, *53*, 497–510.
27. Soroko, K.; Zgrzebski, P.; Stach, R.; Gola, S. Improving underground mine climate conditions using ventilation dams to seal abandoned areas in Polish copper mines. *CIM J.* **2020**, *11*, 188–197. [CrossRef]
28. Xiao, Y.; Deng, H.; Xie, Z.; He, W. Application of Nanoporous Super Thermal Insulation Material in the Prevention and Control of Thermal Hazards in Deep Mining of Metal Mines. *J. Nanomater.* **2022**, *2022*, 2390616. [CrossRef]
29. Szlązak, N.; Obracaj, D.; Swolkień, J. Thermal Insulation of Excavations and Its Effect on Climate Conditions. *Energies* **2021**, *14*, 4170. [CrossRef]
30. Zawiślak, M. *Method of Fluid-Flow Machinery and Systems Design and Modernisation by Computational Fluid Dynamics Application*; Publishing House of the Poznan University of Technology: Poznan, Poland, 2017. (In Polish)
31. Xu, G.; Luxbacher, K.D.; Ragab, S.; Xu, J.; Ding, X. Computational fluid dynamics applied to mining engineering: A review. *Int. J. Min. Reclam. Environ.* **2017**, *31*, 251–275. [CrossRef]
32. Ren, T.; Balusu, R. The use of CFD modelling as a tool for solving mining health andsafety problems. In Proceedings of the 10th Underground Coal Operators' Conference, Wollongong, Australia, 11–12 February 2010.
33. Yi, H.; Kim, M.; Lee, D.; Park, J. Applications of Computational Fluid Dynamics for Mine Ventilation in Mineral Development. *Energies* **2022**, *15*, 8405. [CrossRef]
34. Brodny, J.; Tutak, M. Applying computational fluid dynamics in research on ventilation safety during underground hard coal mining: A systematic literature review. *Process Saf. Environ. Prot.* **2021**, *151*, 373–400. [CrossRef]
35. Peng, S.; Huang, Z.; Dong, D. Numerical Simulation Study On Fire Hazard Of A Coal Mine Transport Roadway. *Min. Sci.* **2022**, *29*, 33–52.
36. Nezarat, H.R.; Jalali, S.M.E.; Khosrotash, M.; Nazari, M. *Reduction of Dead Zones by Improving the Ventilation System Inside the TBM*; Technical Report; EasyChair: Manchester, UK, 2019.
37. Liu, Q.; Nie, W.; Hua, Y.; Jia, L.; Li, C.; Ma, H.; Wei, C.; Liu, C.; Zhou, W.; Peng, H. A study on the dust control effect of the dust extraction system in TBM construction tunnels based on CFD computer simulation technology. *Adv. Powder Technol.* **2019**, *30*, 2059–2075. [CrossRef]
38. Liu, C.; Bao, Q.; Nie, W. The influence of ventilation parameters on dust pollution in a tunnel's environment using the CFD method. *J. Wind Eng. Ind. Aerodyn.* **2022**, *230*, 105173. [CrossRef]
39. Xu, G.; Chang, P.; Mullins, B.; Zhou, F.; Hu, S. Numerical study of diesel particulate matter distribution in an underground mine isolated zone. *Powder Technol.* **2018**, *339*, 947–957. [CrossRef]
40. Chang, P. Comparison of Diesel Particulate Matter Simulation Models and Ventilation Optimisation by Using Computational Fluid Dynamics. Ph.D. Thesis, Curtin University, Bentley, Australia, 2019.
41. Chang, P.; Xu, G.; Zhou, F.; Mullins, B.; Abishek, S.; Chalmers, D. Minimizing DPM pollution in an underground mine by optimizing auxiliary ventilation systems using CFD. *Tunn. Undergr. Space Technol.* **2019**, *87*, 112–121. [CrossRef]
42. Wróblewski, A.; Wodecki, J.; Trybała, P.; Zimroz, R. A Method for Large Underground Structures Geometry Evaluation Based on Multivariate Parameterization and Multidimensional Analysis of Point Cloud Data. *Energies* **2022**, *15*, 6302. [CrossRef]
43. Janus, J. Air Flow Modelling on the Geometry Reflecting the Actual Shape of the Longwall Area and Goafs. *Arch. Min. Sci.* **2021**, *66*, 495–509.
44. Janus, J.; Krawczyk, J. Measurement and Simulation of Flow in a Section of a Mine Gallery. *Energies* **2021**, *14*, 4894. [CrossRef]
45. Aboosaidi, F.; Warfield, M.J.; Choudhury, D. Computational fluid dynamics applications in airplane cabin ventilation system design. *SAE Trans.* **1991**, *100*, 2082–2091.
46. Zhang, T.; Chen, Q.Y. Novel air distribution systems for commercial aircraft cabins. *Build. Environ.* **2007**, *42*, 1675–1684. [CrossRef]
47. Yan, W.; Zhang, Y.; Sun, Y.; Li, D. Experimental and CFD study of unsteady airborne pollutant transport within an aircraft cabin mock-up. *Build. Environ.* **2009**, *44*, 34–43. [CrossRef]

48. Li, M.; Yan, Y.; Zhao, B.; Tu, J.; Liu, J.; Li, F.; Wang, C. Assessment of turbulence models and air supply opening models for CFD modelling of airflow and gaseous contaminant distributions in aircraft cabins. *Indoor Built Environ.* **2018**, *27*, 606–621. [CrossRef]
49. Liu, W.; Mazumdar, S.; Zhang, Z.; Poussou, S.B.; Liu, J.; Lin, C.H.; Chen, Q. State-of-the-art methods for studying air distributions in commercial airliner cabins. *Build. Environ.* **2012**, *47*, 5–12. [CrossRef]
50. Zawiślak, M. Impact of vehicle interior geometry on chosen volatile carcinogens concentration distribution in vehicle cabin. *Proc. ECOpole* **2013**, *7*, 177–184.
51. Rodak, M.; Skrętowicz, M.; Janicka, A.; Zawiślak, M. Numerical analysis of volatile organic compounds concentration in a C segment vehicle interior. *Combust. Engines* **2015**, *54*, 1042–1045. [CrossRef]
52. Dudycz, A.; Górniak, A.; Janicka, A.; Rodak, M.; Skrętowicz, M.; Trzmiel, K.; Włostowski, R.; Woźniak, J.; Zawiślak, M. Numerical analysis of volatile organic compounds concentration in AC segmnet vehicle interior–diesel engine exhoust pollution. *J. Pol. CIMEEAC* **2015**, *10*, 41–48.
53. Zawiślak, M. Application of computational fluid dynamics in the assessment of the spread of toxic compounds emitted by a diesel engine inside a vehicle. *J. Pol. CIMAC* **2014**, *9*, 197–205.
54. Hadi, J.M.; Alturaihi, M.H.; Jasim, N.Y.; Habeeb, L.J. Numerical study of airflow and temperature variations inside car at different solar intensity angles. *Mater. Today Proc.* **2022**, *60*, 1689–1695. [CrossRef]
55. Mao, Y.; Wang, J.; Li, J. Experimental and numerical study of air flow and temperature variations in an electric vehicle cabin during cooling and heating. *Appl. Therm. Eng.* **2018**, *137*, 356–367. [CrossRef]
56. Basciotti, D.; Dvorak, D.; Gellai, I. A novel methodology for evaluating the impact of energy efficiency measures on the cabin thermal comfort of electric vehicles. *Energies* **2020**, *13*, 3872. [CrossRef]
57. Singh, S.; Abbassi, H. 1D/3D transient HVAC thermal modeling of an off-highway machinery cabin using CFD-ANN hybrid method. *Appl. Therm. Eng.* **2018**, *135*, 406–417. [CrossRef]
58. Oh, J.; Choi, K.; Son, G.h.; Park, Y.J.; Kang, Y.S.; Kim, Y.J. Flow analysis inside tractor cabin for determining air conditioner vent location. *Comput. Electron. Agric.* **2020**, *169*, 105199. [CrossRef]
59. Akdemir, S.; Öztürk, S.; Ülger, P. CFD modelling of ambient factors in a tractor cabin for summer conditions. *J. Tekirdag Agric. Fac.* **2016**, *13*, 46–54.
60. Akdemir, S.; Öztürk, S.; Ülger, P. CFD modelling of air velocity in a tractor cabin for winter conditions. In Proceedings of the VII International Scientific Agriculture Symposium, "Agrosym 2016", Jahorina, Bosnia and Herzegovina, 6–9 October 2016; University of East Sarajevo, Faculty of Agriculture: Sarajevo, Bosnia and Herzegovina, 2016; pp. 1829–1835.
61. Hwang, J.H.; Lee, J.S.; Kim, Y.H.; Kim, P.Y. Evaluation Method of Thermal Comfort in Excavator Cabin. In Proceedings of the ASME International Mechanical Engineering Congress and Exposition, San Diego, CA, USA, 15–21 November 2013; Volume 56345, p. V08AT09A031.
62. Ghorpade, U.S.; Patil, M.A.; Shelake, A.S.; Ghodake, H.S. CFD Analysis of Manipulator Cabin by Selecting Proper Air Conditioning System. *J. Eng. Res. Appl.* **2017**, *7*, 116–120. [CrossRef]
63. Cimbala, J.M.; Cengel, Y.A. *Fluid Mechanics: Fundamentals and Applications*; McGraw-Hill Higher Education: New York, NY, USA, 2006.
64. ANSYS, Inc. *ANSYS FLUENT Theory Guide*; ANSYS, Inc.: Canonsburg, PA, USA, 2013; pp. 90311–90312.
65. Shaheed, R.; Mohammadian, A.; Kheirkhah Gildeh, H. A comparison of standard k–ε and realizable k–ε turbulence models in curved and confluent channels. *Environ. Fluid Mech.* **2019**, *19*, 543–568. [CrossRef]
66. KGHM ZANAM. Stationary Rock Breaker Type URB/K. Available online: https://www.kghmzanam.com/wp-content/uploads/2020/06/URB_K_PL.pdf (accessed on 5 December 2022).
67. Schneider, L.W. *Development of Anthropometrically Based Design Specifications for an Advanced Adult Anthropomorphic Dummy Family*; Technical Report; Final Report; Regents of the University of Michigan: Ann Arbor, MI, USA, 1983; Volume 1.
68. Zore, K.; Sasanapuri, B.; Parkhi, G.; Varghese, A. Ansys mosaic poly-hexcore mesh for high-lift aircraft configuration. In Proceedings of the 21st AeSI Annual CFD Symposium, Bangalore, India, 8–9 August 2019.
69. Fanger, P.O. Calculation of thermal comfort: Introduction of a basic comfort equation. *ASHRAE Trans. Part II* **1967**, *73*, III4.1–III4.20.
70. Fanger, P.O. *Thermal Comfort. Analysis and Applications in Environmental Engineering*; Danish Technical Press: Copenhagen, Denmark, 1970; 244p.
71. *ISO7730*; Ergonomics of the Thermal Environment—Analytical Determination and Interpretation of Thermal Comfort Using Calculation of the PMV and PPD Indices and Local Thermal Comfort Criteria. International Organization for Standardization: Geneva, Switzerland, 2006.

energies

Article

# Scale Effect on Hydraulic Properties of Pore-Fissure Deep Rock Formations and Its Importance for the Mining Shaft-Sinking Process

**Maciej Pikuła [1], Krzysztof Chudy [2], Magdalena Worsa-Kozak [2,*] and Mariusz Czop [3]**

1 KGHM Cuprum Research and Development Centre, 53-659 Wrocław, Poland
2 Faculty of Geoengineering, Mining and Geology, Wrocław University of Science and Technology, 50-421 Wrocław, Poland
3 Faculty of Geology, Geophysics and Environmental Protection, AGH University of Science and Technology, 30-059 Kraków, Poland
* Correspondence: magdalena.worsa-kozak@pwr.edu.pl

**Abstract:** The problem of hydraulic parameters estimation lies in the depth: the deeper the rock formation, the more expensive and difficult the field tests and samples acquisition, and the more challenging the technical issues. The article assesses the Triassic sandstone's drainage potential at the stage of shaft sinking. It focuses on parameter analysis in varied scales, from drill-core sample laboratory testing, through a single well drawing test, to long-term pumping and recovery tests in the well with observation piezometers. The obtained results are compared to the values estimated in the past using different methods. Finally, the paper states whether it is reliable to forecast pore-fissure sandstone drainage potential based only on core samples' laboratory tests. This research proved that lab tests underestimate pore-fissure rocks' hydraulic parameters (mean hydraulic conductivity $k = 9.79 \times 10^{-8}$ m/s) tenfold more than long-term pumping tests (mean $k = 4.45 \times 10^{-7}$ m/s). However, it can be concluded that the group of so-called "witness samples", 10% of all core samples with a top value of the hydraulic conductivity tested in the laboratory, can be representative of the aquifer and comparable to the values obtained in pumping tests. With this in mind, we recommend using the highest values of hydrogeological parameters from laboratory tests based on the worst-case scenario. Therefore, it is possible to forecast inflows to the shafts reliably. This methodology is recommended only for rocks of porous and pore-fissure character.

**Keywords:** open porosity; specific yield; hydraulic conductivity; pumping test; Triassic sandstone

check for updates

**Citation:** Pikuła, M.; Chudy, K.; Worsa-Kozak, M.; Czop, M. Scale Effect on Hydraulic Properties of Pore-Fissure Deep Rock Formations and Its Importance for the Mining Shaft-Sinking Process. *Energies* **2023**, *16*, 2263. https://doi.org/10.3390/en16052263

Academic Editors: Sergey Zhironkin and Dawid Szurgacz

Received: 9 December 2022
Revised: 8 February 2023
Accepted: 23 February 2023
Published: 27 February 2023



## 1. Introduction

The reliable hydraulic properties of pore-fissure rocks are crucial to the accurate prognosis of groundwater inflow to the sinking shafts. Moreover, they determine the safety of mining operations and water management strategies. For example, a known unpredicted catastrophic influx to the sinking shaft is the R-XI shaft accessing Poland's "Rudna" copper ore deposit [1]. Groundwater intrusion in 2002/2003 stopped the sinking process at a depth of 600–635 m for nearly a year and showed that more attention must be paid to hydrogeology when accessing deep deposits. Copper ore deposits in Poland's Fore-Sudetic Monocline (FSM) have been exploited for more than 60 years. As a result of many years of mining, the resource base decreases, making it necessary to mine in deeper and less accessible parts of the copper ore body. However, access to mineral resources at greater depths is performed within complicated geological and hydrogeological conditions, causing health and safety challenges. Appropriate solutions for the ore body-opening technology should be based on reliable geological and hydrogeological research results. The best method of assessing these conditions should rely on a factual assessment of the previous exploration methods. Combined with an analysis of the predicted and measured data, weaknesses, and strengths of particular practices have been indicated. Hydrogeological research preceding the design and construction of a shaft, if performed on a limited scale or using low accuracy and reliability methods, provides incomplete and unreliable data [2]. Such imprecise recognition of hydrogeological conditions and geological structure leads to inaccurately estimated inflows, as was proved during the long-term exploitation in FSM (Figure 1). As a result, several critical influxes occured during shafts' sinking and disrupted or even stopped the process [1,3–5].

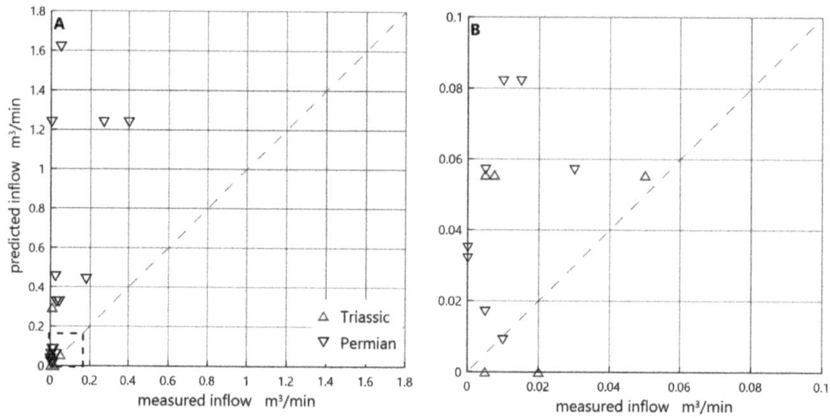


**Figure 1.** Comparison of predicted and measured inflows into the sinking shaft R-XI; (**A**) complete data, (**B**) data in the range up to 0.1 $m^3/min$ (black square in (**A**); [2] modified).

The problem of adequately identifying hydrogeological conditions in the copper mining area in FSM began to be described in the 1970s and 1980s [3,6,7] and remains valid. It became more important with the development of numerical modeling methods and their implementation to prognoses of water inflow to underground mine workings [2,5,8,9]. However, the software development was not followed by the explicit recognition of rock mass filtration parameters. As a result, it caused problems in properly defining boundary conditions or hydrogeological schematization [10] and adequate integration of laboratory and field measurements [11,12]. Similar challenges were noticed in hydrogeological modeling in mining areas worldwide. For example, one of the first review papers on data reliability and quality and the investigation methodology for determining the hydraulic parameters necessary for the initial inflow to shafts estimations for modeling purposes highlighted the importance of field tests in drill holes [13]. Likewise, the application of groundwater flow modeling for mining purposes and data quantity and quality challenges were widely described based on examples from the Czech Republic [14]. However, predictions of groundwater states and flows based on numerical modeling are challenging in mining and every application. Accurate recognition of hydrogeological parameters of rock mass for shaft-sinking purposes is also crucial for proper design of the sinking technology [15,16]. Thus, special attention must be paid to data collection, which was presented in detail based on the case study of the Death Valley regional groundwater flow system [17]. Cases from Poland and other countries mentioned above show that more attention must be paid to the hydrogeological properties of individual water-bearing horizons. For deep multilayered aquifers, it is a considerable challenge mining engineering, hydraulic engineering, and hydrogeology must face minimizing the risk of disaster, specifically at the stage of shafts sinking.

The accuracy and reliability of estimated hydrogeological parameters of rocks and soils vary with the scale of the investigation [18–24]. In most cases, the lowest values are obtained on the test in the smallest point scale, such as laboratory tests of soil or core samples [18,21,23]. In contrast, the largest values that characterize rock mass on a local or regional scale are derived by the filed tests (single- or cross-borehole pumping tests) [19–21]. The novelty presented by our work is comparing different methods of hydraulic properties' estimation and an indication of the most reliable methodology of their evaluating for shaft sinking in the pore-fissure rocks.

Our study provides: (1) reliable and unique data on Triassic pore-fissure sandstone parameters as a referential for numerical modeling, (2) best practice guidance on the usefulness of the results obtained in different scales, (3) hydrogeological parameters of the Middle and Lower Bunter sandstone in FSM are the objectives of this study. It has been done based on field and laboratory tests presented in this article and allowed for comprehensive hydrogeological characteristics of sandstone formations, classified as the Middle and Lower Bunter sandstone, together with an assessment of their drainage potential at the stage of shaft construction. Moreover, we ask whether it is reliable to forecast pore-fissure sandstone drainage potential based only on core samples' laboratory tests. We answer this question by analyzing porosity, specific yield, storage coefficient, and hydraulic conductivity, examined in varied scales from drill-core samples' laboratory testing and a single well-drawing test to long-term pumping and recovery tests in the cross-borehole. The novel approach was implemented based on the methodology for estimating representative values of hydrogeological parameters for pore-fissure rocks, taking into account the often overlooked laboratory tests for samples derived from drill cores. Finally, the obtained results are compared to the values estimated in the past using different methods and matched with real inflows.

## 2. Materials and Methods

### 2.1. Study Area

The study area is situated in southwestern Poland on the FSM, between Gawronki and Gawrony (Figure 2) in the north-eastern part of the Retków copper ore deposit, which is a prospective part deposit designated for future exploitation.

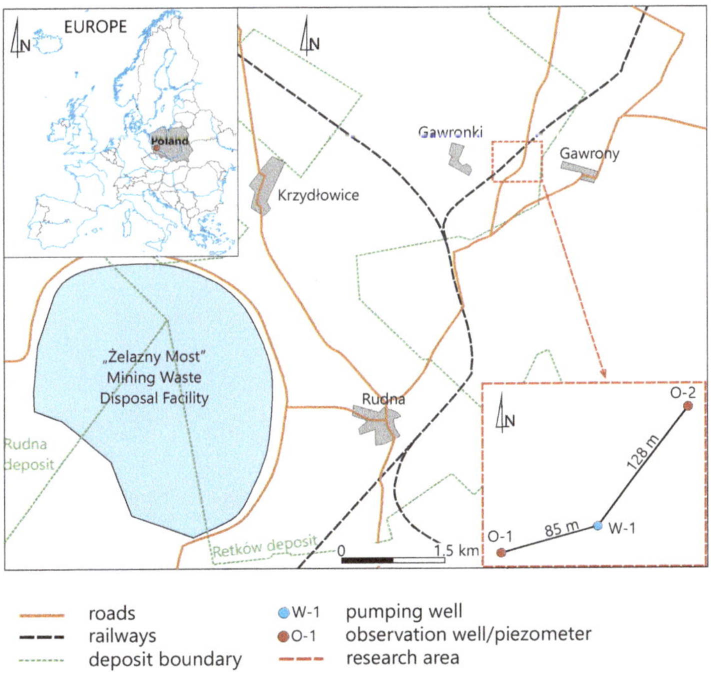


**Figure 2.** Location of the research area within a copper ore deposit.

The bedrock of the FSM is a complex of crystalline rocks formed in the Proterozoic and older Palaeozoic periods (Figure 3) [25]. The Permian formations are represented by the Rotliegendes sediments of red-colored conglomerate, shale, sandstone, and higher volcanic rocks (rhyolite, rhyolitic tuff) [26]. They are overlaid by brownish-red sandstone, which changes into grey and white in the upper part. Zechstein sediments are copper-bearing shale, limestone, dolomite, anhydrite, rock salt, and clay shale [27], and overlay the Rotliegendes part. The Zechstein is divided into four cyclothems: Aller (Z4), Leine (Z3), Stassfurt (Z2), and Werra (Z1) (Figure 3). Triassic deposits, represented by the lower and middle Triassic, lie conformably on the Zechstein formations. In the L=lower Triassic, the Bunter sandstone is represented by sandstone, siltstone, and shale with limestone interbeds in the lower and middle parts. The upper part of the lower Triassic (Röt Formation) is characterized by marl, shale, siltstone, dolomite, limestone, anhydrite, and gypsum with interbeds of marl. The Middle Triassic, represented by Muschelkalk, comprises limestone, dolomite, marl, anhydrite, and gypsum. The Cenozoic formations are characterized by Paleogene, Neogene, and Quaternary formations, which lie unconformably on Triassic deposits. The Palaeogene–Neogene sediments are mainly quartz sand, glauconitic sand, and clay with an interbedding of mud or sand. Quaternary formations consist of sand, gravel, clay, and silt of the Southern and Central Poland glaciations. The Quaternary is characterized by high lithological variability in vertical and horizontal profiles.

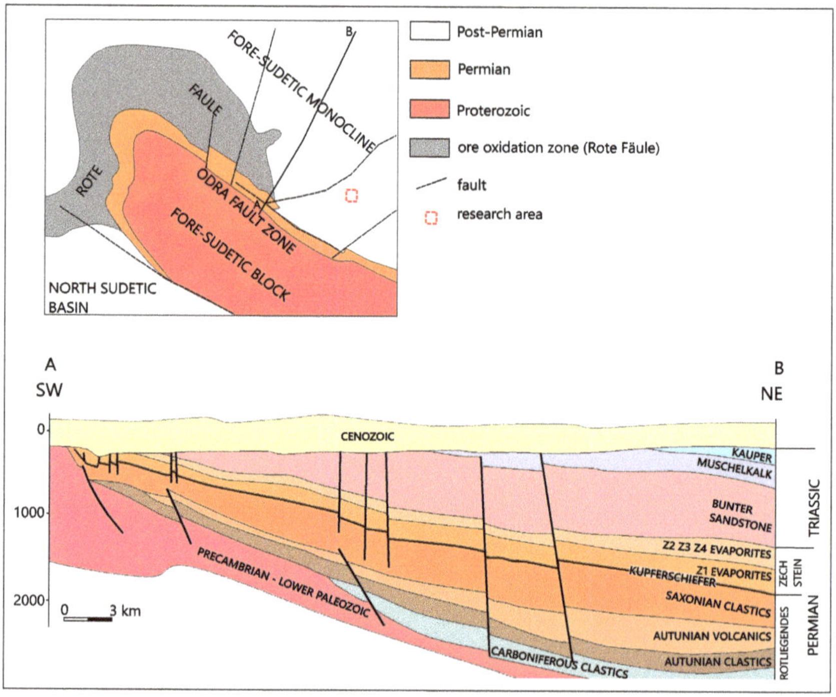


**Figure 3.** Location and geologic setting of the research area. Geologic cross-section along line A–B in upper figure [28].

The hydrogeologic profile of the study area includes four aquifers: Quaternary, Neogene–Paleogene, Triassic, and Permian (Figure 4) [5]. The Quaternary aquifer covers sandy and sandy-gravel water-bearing layers of the Holocene and Pleistocene, which are usually separated from lower layers by Pliocene clays of the Poznan series.

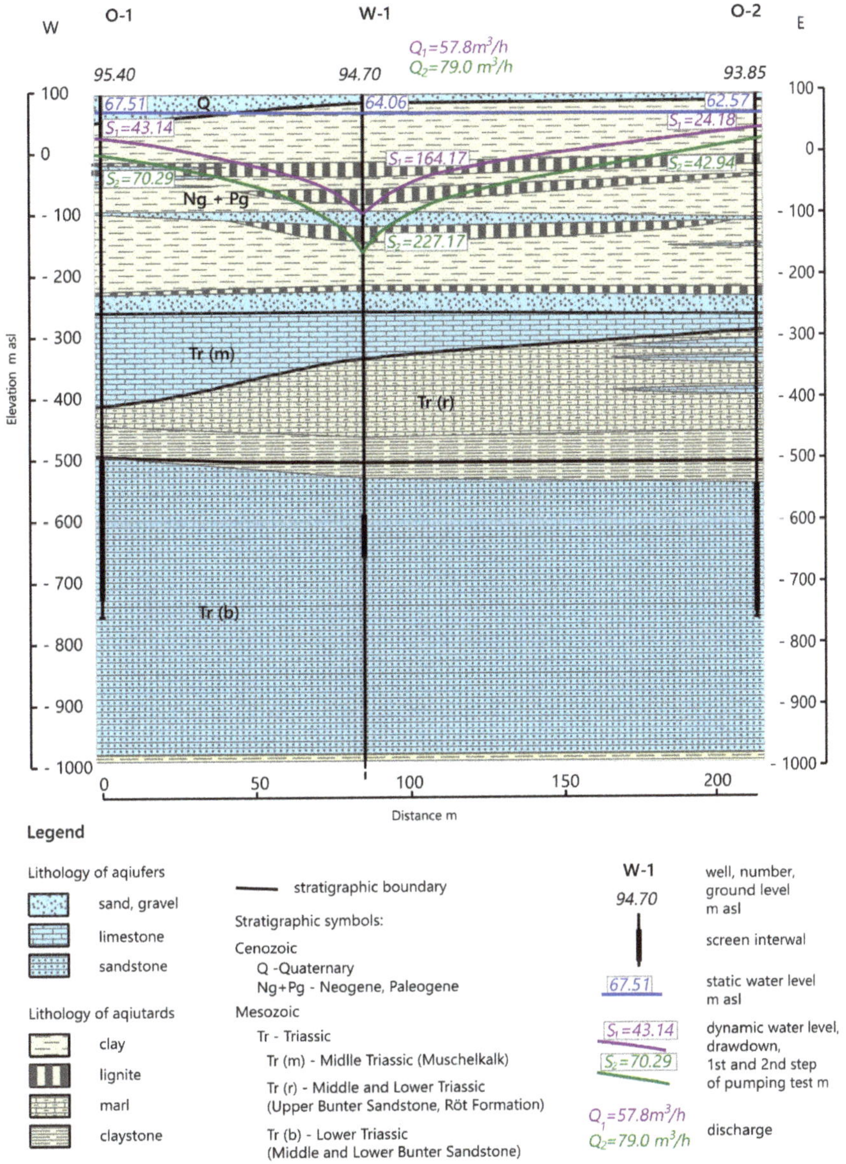


**Figure 4.** Scheme of hydrogeological conditions of the pumping test area.

The Neogene–Paleogene water-bearing horizon includes water in sandy and sandy-gravel layers within isolating layers: clay, silt, and brown coal. Three water-bearing horizons can be distinguished within this horizon: over-coal (Pliocene and upper Miocene), inter-coal (middle and lower Miocene), and under-coal (Oligocene). These three levels of the Neogene–Pliocene aquifer may have hydraulic connectivity locally. The subject of the presented research, the Triassic aquifer, is distinguished by an aquifer of shell limestone, upper Bunter sandstone (Röt Formation), and middle and lower Bunter sandstone. The middle and lower Bunter sandstone aquifer's water occurs in the arkose sandstone, more rarely in quartz sandstone. This sandstone is interlayered with insets of poorly permeable and impermeable sediments—shale and gypsum. The studied sandstone is characterized by diversified grain sizes—from fine- to coarse-grained. Water-bearing formations of the discussed zone occur in most of the FSM area. Sub-Cenozoic outcrops of this aquifer in the area of the monocline spread in a wide belt, several kilometers wide, located to the northeast of outcrops of older, lower Zechstein formations (cyclothem Z4). Paleogene sediments are present above the sub-Cenozoic outcrops of this aquifer in unconsolidated sandy formations and isolating layers of silt and clay. To the north and north-west of the sub-Cenozoic outcrops of the middle and lower sharp sandstone, the aquifer is covered by isolating sediments (mudstones) overlain by carbonate water-bearing sediments of the upper Bunter sandstone and Muschelkalk. The middle and Bunter sandstone sediments are underlaid by separating rocks of Zechstein's siltstone and anhydrite. In the copper-bearing area, the thickness of the Middle and Lower Bunter sandstone completely disappears in the southwestern region of its sub-Cenozoic outcrops. In the northern areas of the documented copper ore deposits, the maximum thickness reaches approx. 600 m. In the study area, the thickness of sediments of this aquifer ranges about 450 m. The hydraulic conductivity of the discussed aquifer in the copper-bearing area of the FSM ranges between $10^{-9}$ m/s ÷ $10^{-6}$ m/s [29]. The Permian aquifer comprises the Zechstein and Rotliegendes aquifers and includes sedimentary and relict water. It is built of dolomite, limestone, and gray and red sandstone separated by impermeable layers.

### 2.2. *Methods and Calculations*

#### 2.2.1. Field Test

Field hydrogeologic investigations in the middle and lower Bunter Sandstone aquifer were based on pumping tests in the well (W-1) with two observation wells (O-1, O-2; Figure 2). The pumping test ran between 08.07 and 02.09.2016 at a depth from 684.7 m to 755.0 m. It covered purge pumping and measurement pumping. Each was completed by recovery of the water level. Purge pumping was conducted to remove the drilling mud and decontaminate the near-borehole zone [30,31]. A submersible electronic hydrostatic pressure sensor—APLISENS SG-25, and a backup Solinst Levelogger Edge water level sensor were used for constant, automated observation of the water level in the wells. Additionally, control manual measurements were taken with the SEBA Electric Contact Meter Type KLL. The pumping test flow rate was measured using an electromagnetic flow sensor Siemens MAG 3100. The pumping test was performed as a two-drawdown step procedure (Figure 5) [30]. The first drawdown step was carried out with the pumping rate of 60.0 $m^3/h$ for 9 days and 23 h, and the second was carried out with the rate of 80.0 $m^3/h$ for 40 days and 1 h 30 min. After the pumping test was completed, observations of water table recovery in the boreholes were conducted for 74 days in the pumping borehole and 56 days in the piezometers. Interpreting the pumping test with the determination of the hydraulic conductivity was carried out using methods for non-steady flow. The hydraulic conductivity and storage coefficient was determined based on measurements from a pumping borehole (W-1) and two piezometers (O-1 and O-2) using the AquiferTest software. For the W-1 borehole, both for the period of pumping and stabilization of the water table, methods were applied which consider the influence of so-called "well effects" on the obtained measurement results. For measurements from pumping, "Agarwal's method" was used, and for measurements from stabilization, "Agarwal's solution" was applied, taking into account the assumptions of the above method [32–34]. Measurement data from piezometers O-1 and O-2 were interpreted using the Theis method, and for data from the stabilization period, the Theis method, Agarwal's solution [32,34,35]. Data from the complete pumping and stabilization period were included in the interpretation. Only the initial 10 min of the 2nd pumping step were omitted because of the significant stepwise variability in the magnitude of the pumping rate. For the 1st and 2nd pumping test steps, the variability in pumping rates during the tests was included in the calculations.

For comparison of parameters estimated with the software storage coefficient was also calculated using the formula [36]:

$$S = 3 \times 10^{-6} b \quad (1)$$

where $b$ is saturated aquifer thickness in meters.

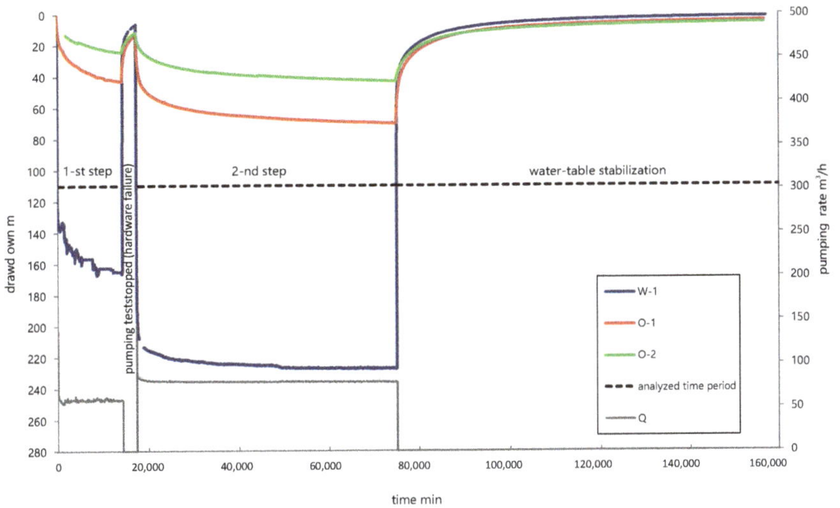


**Figure 5.** Pumping test progress.

The rock quality designation (RQD) index, developed by Deere et al. (1967) [37], was also implemented to analyze hydrogeological parameters. RQD is defined as the borehole core recovery percentage or ratio incorporating only pieces of a solid core longer than 100 mm in length measured along the core's centerline of the core [38]. It is distinguished for selected structural domains or specific sizes of core [39]. It was estimated in the field while the core was logged during the drilling operations.

### 2.2.2. Laboratory Test

Laboratory studies were also significant in determining aquifer parameters. After the drilling of boreholes, rock samples were taken from drill cores to test hydrogeological parameters in terms of open porosity ($p_o$), specific yield ($S_y$), and hydraulic conductivity ($k$). The hydrogeological parameters of sandstone were studied on 36 samples at the AGH University of Science and Technology in Krakow. Samples were taken from those sections of the core that allowed a smaller diameter sample to be cut. The samples ready to test were cylindrical (diameter 50 mm; length 55 mm, (Figure 6A)). First, the open porosity was determined. The next step was the determination of specific yield using a high-speed centrifuge. Finally, hydraulic conductivity was determined.

Open porosity ($p_o$) (interconnected porosity) is one of the elementary microstructural characteristics of rocks. It determines the proportion of interconnected pores regardless of their size in the volume of the rock sample. Tests were performed based on the method described in the literature and saturating samples in water, called the Archimedes' method [10,40–43]. The value of the open porosity was calculated from the formula [43,44]:

$$p_o = (G_a - G_d)/(G_a - G_w) \times 100\% \quad (2)$$

where $G_a$ is the weight of the sample saturated with water (24 h) and weighed in air, $G_d$ is the weight of the sample dried at 110 °C for 24 h, $G_w$ is the weight of the sample saturated with water (24 h) and weighed in water.

To determine the ability of rocks to drain free water under the effect of gravity, we estimated the specific yield ($S_y$), which defines the volume of water that can drain away from a unit volume of rock [36,45,46]. The method used to determine the rock's $S_y$ is based on a laboratory centrifuge (Figure 6B) [47,48].

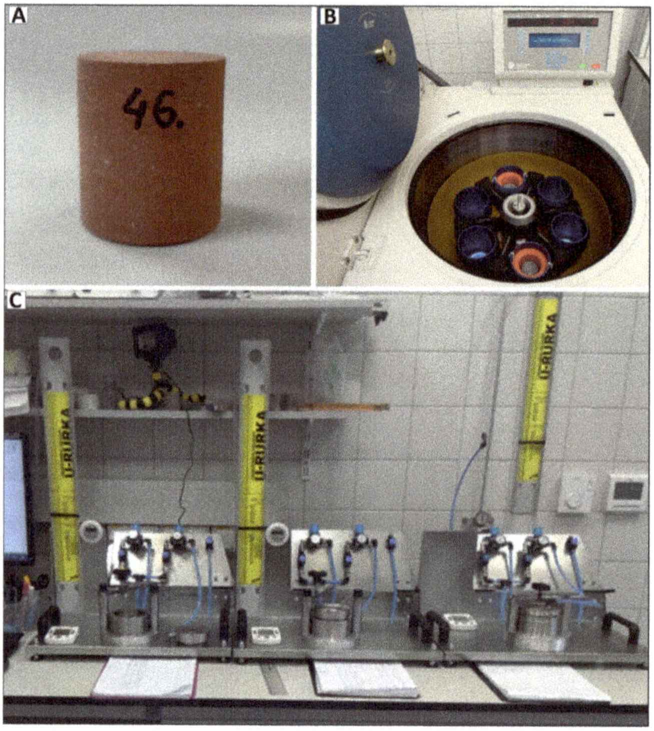


**Figure 6.** Core prepared for parameter determination (**A**), samples placed in a high-speed centrifuge (**B**), test equipment for measurement of $k$ (**C**).

The centrifuge's speed is adjusted to the height of the sample to simulate a negative pressure of 10 m of the water column, which is assumed to be the maximum value occurring in nature and simulates natural drainage lasting 5–20 years. This value is used in international research [10,42–44,49]. The $S_y$ was calculated from the formula:

$$S_y = V_w / V_r \tag{3}$$

where $V_w$ is the volume of drained water released by a suction pressure equivalent to a water column 10 m high (cm$^3$), $V_r$ is the volume of the sample/rock (cm$^3$).

All the samples were centrifuged for 30 min, equivalent to the percolation time under natural conditions from 2 to 2.5 years. This interval was calculated according to Prill et al. [48] as the relationship between percolation time ($T_n$) of gravitational water in nature and centrifugation time ($t$):

$$(T_n / t) = (a / g)^2 \tag{4}$$

where: $a$—centrifugal acceleration, $g$—gravity.

Hydraulic conductivity ($k$) is one of the most important parameters used in hydrogeology. It is determined from the measured intrinsic permeability ($k_p$). Intrinsic permeability was determined in Dulinski's [50] apparatus, the gas permeameter (Figure 6C). This works on forcing the flow of compressed gas (liquid) through the dried sample. The absolute pressure "before" and "after" the sample and the amount of gas flow is measured. Formula (5) [42,43] is used for calculations:

$$k_p = 2Qpl\eta / F(p_1^2 - p_2^2) \quad (5)$$

where: $k_p$ is intrinsic permeability (Darcy), $Q$ is the volume of flowing gas (cm$^3$/s), $p$ is atmospheric pressure (at), $l$ is sample length (cm), $\eta$ is dynamic gas viscosity coefficient (cP), $F$ is sample cross-sectional area (cm$^2$), $p_1$ is the pressure of gas before sample (at), $p_2$ is the pressure of gas behind sample (at).

The Formula (6) describes the relationship between intrinsic permeability and hydraulic conductivity [43]:

$$k = k_p(\gamma_w / \eta_w) \quad (6)$$

where: $k$—hydraulic conductivity (cm/s), $k_p$—intrinsic permeability (Darcy), $\gamma_w$—water specific gravity (g/cm$^3$), $\eta_w$—water dynamic viscosity coefficient (cP).

From (6) we find that (at temp = 10 °C):

$$k = 7.66 \times 10^{-6} k_p \quad (7)$$

## 3. Results

### 3.1. Field Test

The value of hydraulic conductivity $k$ determined on measurements from the first step of test pumping for the W-1 borehole was $3.31 \times 10^{-7}$ m/s. For the observation wells, it was $3.71 \times 10^{-7}$ m/s (O-1) and $5.80 \times 10^{-7}$ m/s (O-2). The arithmetic mean of the whole system (pumping well with observation wells) was $4.27 \times 10^{-7}$ m/s.

Based on measurement data obtained in the second step of the pumping test, the calculated value of $k$ for the test borehole (W-1) was $3.82 \times 10^{-7}$ m/s, and for the observation wells—$3.92 \times 10^{-7}$ m/s (O-1) and $5.62 \times 10^{-7}$ m/s (O-2). As a result, the arithmetic mean value of the $k$ equal $4.45 \times 10^{-7}$ m/s was calculated for the whole system (pumping well with observation wells) (Figures 7–10).

Measurements made during water table recovery in a pumping well (W-1; (Figure 11) determined $k$ of $3.88 \times 10^{-7}$ m/s and in observation wells—$3.71 \times 10^{-7}$ m/s (O-1) and $4.58 \times 10^{-7}$ m/s (O-2) (Figure 12), on average for the whole system (pumping well with observation wells)—$4.06 \times 10^{-7}$ m/s.

Matching calculated depression with measured depression (Figure 13), as well as matching statistics (Table 1), indicate that the above parameters calculations are reliable.

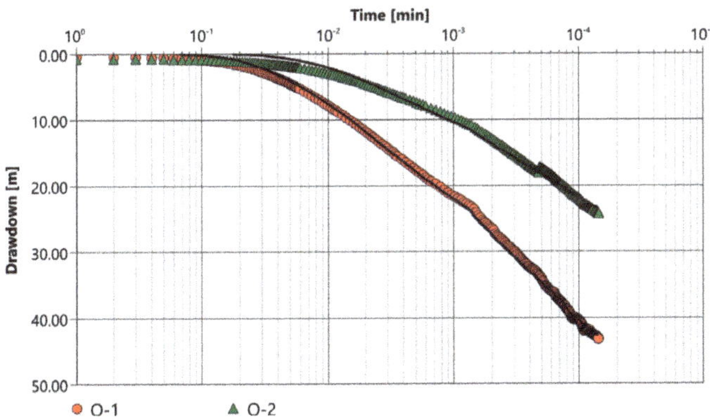


**Figure 7.** Analytical diagram for measurements in O-1 and O-2 during the first step of the pumping test (Theis method).

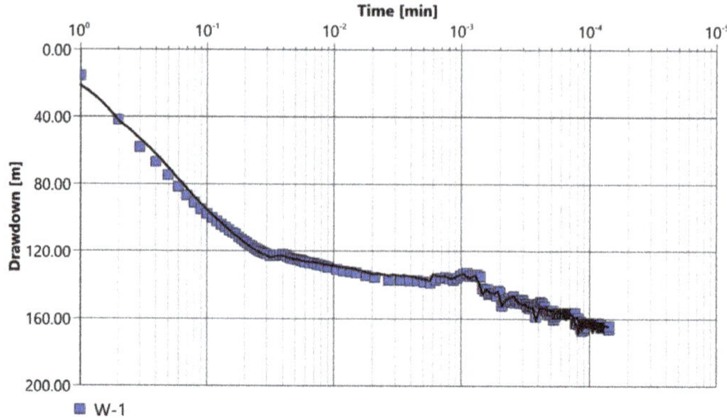


**Figure 8.** Analytical diagram for measurements in W-1 well during the first step of the pumping test (Agarwal's method).

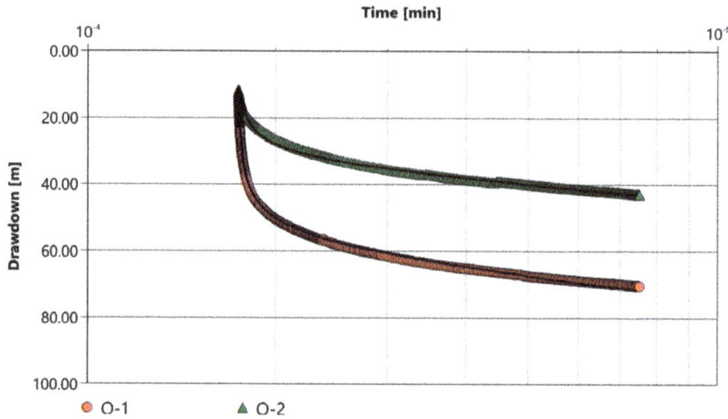


**Figure 9.** Analytical diagram for measurements in O-1 and O-2 during the second steps of the pumping test (Theis method).

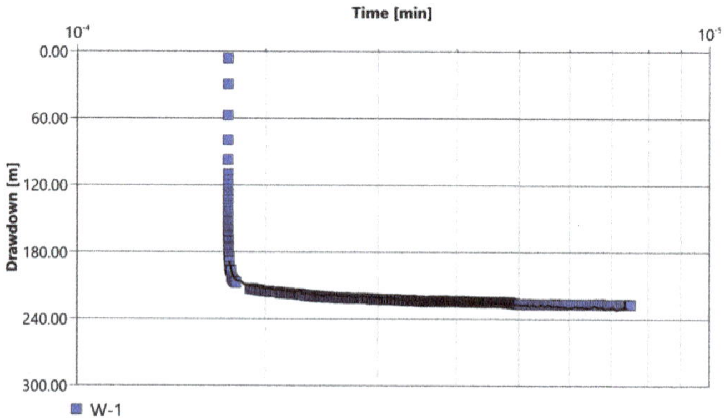


**Figure 10.** Analytical diagram for measurements in W-1 well during the second step of the pumping test (Agarwal's method).

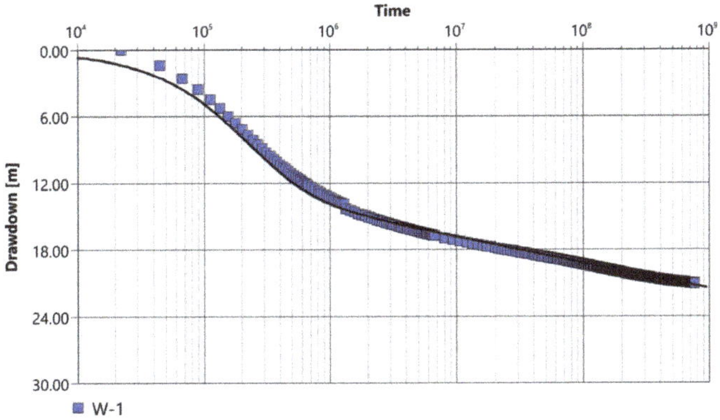


**Figure 11.** Analytical diagram for measurements during recovery in the W-1 well (Agarwal's method considering well effects + Agarwal's solution).

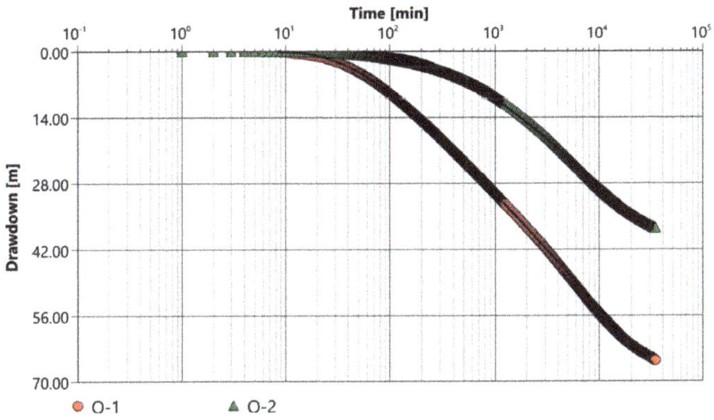


**Figure 12.** Analytical diagram for measurements during recovery in observation wells O-1 and O-2 (Theis method + Agarwal's solution).

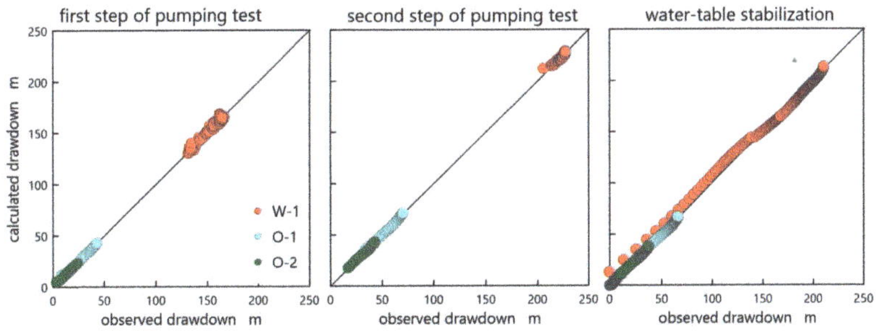


**Figure 13.** The plot of measured borehole depression vs. value calculated by the software

The results obtained from the pumping test classify the rocks into semi-permeable [51] or low-permeable rocks [52]. The water storage coefficient *S* value was also determined based on the analyses performed. Averaged values of this parameter for observation wells O-1 and O-2 of the first and second step of the pumping test are equal to $2.27 \times 10^{-4}$ and $2.18 \times 10^{-4}$, respectively, and the average value for measurements during recovery is $4.02 \times 10^{-4}$. A similar result, $4.49 \times 10^{-4}$, was obtained from Formula (1) calculations by which the *S*-value is estimated from the thickness of the aquifer.

**Table 1.** Matching statistics between measurements taken during the pumping test and model values calculated with the software.

| Pumping Test | Well No | Mean Δs [m] | Sum of Squared Errors [m$^2$] | Variance [m$^2$] | Standard Deviation [m] |
|---|---|---|---|---|---|
| 1st step | W-1 | 0.014 | 799.087 | 3.386 | 1.84 |
| 1st step | O-1 | 0.054 | 59.743 | 0.234 | 0.484 |
| 1st step | O-2 | 0.108 | 42.545 | 0.212 | 0.460 |
| 2nd step | W-1 | 0.003 | 795.837 | 0.846 | 0.920 |
| 2nd step | O-1 | 0.007 | 173.542 | 0.496 | 0.704 |
| 2nd step | O-2 | 0.014 | 81.773 | 0.224 | 0.473 |
| Recovery after 2nd step | W-1 | −0.097 | 4578.936 | 3.134 | 1.770 |
| Recovery after 2nd step | O-1 | −0.003 | 1367.277 | 0.522 | 0.722 |
| Recovery after 2nd step | O-2 | 0.074 | 1038.755 | 0.396 | 0.630 |

### 3.2. Laboratory Test

#### 3.2.1. Characteristics of the Rock Samples

Macroscopic analysis of the rock samples identified three groups of sandstone: very fine-grained, fine-grained, and medium-grained.

The very fine-grained sandstone samples (seven samples) occur in the depth interval of 604–687 m in single layers of small thickness. These are samples of red or brownish-red quartz sandstone with gray layers added. They are layered flat-parallel or have wavy lamination. The mineral composition is dominated by transparent and semitransparent quartz grains, mainly pink, of spherical or ellipsoidal shape. The shape of the grains is semi-sharp to rounded. The lithic components are light and dark micas. Ferrosilicate binder predominates.

Fine-grained sandstone was the most popular in the analyzed samples and occurred in the complete profile (24 samples) as red or brownish-red quartz sandstone. In the depth interval of 640.0–644.0 m and 697.0–730.0 m, light gray and light green color dominates. Flat-parallel lamination up to 1mm thick predominates. Quartz dominates the mineral composition with semitransparent, pink, or grey grains. Grains are spherical and semi-rounded to rounded. The lithic constituents are light and dark micas. Silica binder predominates.

Medium-grained sandstone represents the minor group of four analyzed samples. They occur locally below 700.0 m and comprise red-brown to reddish quartz sandstone. Flat-parallel, continuous lamination dominates here, and laminae are light red and brownish red. The main mineral component is transparent and semitransparent Quartz, locally pink or yellow. Grains are semi-rounded and shaped spherical or ellipsoidal. The lithic components are light and dark micas and clay minerals. The sandstone is brittle and porous. Iron–silica or clay–silica binder dominates.

#### 3.2.2. Interconnected Porosity ($p_o$)

The studied sandstone's open porosity ($p_o$) is relatively high. Values are between 0.0098 (Table 2) and 0.1894, with a mean of 0.0918 and a standard deviation of 0.0105. Very fine-grained and fine-grained sandstone showed similar $p_o$ values. However, the highest $p_o$ values were recorded for medium-grained sandstone (mean 0.1308).

The distribution of $p_o$ is not homogeneous. Two subgroups with a similar parameter distribution can be distinguished but shifted to each other (Figure 14). In the first subgroup, we have fine-grained sandstone with low $p_o$ (up to about 0.06). Sandstone with much higher $p_o$ values (0.12–0.19) is placed in the second subgroup. Lower values belong to fine-grained sandstone ($p_o < 0.15$), and higher $p_o > 0.15$ to very fine-grained and medium-grained sandstone.

**Table 2.** Statistics of the measurement date set.

| Rock Type | Parameter | Very Fine-Grained Sandstone | Fine-Grained Sandstone | Medium-Grained Sandstone | Sandstone in Total |
|---|---|---|---|---|---|
| $p_o$ | Number of samples | 6 | 24 | 4 | 34 |
| | Min. | 0.0098 | 0.0136 | 0.0614 | 0.0098 |
| | Max. | 0.1893 | 0.1829 | 0.1820 | 0.1894 |
| | Arithmetic Mean | 0.0807 | 0.0881 | 0.1308 | 0.0918 |
| | Geometric Mean | 0.0514 | 0.0219 | 0.1211 | 0.0683 |
| | Median | 0.0401 | 0.0583 | 0.1398 | 0.0624 |
| | Standard Deviation | 0.0771 | 0.0583 | 0.0521 | 0.0611 |
| | Variance | 0.0059 | 0.0034 | 0.0027 | 0.0037 |
| $S_y$ | Number of samples | 1 | 13 | 3 | 17 |
| | Min. | - | 0.0017 | 0.0160 | 0.0018 |
| | Max. | - | 0.0849 | 0.0495 | 0.1059 |
| | Arithmetic Mean | - | 0.0422 | 0.0374 | 0.0451 |
| | Geometric Mean | - | 0.0219 | 0.0333 | 0.0259 |
| | Median | - | 0.0389 | 0.0466 | 0.0466 |
| | Standard Deviation | - | 0.0322 | 0.0185 | 0.0327 |
| | Variance | - | 0.0010 | 0.0003 | 0.0011 |
| $k$ | Number of samples | 5 | 23 | 3 | 31 |
| | Min. | $7.00 \times 10^{-12}$ | $4.25 \times 10^{-12}$ | $3.67 \times 10^{-10}$ | $4.25 \times 10^{-12}$ |
| | Max. | $1.65 \times 10^{-6}$ | $4.89 \times 10^{-7}$ | $1.85 \times 10^{-7}$ | $1.65 \times 10^{-6}$ |
| | Arithmetic Mean | $3.31 \times 10^{-7}$ | $5.10 \times 10^{-8}$ | $6.90 \times 10^{-8}$ | $9.79 \times 10^{-8}$ |
| | Geometric Mean | $1.22 \times 10^{-10}$ | $4.58 \times 10^{-10}$ | $1.22 \times 10^{-10}$ | $5.24 \times 10^{-10}$ |
| | Median | $9.73 \times 10^{-12}$ | $5.31 \times 10^{-10}$ | $2.17 \times 10^{-8}$ | $3.67 \times 10^{-10}$ |
| | Standard Deviation | $7.40 \times 10^{-7}$ | $1.14 \times 10^{-7}$ | $1.01 \times 10^{-7}$ | $3.06 \times 10^{-7}$ |
| | Variance | $5.47 \times 10^{-13}$ | $1.29 \times 10^{-14}$ | $1.02 \times 10^{-14}$ | $9.39 \times 10^{-14}$ |

### 3.2.3. Specific Yield ($S_y$)

$S_y$ values were measured for 17 samples of fine-grained sandstone. The remaining 17 samples were mechanically disintegrated at the centrifugation stage. Unfortunately, this does not allow a comparison of results for the three distinguished sandstone groups. The $S_y$ value of tested samples ranges from 0.0018 to 0.1059 (Table 2). The arithmetic mean is 0.0451, with a standard deviation of 0.0079.

### 3.2.4. Hydraulic Conductivity ($k$)

The last measured parameter in the laboratory tests was the hydraulic conductivity ($k$), determined for 31 samples. Measurements were made three times, and the average of these three measurements was taken for analysis. The $k$ of the sandstone matrix shows variation from $4.25 \times 10^{-12}$ m/s to $1.65 \times 10^{-6}$ m/s; the geometric mean is $9.79 \times 10^{-8}$ m/s, and the standard deviation is $3.06 \times 10^{-7}$ m/s. Very fine-grained and fine-grained sandstone show very similar $k$ values (Table 2). In medium-grained sandstone, the lower limit of the $k$ value is two orders higher than in fine-grained.

The distribution of $k$ values is logarithmic but not homogeneous (Figure 14). Three subgroups can be distinguished. The first is sandstone with a very low $k$ value ($7.37 \times 10^{-12}$ m/s to $9.73 \times 10^{-12}$ m/s), mainly fine-grained and very fine-grained, probably with small pore sizes. The second subgroup is very fine-grained, fine-grained, and medium-grained sandstone with $k$ oscillating between $1.58 \times 10^{-11}$ m/s to $1.39 \times 10^{-9}$ m/s). In this case, the slope of the approximating line is significantly lower, which indicates a more significant variation in the pore space distribution in the samples. The third group consists of sandstone with higher permeability values ($1.26 \times 10^{-8}$ m/s to $1.65 \times 10^{-6}$ m/s)) and well-developed pores where individual pores connect and enable water flow. The macroscopic observations show that the samples with clay interlayers tended to have lower $k$ values than those without clay interlayers.

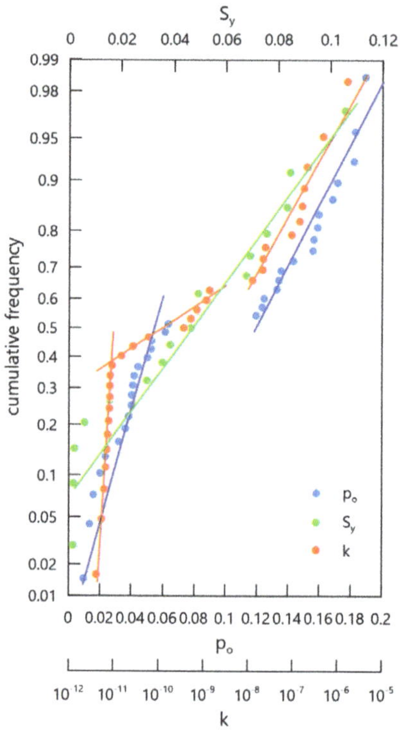


**Figure 14.** Cumulative frequencies of $p_o$, $S_y$, and $k$ [m/s].

### 3.2.5. Variation of the RQD and Hydrogeological Properties with Depth

The analysis of changes in the described parameters should start with assessing the quality (strength) of the drill core since it describes the mechanical strength of rocks and indirectly tells about the binder that holds grains together. The RQD index was used for this purpose, and its analysis indicates that the studied rock mass has different mechanical strengths depending on the depth (Figure 15). Three zones with different RQD values can be distinguished. The first zone begins at 595.0 m and extends to a depth of 723 m. In this zone, the rock mass shows high mechanical strength, RQD > 80 for most of this zone (Figure 15). The exception is the interval from the depth of 690.0 m to 700.0 m, where locally, the RQD drops to the value of 60–63. Below the depth of 730 m, the zone of weakened rock mass begins, where the RQD index drastically drops to the value from 20 to 50. In an interval of 735–737.5 m, it reaches a minimum value of RQD = 0, and at a depth between 747.0 and 750.2 m, RQD equals 6.25. This zone ends at a depth of 796.0 m, where there is a significant increase in the strength of the rock mass RQD > 80, with local zones of weakness in the range of 804.0 m–808.0 m and 810.0 m–814.0 m. Such a distribution of RQD indicates the mechanical strength of the rock mass translated into the ability to collect samples for laboratory testing. Therefore, samples were taken from those strong enough fragments to be mechanically processed (cutting smaller core fragments of a specific, smaller diameter). In addition, the samples had to have sufficient mechanical strength to survive centrifugation in a high-speed centrifuge. For this reason, there are no samples from core fragments with

RQD < 50. There is no evident variability of hydrogeological parameters with depth. It is possible to distinguish zones in the profile with low values interspersed with zones with high values of measured parameters. The zones with low values are the depth zones 625.0 m–640.0 m, 680.0 m–700.0 m, and below 810 m (Figure 15). A slight correlation between their occurrence with local zones of reduced RQD can be noticed.

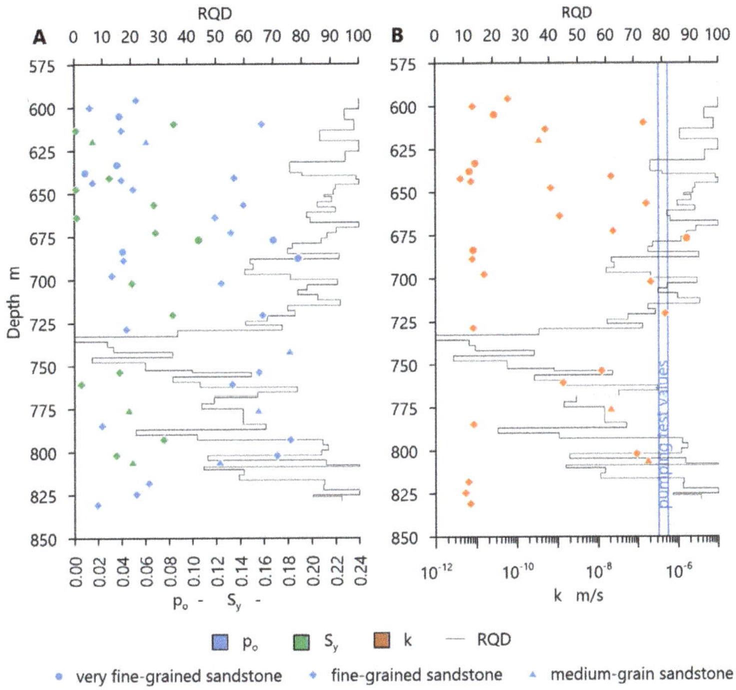


**Figure 15.** Distribution of $p_o$, $S_y$ (**A**) and $k$ (**B**) in the vertical profile.

### 3.2.6. Correlation between Hydrogeological Parameters

The arrangement and geometry of the matrix determine the values of $p_o$, $S_y$, and $k$. The homogeneity of the grain size, the shape of the grains, and the degree of grain cementation influence this. The degree of grain cementation will be indirectly indicated by the RQD parameter, the variation of which with depth was discussed earlier. The correlation between RQD and depth is moderately negative. Negative moderate correlation between RQD and $p_o$ is evident (Figure 16, Table 3).

The analysis shows a fairly strong correlation between $p_o$ and $S_y$ (R = 0.70) (Figure 17). Correlation analysis for $k$ shows its moderate correlation with $S_y$ (R = 0.67) and $p_o$ (R = 0.45). On the other hand, there is no visible correlation between depth and $S_y$ and $k$, as well as between RQD and $S_y$ and $k$.

**Table 3.** Correlation matrix between measured parameters.

| | Depth | $p_o$ | $S_y$ | $k$ | RQD |
|---|---|---|---|---|---|
| depth | 1.00 | 0.27 | 0.13 | 0.00 | −0.62 |
| $p_o$ | 0.27 | 1.00 | 0.70 | 0.45 | −0.49 |
| $S_y$ | 0.13 | 0.70 | 1.00 | 0.67 | −0.08 |
| $k$ | 0.00 | 0.45 | 0.67 | 1.00 | −0.03 |
| RQD | −0.62 | −0.49 | −0.08 | −0.03 | 1.00 |

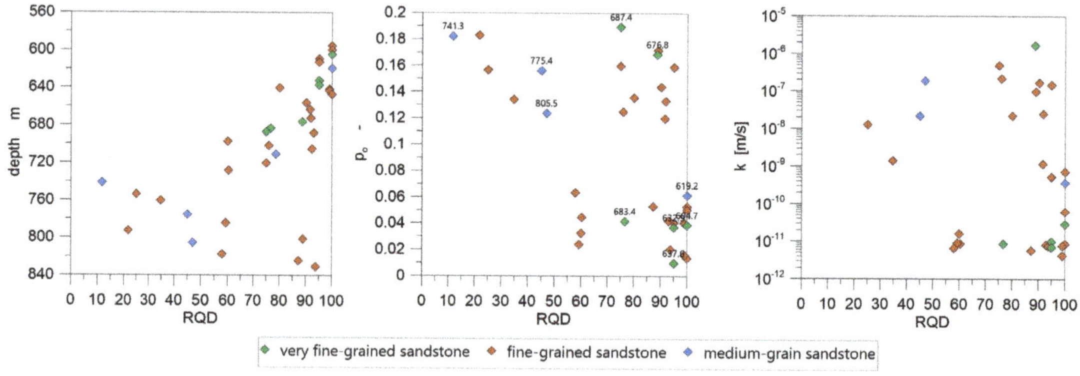


**Figure 16.** RQD vs. depth, $p_o$ and $k$ in the sandstone.

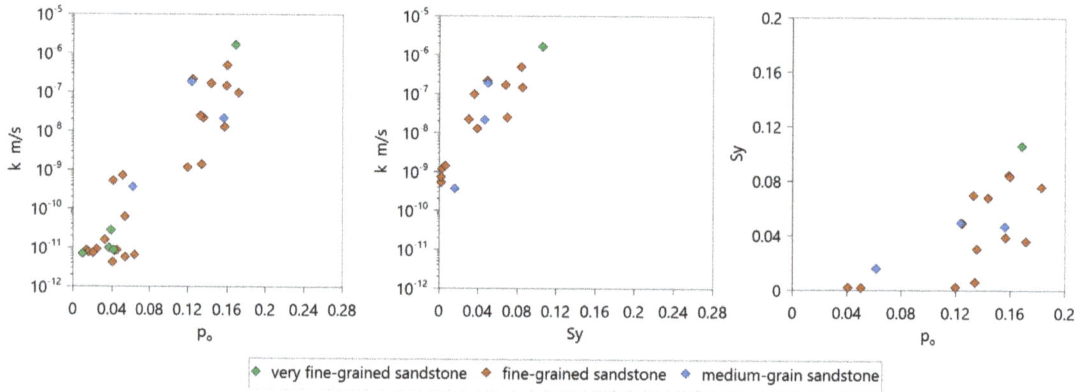


**Figure 17.** $k$ vs. $p_o$ and $S_y$, $p_o$ vs. $S_y$.

## 4. Discussion

The middle and lower Bunter sandstone horizon is characterized by high horizontal and vertical variability in terms of water flow. It is formed in fine- and medium-grained, medium-bedded sandstone, locally brittle and multi-directionally fractured. Core recovery was observed in a wide range of 10–90%. Additionally, mudflow escapes were common while drilling this horizon.

RQD parameter analysis indicates the presence of a zone of mechanically strong, highly cemented sandstone and a zone of mechanically weak sandstone. In the weak zone, there are numerous clay inserts and delaminations. This results in privileged groundwater flow paths, reflected in mudflow escapes and large water inflows to the borehole. In the discussed borehole, such a zone occurs below a depth of 700 m. The negative correlation between RQD and $p_o$ indicates that binder influences the level of permeable fractures. We observe lower interconnected porosity in rocks with higher strength and more binder. The second factor determining interconnected porosity is sandstone grain sorting. Samples with good sorting (medium-grained sandstone) have higher $p_o$ values than samples with worse sorting (finer material fills the pore spaces).

The good correlation of $p_o$ with $S_y$ indicates a close relationship between these parameters, which is typical for sandstone where water fills the pore spaces. If a depression cone is formed, it is drained according to local pressure gradients and determines the long-term inflow to the mine workings.

One of the most important parameters determining the possibility of groundwater flow is hydraulic conductivity ($k$). The studied samples show three subgroups of results with different water filtration properties, which depend mainly on pore formation and connections between them. Low values of *k* are found in local weak zones of the rock mass, which may indicate admixtures of clay minerals that fill the pore space and restrict water flow. Analysis of the results from the laboratory measurements shows that we obtain much lower *k* values in these studies than in the pumping tests. Only the six highest *k* values from laboratory measurements were similar to the *k* result from the pumping test. It means that direct adopting the results of laboratory measurements to calculate groundwater inflow gives the underestimated ability of rock mass drainage.

Information from archival hydrogeologic documentation was collected to evaluate the results on a regional scale. These are data from pumping tests performed in the middle and lower Bunter sandstone horizon since the 1970s. One hundred verified field tests were used to create a probability distribution plot of *k*. Results show hydraulic conductivity calculated from the single well (borehole) pumping tests ranging from $2.30 \times 10^{-8}$ m/s to $2.60 \times 10^{-6}$ m/s (Figure 18, blue points), average $6.48 \times 10^{-7}$ m/s. The results obtained during the pumping tests classify the rocks into semi-permeable [51] or low-permeable rocks [52].

The *k*-values estimated based on current laboratory tests and pumping well W-1 are plotted on this graph for comparison to archival results from pumping tests. The W-1 pumping test results indicate that the rocks are low permeable and included in the regional values' upper zone. The laboratory results are shifted towards lower values and indicate that very low and low permeability rocks dominate. Such laboratory and pumping *k* values distribution suggest that the inflow to the mine will have two components (short-term and long-term), as described below, that are related to the scale of water-bearing voids.

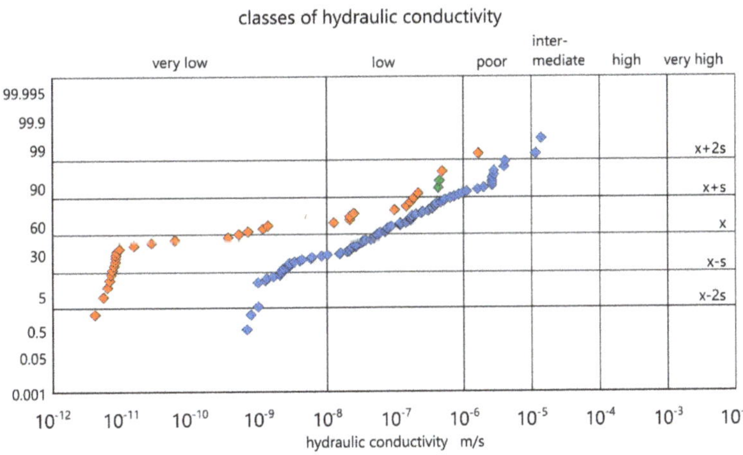


**Figure 18.** Values of the hydraulic conductivity of the middle and lower Bunter sandstone horizons determined from a single well pumping test (blue) ([53]), W-1 well-pumping test (green), and laboratory tests (orange) within the current study. x ± s: mean ± standard deviation, x ± 2s: mean ± 2 times standard deviation. Classification according to the *Hydrogeological Dictionary* [45].

The tested volume of the aquifer or sample strongly influences the obtained results, as was also proved by studies conducted in various laboratories (e.g., [18,54,55]). Different k values are obtained under laboratory conditions and during field measurements for the same sampled lithological interval (Figure 19). However, hydraulic conductivity increases with the scale of the tested pore-fissure rock mass (Figure 19). The same rule was observed in a described case; k measured in a laboratory is tenfold lower than in the pumping test in the well W-1. Moreover, such a distribution of results indicates that the pumping test shows the drainage ability of large rock intervals and gives information on water stored in privileged regional structures (tectonic zones, rock faults, connected voids, etc.). These structures will be drained at first during the opening of the deposit and will provide a large and fast inflow (short-term component). They will also be responsible for the inflow to the shaft during the sinking operation. The laboratory measurements indicate values typical for the local rock matrix that will determine the inflow conditions to the mine long after the drained privileged zones. It will shape the long-term and smaller-intensity inflow.

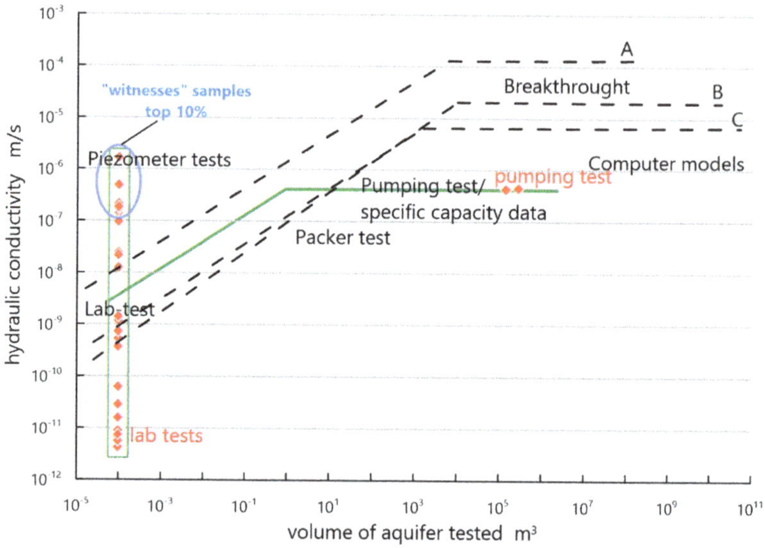


**Figure 19.** Dependence of the hydraulic conductivity on the volume of the tested rock mass. Green line: Fore-Sudetic Monocline Triassic sandstone, Black dashed lines: carbonate series rocks A: Thiensville Formation, B: Mayville carbonate rocks, C: Romeo carbonate rocks [56] after [55].

While discussing scale influence on hydraulic conductivity, we should also consider the thickness of the studied aquifer, which plays an important role in shaping the inflow to the sinking shaft. Parametrically it is defined by water transmissivity index $T$ [$m^2/s$], which is a function of the properties of the liquid, the aquifer, and the thickness of the porous media [57]. In the study area, the average thickness of the middle and lower Bunter sandstone water-bearing formations reaches about 450 m. Consequently, the water transmissivity $T$ index determined from the averaged pumping test results ranges from $1.82 \times 10^{-4}$ $m^2/s$ to $2.00 \times 10^{-4}$ $m^2/s$ (15.72–17.28 $m^2/d$). Therefore, according to the VI step classification proposed by J. Krásný (1993) [58], we classify the hydraulic transmissivity of the middle and lower Bunter sandstone horizon as class III: intermediate water transmissivity.

## 5. Conclusions

The process of sinking a mine shaft is a technologically challenging operation, especially in complex geological conditions. As described in the article, the Fore-Sudetic Monocline is precisely one of the areas with challenging geology. In order to reach a copper ore deposit at present, it is necessary to mine more than 1200 m deep and to pass through many aquifers. To date, the commonly used methodology for predicting inflows to the shaft has relatively often proved inaccurate and unreliable due to unrepresentative values of the hydraulic conductivity obtained from simple or inadequate estimates. An accurate assessment of the hydrogeological conditions and estimating the expected inflows to the sunken shaft determines its construction technology and influences the crew's safety and investment.

Our work and other scientific articles prove it [10,56,59] that to obtain the most realistic inflow predictions, it is necessary to rely on hydrogeological parameters from long-term pumping tests at nodes. However, such surveys for shafts deeper than 1000 m are extremely costly and time-consuming, and hence there is often great resistance to their use. Performing a long-duration pumping test also delays the mining works associated with the shaft excavation, which is also an unfavorable factor for performing this testing. New to our work were hydraulic conductivity measurements obtained from laboratory tests for samples obtained from drill cores. These types of samples are easy to collect in the course of drilling works related to the rock mass's reconnaissance and the subsequent shaft excavation. It is obvious that these samples characterize hydrogeological parameters on a point scale. Still, their sufficiently large population makes it possible to estimate a representative value of the hydraulic conductivity of the whole rock mass.

Our research showed that lab tests underestimate hydraulic parameters of the pore-fissure Triassic sandstone (mean hydraulic conductivity $k = 9.79 \times 10^{-8}$ m/s) tenfold more than long-term pumping tests (mean $k = 4.45 \times 10^{-7}$ m/s). This situation is consistent with cases known from the literature of the influence of the test scale on the values of the hydraulic conductivity in porous and pore-fissure aquifers with low rock matrix porosity. In our study for the Triassic sandstone, we found that this underestimation, i.e., the ratio of the hydraulic conductivity value obtained from the pumping test to the mean value of this parameter from the laboratory method, is relatively small and amounts to less than an order of magnitude ("underestimation" ratio = 4.54). The results published in the literature for fractured karst rocks show a difference in the results from laboratory tests and pumping tests of up to several orders of magnitude.

Individual laboratory samples cannot indicate the representative value of hydraulic conductivity, as it depends strongly on the research scale. Hydraulic conductivity typically increases with the test scale and then approaches an asymptote (Figure 19). Only if the tests cover the rock mass volume of at least 10,000–100,000 $m^3$ is it possible to obtain representative values of the hydrogeological parameters, including hydraulic conductivity. In such a volume, it is highly probable to capture most of the privileged zones affecting the groundwater flow conditions, e.g., those associated with higher porosity or characterized by a higher density of cracks.

For porous and pore-fissure rocks, such as the studied sandstone, collecting a large population of samples with the most diverse porosity makes it possible to find "witnesses" samples with parameter values close to a representative. This situation does not occur in the case of published studies of carbonate rocks where the permeability is related only to cracks and not to the porosity of the matrix (e.g., [18,39,54]. Based on our research, it can be concluded that the group of these "witness samples", for which the values of the hydraulic conductivity correspond to the representative value from the pumping test, comprises 10% of all core samples tested in the laboratory. With this in mind, we recommend using the highest values of hydrogeological parameters from laboratory tests based on the worst-case scenario. Therefore, it is possible to forecast inflows to the shafts reliably. This methodology is recommended only for rocks of porous and pore-fissure character. We recommend continuing research in this area to confirm whether the relationship found is valid only for the studied Triassic sandstone from the area of the Fore-Sudetic Monocline or also for all sandstone types from other locations.

The main conclusion of practical importance arising from our study is the confirmation of the relatively very low values of the hydraulic conductivity for Triassic sandstone from the area of the Fore-Sudetic Monocline. This is confirmed by results obtained by two methods at different scales, not only from drill core samples but from long-term and much more representative pumping tests.

The very low values of the hydraulic conductivity of the Triassic sandstone mean that, consequently, the expected inflows to the deep shaft will also be relatively small and technically feasible to drain using pumps. We, therefore, recommend considering the use of direct dewatering techniques for the rock mass without the need to freeze the ground, which for deeper parts of the rock mass above 1 km is costly and technically difficult due to the high temperature of the rock and the presence of highly saline groundwater (brines).

**Author Contributions:** Conceptualization, M.P., M.W.-K., K.C. and M.C.; methodology, M.P., M.W.-K. K.C. and M.C.; software, M.P.; validation, M.P. and M.C.; formal analysis, M.P. and K.C.; investigation, M.P. and K.C.; resources, M.P., M.W.-K. and K.C.; data curation, M.W.-K. and K.C.; writing—original draft preparation, M.P., M.W.-K. and K.C.; writing—review and editing, M.W.-K. and K.C.; visualization, M.P. and K.C.; supervision, M.W.-K. and M.C.; project administration, M.W.-K.; funding acquisition, M.W.-K. and K.C. All authors have read and agreed to the published version of the manuscript.

**Funding:** The research work was partially funded with the grant co-financed by NCBiR and KGHM Polska Miedź S.A., entitled "IMore—Innovative methods of accessing deep deposits" (No. CuBR/I/1/NCBiR/2014) under the CuBR program and Wrocław University of Science and Technology research fund from the Polish Ministry of Education and Science granted for 2023.

**Conflicts of Interest:** The authors declare no conflict of interest.

## References

1. Kalisz, M.; Niedbał, M. Wpływ odwadniania utworów triasowych w trakcie głębienia szybu R–XI na warunki hydrodynamiczne i powierzchniowe w północnej części OG "Rudna". *Mater. Symp. Nauk. Probl. Hydrogeol. Górnictwa Rud Miedzi* **2004**, 148–160.
2. Chudy, K.; Worsa-Kozak, M.; Pikuła, M. Rozwój metod rozpoznania warunków hydrogeologicznych na potrzeby wykonywania pionowych wyrobisk udostępniających złoże-przykład LGOM [Evolution of methods for hydrogeological condition recognition for the need of shaft sinking: An example from the Legnica-Glogow Copper District]. *Przegląd Geol.* **2017**, *65*, 1035–1043. Available online: https://geojournals.pgi.gov.pl/pg/article/view/27216/18934 (accessed on 2 December 2022).
3. Bocheńska, T.; Leśniak, L.; Tomaszewski, J. Związek między zawodnieniem wyrobisk górniczych a cechami litologicznymi i strukturalnymi górotworu. *Przegląd Geol.* **1971**, *19*, 278–284. Available online: https://geojournals.pgi.gov.pl/pg/article/view/23497/16362 (accessed on 2 December 2022).
4. Bocheńska, T.; Fiszer, J.; Kalisz, M. Weryfikacja prognoz dopływów do kopalń w Legnicko—Głogowskim Obszarze Miedzionośnymna podstawie badań modelowych. *Współczesne Probl. Hydrogeol.* **1995**, *VII*, 39–46.
5. Bocheńska, T.; Fiszer, J.; Kalisz, M. Prediction of groundwater inflow into copper mines of the Lubin Glogow Copper District. *Environ. Geol.* **2000**, *39*, 587–594. [CrossRef]
6. Bocheńska, T.; Bieniewski, J. Dopływ wód podziemnych do kopalni rudy miedzi na monoklinie przedsudeckiej. *Geol. Sudet.* **1978**, *XIII*, 133–141. [CrossRef]
7. Bocheńska, T.; Borysow, M. Zastosowanie metod modelowania elektrohydrodynamicznego do prognozy zawodnienia szybów [Application of electrohydrodynamic modeling methods to predict shaft inflows]. *Tech. Posz. Geol. Geosyn. Geoter.* **1982**, *3*, 35–38.
8. Sawicki, J.; Derkowska-Sitarz, M. Methods of definition of parts of water inflow from groundwater storage and groundwater renewable resources in total water inflow to mines. *Min. Sci.* **2006**, *8*, 167.
9. Pikuła, M.; Worsa-Kozak, M.; Chudy, K.; Kaczorek, K.; Szymański, A. Zastosowanie zmodyfikowanej metody analogii hydrogeologicznej z wykorzystaniem GIS w prognozowaniu dopływu wód do podziemnych wyrobisk górniczych [Application of modified hydrogeological analogy method using GIS in forecasting water inflow to underground mine workings]. *Górnictwo Odkryw.* **2018**, *2*, 67–75.
10. Zuber, A.; Motyka, J. Matrix porosity as the most important parameter of fissured rocks for solute transport at large scales. *J. Hydrol.* **1994**, *158*, 19–46. [CrossRef]
11. Gurwin, J.; Staśko, S.; Wcisło, M. Dokładność odwzorowa nia wielowarstwowych systemów hydrogeologicznych na szczegółowych modelach numerycznych—Analiza dla obszaru LGOM [Hydrogeological regional model as a permanent tool in solving various-scale tasks—Experience from the Legnica–Glogow Copper Region (LGOM)]. *Modele Mat. Hydrogeol.* **2014**, *471*, 45–50.
12. Staśko, S.; Gurwin, J.; Wcisło, M.; Modelska, M.; Kryza, H.; Kryza, J.; Olichwer, T.; Buczyński, S.; Tarka, R.; Wąsik, M.; et al. Model koncepcyjny systemu hydrogeologicznego obszaru oddziaływania Lubińsko-Głogowskiego Obszaru Miedzionośnego (LGOM) [Conceptual model of hydrogeological system of Lubin-Głogów Ore District impact area]. *Biul. Państwowego Inst. Geol.* **2012**, *451*, 203–210.
13. Lloyd, J.W.; Edwards, M.G. Estimation of groundwater inflow to an underground mining operation. *Int. J. Mine Water* **1988**, *7*, 25–47. [CrossRef]
14. Rapantova, N.; Grmela, A.; Vojtek, D.; Halir, J.; Michalek, B. Ground Water Flow Modelling Applications in Mining Hydrogeology. *Mine Water Environ.* **2007**, *26*, 264–270. [CrossRef]
15. d'Obyrn, K.; Kamiński, P.; Motyka, J. Influence of Hydrogeological Investigation's Accuracy on Technology of Shaft Sinking and Design of Shaft Lining—Case Study from Southern Poland. *Energies* **2021**, *14*, 2050. [CrossRef]

16. Beattie, D.; Bashir, R.; Hatley, J. Impact of Geotechnical and Hydrogeological Parameters upon Shaft Sinking Performance in the Athabasca Basin Canada. In *Mine Water and the Environment*; Rapantova, N., Hrkal, Z., Eds.; VSB—Technical University of Ostrava: Ostrava, Czech Republic, 2008; pp. 3–6. Available online: https://www.imwa.info/docs/imwa_2008/IMWA2008_141_Beattie.pdf (accessed on 2 December 2022).
17. Tiedeman, C.R.; Hill, M.C.; D'Agnese, F.A.; Faunt, C.C. Methods for using groundwater model predictions to guide hydrogeologic data collection, with application to the Death Valley regional groundwater flow system. *Water Resour. Res.* **2003**, *39*, 1010. [CrossRef]
18. Pinto da Cunha, A. *Scale Effects in Rock Masses 93*; CRC Press: Boca Raton, FL, USA, 1993. [CrossRef]
19. Nastev, M.; Savard, M.M.; Lapcevic, P.; Lefebvre, R.; Martel, R. Hydraulic properties and scale effects investigation in regional rock aquifers, south-western Quebec, Canada. *Hydrogeol. J.* **2004**, *12*, 257–269. [CrossRef]
20. Dausse, A.; Leonardi, V.; Jourde, H. Hydraulic characterization and identification of flow-bearing structures based on multi-scale investigations applied to the Lez karst aquifer. *J. Hydrol. Reg. Stud.* **2019**, *26*, 100627. [CrossRef]
21. Sánchez-Vila, X.; Carrera, J.; Girardi, J.P. Scale effects in transmissivity. *J. Hydrol.* **1996**, *183*, 1–22. [CrossRef]
22. Le Borgne, T.; Bour, O.; Paillet, F.; Caudal, J.P. Assessment of preferential flow path connectivity and hydraulic properties at single-borehole and cross-borehole scales in a fractured aquifer. *J. Hydrol.* **2006**, *328*, 347–359. [CrossRef]
23. Godoy, V.A.; Zuquette, L.V.; Gómez-Hernández, J.J. Scale effect on hydraulic conductivity and solute transport: Small and large-scale laboratory experiments and field experiments. *Eng. Geol.* **2018**, *243*, 196–205. [CrossRef]
24. Ghanbarian, B. Estimating the scale dependence of permeability at pore and core scales: Incorporating effects of porosity and finite size. *Adv. Water Resour.* **2022**, *161*, 104123. [CrossRef]
25. Konstantynowicz, E. *Monografia Przemysłu Miedziowego w Polsce*; Wydawnictwa Geologiczne: Warszawa, Poland, 1971.
26. Kiersnowski, H.; Petecki, Z. Budowa geologiczna podcechsztyńskiego podłoża Legnicko-Głogowskiego Okręgu Miedziowego (LGOM) i jego otoczenia: spojrzenie krytyczne [Geology of the Zechstein basement of the Legnica-Głogów Copper] District (LGOM) and its surroundings: A critical overview. *Biul. Państwowego Inst. Geol.* **2017**, *468*, 175–198. Available online: http://yadda.icm.edu.pl/baztech/element/bwmeta1.element.baztech-292ac488-6bc9-4b25-a9fc-dba25f56a442 (accessed on 2 December 2022).
27. Kaczmarek, W.; Kaczmarek, W.; Twardowski, M.; Wasilewska-Błaszczyk, M. Praktyczne aspekty modelowania litologicznych typów rud w złożach cu-ag LGOM(Legnicko-Głogowskiego Okręgu Miedziowego). *Biul. Państwowego Inst. Geol.* **2016**, *468*, 209–226.
28. Wodzicki, A.; Piestrzyński, A. An ore genetic model for the Lubin—Sieroszowice mining district, Poland. *Miner. Depos.* **1994**, *29*, 30–43. [CrossRef]
29. Piestrzyński, A.; Banaszak, A.; Zaleska-Kuczmierczyk, M. *Monografia KGHM Polska Miedź S.A.*, 2nd ed.; KGHM Polska Miedz S.A.: Lubin, Poland, 2007; p. 1086.
30. Dąbrowski, S.; Przybyłek, J. *Metodyka Próbnych Pompowań w Dokumentowaniu Zasobów wód Podziemnych. Poradnik Metodyczny*; Ministerstwo Środowiska: Warszawa, Poland, 2005; p. 292.
31. Gonet, A.; Macuda, J.; Zawisza, L.; Duda, J.; Porwis, J. *Instrukcja Obsługi Wierceń Hydrogeologicznych [Hydrogeological Drilling Operations Manual]*; Wydawnictwa AGH: Kraków, Poland, 2011; p. 134.
32. Agarwal, R.G. A New Method To Account For Producing Time Effects When Drawdown Type Curves Are Used To Analyze Pressure Buildup And Other Test Data. In Proceedings of the SPE Annual Technical Conference and Exhibition, Dallas, TX, USA, 21–24 September 1980. [CrossRef]
33. Agarwal, R.G.; Al-Hussainy, R.; Ramey, H. An Investigation of Wellbore Storage and Skin Effect in Unsteady Liquid Flow: I. Analytical Treatment. *Soc. Pet. Eng. J.* **1970**, *10*, 279–290. [CrossRef]
34. Röhrich, T. *AquiferTest v.4.2 User's Manual*; Schlumberger Water Services: Waterloo, ON, Canada, 2005.
35. Theis, C.V. The relation between the lowering of the Piezometric surface and the rate and duration of discharge of a well using ground-water storage. *Trans. Am. Geophys. Union* **1935**, *16*, 519. [CrossRef]
36. Mays, L.W.; Todd, D.K. *Groundwater Hydrology*; Wiley & Sons: Hoboken, NJ, USA, 2004; p. 656.
37. Deere, D.; Hendron, A.; Patton, F.; Cording, E. Design of Surface and Near Surface Construction in Rock. In Proceedings of the 8th U.S. Symposium on Rock Mechanics (USRMS), Minneapolis, MN, USA, 15–17 September 1966; pp. 237–302.
38. Schön, J. Geomechanical Properties. In *Handbook of Petroleum Exploration and Production*; Elsevier: Amsterdam, The Netherlands, 2011; Volume 8, pp, 245–271. [CrossRef]
39. Carmichael, R.S. *Practical Handbook of Physical Properties of Rocks and Minerals (1988)*; CRC Press: Boca Raton, FL, USA, 2017. [CrossRef]
40. Borczak, S.; Motyka, J.; Pulido-Bosch, A. The hydrogeological properties of the matrix of the chalk in the Lublin coal basin (southeast Poland). *Hydrol. Sci. J.* **1990**, *35*, 523–534. [CrossRef]
41. James, O.D.; Lee, P.C.Y. Flow in fractured porous media. *Water Resour. Res.* **1977**, *13*, 558–566. [CrossRef]
42. Motyka, J. A conceptual model of hydraulic networks in carbonate rocks, illustrated by examples from Poland. *Hydrogeol. J.* **1998**, *6*, 469–482. [CrossRef]
43. Motyka, J.; Pulido-Bosch, A.; Borczak, S.; Gisbert, J. Matrix hydrogeological properties of Devonian carbonate rocks of Olkusz (Southern Poland). *J. Hydrol.* **1998**, *211*, 140–150. [CrossRef]

44. Motyka, J.; Pulido-Bosch, A. Karstic phenomena in calcareous-dolomitic rocks and their influence over the inrushes of water in lead-zinc mines in Olkusz region (South of Poland). *Int. J. Mine Water* **1985**, *4*, 1–11. [CrossRef]
45. Dowgiałło, J.; Kleczkowski, A.; Macioszczyk, T.; Różkowski, A. *Słownik Hydrogeologiczny [Hydrogeological Dictionary]*; Państw. Inst. Geol.: Warszawa, Poland, 2002; p. 461.
46. Agarwal, V.C. *Groundwater Hydrology*; PHI Learning Pvt. Ltd.: Delhi, India, 2012; p. 372.
47. Motyka, J.; Szczepańska, J.; Witczak, S. Zastosowanie wirówki do badania współczynnika odsączalności i dynamiki oddawania wody przez skałę. *Tech. Poszuk.* **1971**, *37*, 38–43.
48. Prill, R.C.; Johnson, A.I.; Morris, D.A. *Specific Yield—Laboratory Experiments Showing the Effect of Time on Column Drainage*; U.S. Geological Survey: Reston, VA, USA, 1965; p. 55.
49. Motyka, J.; Postawa, A. Influence of contaminated Vistula River water on the groundwater entering the Zakrzowek limestone quarry, Cracow region, Poland. *Environ. Geol.* **2000**, *39*, 398–404. [CrossRef]
50. Duliński, W. Aparat do badania przepuszczalnosci z uszczelnieniem pneumatycznym. *Wiadomości Naft.* **1965**, *7*, 117–118.
51. Pazdro, Z. *Hydrogeologia Ogólna [Hydrogeology]*; Wydaw. Geologiczne: Warszawa, Poland, 1983; p. 575.
52. Head, K.; Epps, R. Manual of Soil Laboratory Testing—Volume 2: Permeability, Shear Strength and Compressibility Tests. *Q. J. Eng. Geol. Hydrogeol.* **2014**, *47*, 191–191. [CrossRef]
53. Bezkorowajny, A.; Stochel, B.; Chudy, K.; Worsa-Kozak, M.; Cygan, S.; Janicki, D.; Kalisz, M.; Konsencjusz, D.; Merta, A.; Szumilas, P. Dokumentacja Hydrogeologiczna Określająca Warunki Hydrogeologiczne w Związku z Zamierzonym Wykonywaniem Odwodnień w celu Wydobywania rud Miedzi ze złoża Rudna. 2013, *unpublished*.
54. Zhan, S.S.; Wang, T.T.; Huang, T.H. Variations of hydraulic conductivity of fracture sets and fractured rock mass with test scale: Case study at Heshe well site in Central Taiwan. *Eng. Geol.* **2016**, *206*, 94–106. [CrossRef]
55. Schulze-Makuch, D.; Cherkauer, D.S. Variations in hydraulic conductivity with scale of measurement during aquifer tests in heterogeneous, porous carbonate rocks. *Hydrogeol. J.* **1998**, *6*, 204–215. [CrossRef]
56. Drożdżak, R.; Twardowski, K. Wpływ efektu skali na wyniki badań przepuszczalności porowatych ośrodków gruntowo-skalnych [Influence of scale effect on the results of permeability tests on rock-ground porous media]. *Wiert. Naft. Gaz* **2009**, *26*, 599–611.
57. Şen, Z. Chapter 2—Basic Porous Medium Concepts. In *Practical and Applied Hydrogeology*; Şen, Z., Ed.; Elsevier: Oxford, UK, 2015; pp. 43–97. [CrossRef]
58. Krasny, J. Classification of Transmissivity Magnitude and Variation. *Ground Water* **1993**, *31*, 230–236. [CrossRef]
59. Ziemianin, K.; Jędrzejowska-Tyczkowska, H. Efekt skali w interpretacji geologicznej danych geofizycznych i analiz laboratoryjnych—Wprowadzenie do tematyki oraz przykłady obliczeń na danych modelowych [Scale effect in geological interpretation of geophysical and laboratory data—Introduction to the topic and model data calculations examples]. *Nafta-Gaz* **2013**, *69*, 719–734.

energies

Article

# Prediction of NOx Emission Based on Data of LHD On-Board Monitoring System in a Deep Underground Mine

Aleksandra Banasiewicz [1], Paweł Śliwiński [2], Pavlo Krot [1,*], Jacek Wodecki [1] and Radosław Zimroz [1]

1 Faculty of Geoengineering, Mining and Geology, Wroclaw University of Science and Technology, Na Grobli 15, 50-421 Wroclaw, Poland
2 KGHM Polska Miedz S.A., ul. Marii Skłodowskiej-Curie 48, 59-301 Lubin, Poland
* Correspondence: pavlo.krot@pwr.edu.pl

**Abstract:** The underground mining industry is at the forefront when it comes to unsafe conditions at workplaces. As mining depths continue to increase and the mining fronts move away from the ventilation shafts, gas hazards are increasing. In this article, the authors developed a statistical polynomial model for nitrogen oxide (NOx) emission prediction of the LHD vehicle with a diesel engine. The best-achieved prediction accuracy by the 4th order polynomial model for 11 and 10 input variables is about 8% and 13%, respectively. It is comparable with the sensors' accuracy of 10% at a stable regime of loading and 20% in the transient periods of operation. The obtained results allow planning of ventilation system capacity and power demand for the large fleet of vehicles in the deep underground mines.

**Keywords:** NOx emission; LHD machines; deep underground mine; statistical model; ventilation; prediction

check for updates

**Citation:** Banasiewicz, A.; Śliwiński, P.; Krot, P.; Wodecki, J.; Zimroz, R. Prediction of NOx Emission Based on Data of LHD On-Board Monitoring System in a Deep Underground Mine. *Energies* **2023**, *16*, 2149. https://doi.org/10.3390/en16052149

Academic Editors: Sergey Zhironkin and Dawid Szurgacz

Received: 20 January 2023
Revised: 14 February 2023
Accepted: 21 February 2023
Published: 23 February 2023



## 1. Introduction

One of the main tasks of an employer is to ensure safe working conditions for its employees regardless of the industry in which they work. According to [1], the underground mining industry is at the forefront when it comes to unsafe conditions at workplaces. As mining depths continue to increase, and the mining fronts move away from the ventilation shafts, gas hazards are increasing. This article will analyze the danger of nitrogen oxides (NOx). Sources of NOx emissions into the mine atmosphere can be divided into natural and technological. It is assumed that diesel mining machinery has the greatest impact on air pollution. It is important to properly control gas concentrations in mine workings. Unfortunately, not all mining machines have NOx measurement sensors installed—making it difficult to control the concentrations of these gases in workings. Rapidly changing standards for reducing gas concentration limits at workplaces are forcing the use of new methods in assessing workers' exposure to harmful gases. Not only is the lack of machine-mounted sensors a major problem in assessing the work environment, but also the problem of variable selection or data recording errors.

Monitoring, analyzing, and now also predicting parameters to ensure safe and trouble-free continuous operation of underground crews has been carried out for years [2–5]. Due to the difficult working conditions and unreliability of electronic equipment in underground conditions, the use of prediction sometimes turns out to be the only option for assessing the state of the environment or the condition of the equipment/vehicle.

The article uses measurement data from the SYNAPSA system, which data is obtained from the monitoring system mounted on an LHD. Authors selected from the full list of parameters those ones, which affect the value of NOx concentrations. Based on them, a model for the prediction of the value of these concentrations was created. To ensure safe working conditions for underground crews, it is important to know the values of emissions of harmful compounds from mining machinery. The created prediction model can be used to estimate the values that are generated by mining machines that do not have a nitrogen oxide sensor installed.

Nitrogen oxides are currently a critical problem. Their negative impact on human health results in continuous changes and lowering of nitrogen oxide concentration limits in the working environment of underground crews. The solution presented in the article will help improve working conditions. In further research, the proposed statistical models can provide valuable assistance in determining further parameters that can affect NOx from vehicle emissions, like measurements of other important engine variables such as cylinder pressure. In addition, statistical modeling can be very useful for predicting emissions under transient conditions of engine operation, where physical models still need significant improvement.

The main difference is that we used a new multi-polynomial statistical model to predict the NOx emissions and verified it by the unique data of LHD working in the underground mine with harsh environmental conditions. The majority of previous studies were conducted either for different types of diesel vehicles on-surface or in-lab conditions The obtained accuracy of NOx emission prediction allows applying the developed model for practical needs.

*Problem Formulation*

Following the own observations of the ventilation systems in the deep underground mines and taking into account the capabilities of onboard monitoring systems in Load-Haul-Dump (LHD) vehicles, the following problems can be distinguished:

- Although all manufacturers of diesel engines for heavy-duty vehicles include the exhaust gas aftertreatment systems, NOx sensors are not present in the LHD machines working in deep underground mines.
- To solve the urgent issues of ventilation, especially in the deep underground mines, and provide safe working conditions, the engineering and management staff would like to know immediately the NOx emissions from every vehicle working in different geological conditions. This is practically more efficient to realize by the on-board monitoring systems instead of permanent numerous sensors installation in the mines with constantly changing the configuration of tunnels. However, all recorded data, including signals from onboard NOx sensors, are uploaded to the server via wireless connections only once per working shift (about 6 h). Therefore, mathematical models and software tools are required for the post-processing of big data sets offline.
- Since the onboard monitoring system can record a huge number of working parameters, the problem arises of their optimal selection from the whole set. Moreover, due to different reasons, data are not always correctly stored (NaN values, missed data, etc.); therefore, data pre-processing and cleaning procedures are required for correct calculations.
- While designing a NOx prediction model, the problem exists of the balance between fit quality and overfitting related to smoothing of initial data (already after pre-processing and invalid data cleaning). Also, having an even enough accurate model tested over certain working conditions and machine technical state, its robustness should be provided for other working locations, operators' experience, and critical elements deterioration (engine, turbine, exhaust system, tires).

## 2. State of the Art

The safety of a worker while at work these days should be the most important criterion for an employer. However, there are industries where the health and safety of workers are at risk more than others. One such process is raw materials mining - mainly in underground mines. The ever-increasing demand for mineral resources results in the exploitation of deposits from ever deeper. The great depth of mining (up to 1000–1500 m) is associated with an increase in the exposure of workers to natural hazards prevailing underground [6–8]. The most dangerous at present is the climatic hazard associated with the constantly increasing primary temperature of the rocks, which in Polish copper ore mines is almost 50 °C [9]. Another critical problem is the gas hazard. As the mining depth increases and the mining fronts move away from the ventilation shafts, the rate of rarefying of harmful gases and ventilation of the workings decreases. According to Struminski [10] and Szlazak [11], the most harmful gases are carbon monoxide (CO), hydrogen sulfide ($H_2S$), methane ($CH_4$), and nitrogen oxides (NOx). In the recent studies of Yin and Linga [12], it has been proposed to use hydrogen or natural gas hydrate as a source of primary fuel to eliminate NOx and SOx.

As reported by Shaw et al. [13], NOx is understood as the sum of nitrogen oxide (NO) and nitrogen dioxide ($NO_2$) compounds. In underground mines, NOx gas hazards can have a natural or technological source. Natural sources include the oxidation of nitrogen from the atmosphere or the natural outflow of nitrogen oxides from the rock mass. The most significant, however, are nitrogen oxides that are generated by technological processes—these include nitrogen oxides that originate in the blasting process, those from welding processes, and, above all, those from diesel engines [14–16].

According to Kampa [17], nitrogen oxides are gases that are harmful to a living organism. Both NO and $NO_2$ are odorless gases. Nitrogen oxide is additionally a colorless gas, while nitrogen dioxide in higher concentrations can take on a brown color [18–20]. According to Galbreath et al. [21], in the exhaust of a diesel engine, the percentage of nitrogen oxides is about 90% NO and about 10% $NO_2$. As reported by Hori et al. in [22], it is nitrogen dioxide that is more toxic. $NO_2$ causes respiratory problems as low as 1.5 ppm, while at 5 ppm, it causes a drop in blood pressure. Death occurs at concentrations near 200 ppm of $NO_2$ [23].

Given the chemical and physical properties of nitrogen oxides and how they affect the human body working in the underground mine, continuous monitoring of the values of these concentrations is being introduced. Measurements are made of the values of NOx concentrations in the mine atmosphere and at the exhaust of the internal combustion engine. The limit values in the exhaust gases are 500 ppm for $NO_2$ and 750 ppm for NO.

Due to the harmfulness of the compounds in the exhaust, numerous studies are being conducted on predicting the emissions of harmful compounds, including nitrogen oxides, into the atmosphere from diesel vehicles [24–26].

This article represents the research results related to NOx emission from load-haul-dump (LHD) vehicles driven by diesel engines.

The articulated load-haul-dump (LHD) machine (LKP-1701), which was under investigation (see Figure 1), is designed for underground application in a confined space of low transportation tunnels. The main parameters of its diesel engine (DEUTZ TCD 12.0 V6) are given in Table 1. These LHDs are equipped with an onboard system for machine working parameters monitoring via CAN bus. The exhaust gas NOx concentration signal from the sensor is stored in the database among other signals of the diesel engine and operator actions (gear selection, torque converter locking, acceleration, and braking). This type of machine is characterized by the continuous reverse motion for blasted bulk material (copper ore) taking and haul truck loading. Due to that, diesel engine exhibits excessive exhaust gas emissions. The most intensive mode of engine loading determined based on working cycles analysis [27] and dynamical model [28] is the bucket digging in the hill.

To reduce the harmful gas emission due to lower combustion temperature, the diesel engine is equipped with the exhaust aftertreatment (EAT) or exhaust gas recirculating (EGR) systems. The exhaust gas is typically routed through a Diesel Oxidation Catalyst (DOC) where a chemical reaction is induced to convert hydrocarbons, NOx, and other pollutants of diesel exhaust to less harmful compounds like carbon dioxide. The remaining particles (soot) are reduced by the Diesel Particulate Filter (DPF).

**Figure 1.** The LHD vehicle under investigation: LKP-1701 (KGHM ZANAM) with a powerful diesel engine for the underground mines [29].

**Table 1.** Parameters of DEUTZ TCD 12.0 V6 diesel engine [30].

| Parameter | Value | Units |
|---|---|---|
| Power output as per ISO 14396 | 390 | kW |
| at speed | 2100 | rpm |
| Max. torque | 2130 | Nm |
| at speed | 1400 | rpm |
| Min. idling speed | 600 | rpm |
| Max. nominal speed | 1800–2100 | rpm |

CanmetMINING Diesel Research Laboratory (Canada) conducted a progressive load test (PLT) and vehicle transient test (VTT) to estimate the contribution in NOx emission of Diesel Oxidation Catalysts (DOC) [31]. The VTT simulated operation of a load-haul-dump (LHD) vehicle's working cycle. Three groups of DOCs are tested: (1) platinum; (2) base metal/palladium; and (3) the "advanced" group. All groups showed a good reduction of carbon monoxide (CO) and total hydrocarbon emissions. However, the change in $NO_2$ (g/kWh) emissions varied from an increase of 446% to a reduction of 47% for groups 1 and 2 while group 3 showed NOx reduction in any mode of operation.

Those systems provide Tier 4 Interim (Stage IIIB) emissions in accordance with EU regulations. For this class of engine power (up to 560 kW), the allowed maximum amount of nitrogen oxides (NOx) is 3.3 g/kWh; non-methane hydrocarbons (NMHC) —0.19 g/kWh; particulate matter (PM) 0.025 g/kWh. In other types of vehicles and EU regulations, these emission parameters are given per kilometer regardless of engine power. Although advanced technologies for emissions reduction are proposed and applied in civil cars [32], e.g., Lean NOx Trap (LNT), Selective Catalytic Reduction (SCR), they have not been yet widely implemented in underground mining vehicles. Moreover, the Common Rail Direct Injection (CRDI) system stabilizes the output power and reduces the fuel consumption of the turbocharged diesel engine under transient modes of loading and speed.

The important role in exhaust gas emission is the setting made by the machine producer in the Electronic Control Unit (ECU), which controls the whole process of machine operation. Depending on certain operator experience, the machine can be operated at different engine rotations and motion speeds. The real power and torque characteristics are given in Figure 2. For this engine, the maximum torque is provided at about 1400 rpm. Although the operators intuitively try to work around this point by gear selection and acceleration regulation; it could not be the optimum by the minimum of exhaust gas emission.

As mentioned earlier, nitrogen oxides from combustion engines are the largest contributor to pollution of the mine atmosphere. Therefore, it is important for the safety of the underground crew to monitor them constantly. Since not all machines are equipped with appropriate sensors, this problem can be solved by modeling. Once the factors and their influence on the increase in the value of NOx concentrations on the engine exhaust have been determined, it is possible to determine the top-down operating parameters of the machine, which the machine operator can control independently, e.g., machine speed, and engine rotation.

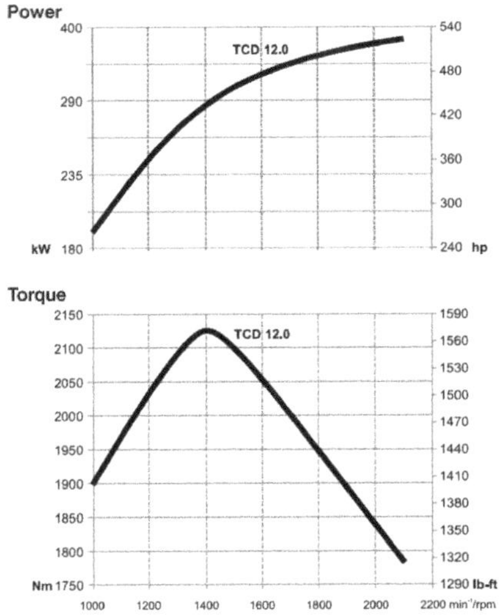


**Figure 2.** The characteristics of the output power and torque by the rotation speed for DEUTZ TCD 12.0 V6 diesel engine [30].

In the area of exhaust gas emission prediction, there are several approaches to get a relation based both on physical and empirical models [33,34]. The limitations of physical models are, in fact, that they require some not measured parameters, e.g., in-cylinder burned gas temperature, the ambient gas-to-fuel ratio, the mass of injected fuel, etc. [35]. Engine emissions due to components' aging, parameters drift, and tolerances violation pose serious problems in meeting emission regulations. To meet practical demand, some authors [36] proposed an optimal linear output feedback controller and a set-point adaption loop on the exhaust gas recirculating rate. Low accuracy restricts the application of physical models in practice. Therefore, in the paper [37], the authors used a statistical approach and correlation analysis to study the main influencing factors of engine torque and NOx emission. They obtained accurate regression models and discovered that ambient temperature in the range 5–30 °C has a great influence both on torque and NOx prediction. The experimental research of intake air humidity influence on the emissions of a turbocharged diesel engine has been conducted in [38]. The relative air humidity was varied from 31 to 80% at a fixed ambient air temperature of 26 °C. The results of tests under three levels of load and rotation speed showed that increasing the intake of air moisture causes less by 3–14% of the NOx emissions. However, since the ambient temperature and humidity in certain underground mines do not variate significantly (+35 °C and 60%), these factors can be neglected in the prediction model.

In general, ANN-based engine models offer a multi-dimensional, adaptive, and learning tool, which does not require knowledge of the governing equations for engine combustion kinetics for emissions prediction [39]. However, this approach requires model training and is difficult to implement in the vehicle onboard monitoring systems due to restricted computing resources. For earth-moving operations with wheel loaders, authors in [40] analyzed energy use and emissions ($CO_2$, CO, NOx, CH4, VOC, PM) based on the criterion of the fuel consumption per cubic meter of hauled material. Using Artificial Neural Networks (ANN) and discrete event simulations, they showed that the fuel consumption and emissions of wheel loaders are mostly dependent on engine load, utilization rate (idle time), and bucket payload.

In the paper [41], three nonlinear models were evaluated: ANN, the split and fit algorithm, and a polynomial NARX model with linear parameters. In the transient mode of the automotive diesel engine, each algorithm showed good prediction accuracy and a short time (0.3 ms) of calculation. By the training time, the split and fit algorithm was the quickest (50 s). The authors concluded that such models are much more accurate than the frequently used engine map and the linear fit models, moreover, in the transient mode. Authors in [42] developed fast one-dimensional models for NOx emission prediction based on the Extended Zeldovich mechanism and different calibration multiplier maps. It is shown that turbine inlet temperature, in-cylinder maximum temperature, maximum pressure, load, CA50 (Crank Angle position where 50% of the heat is released), exhaust gas recirculating rate, and fuel-air ratio are the most critical map parameters for accurate NOx prediction The problem of input parameter selection for the AI-based NOx emission prediction models is considered in [43]. The gradient boosting regression (GBR) model was used to train based on 10 input features. The coefficient of determination ($R^2$) values is within 0.88–0.99 for different driving routes. The most important features for the NOx prediction are mass air flow rate (g/s), exhaust flow rate ($m^3$/min), and $CO_2$ (ppm).

## 3. Measurement Method

The NOx sensor is permanently installed on the underground articulated LHD vehicle with the diesel engine DEUTZ TCD 12.0 V6 (see Figure 3). This is a 6-cylinder in-line engine with a charge air cooling and exhaust turbocharger. The engine manufacturer declares a lifespan of about 1 million km. By official information, the engine copes well with sharply increasing loads providing 90% of the maximum torque already at 1300 rpm. Additional parameters of the engine are given in Table 2. Best point consumption refers to diesel with a density of 0.835 kg/dm$^3$ at 15 °C.

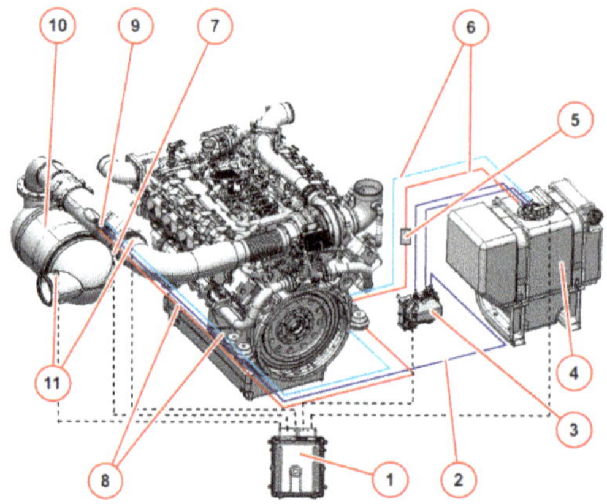


**Figure 3.** The diesel engine DEUTZ TCD 12.0 V6 installed on the LHD-1701 vehicle and its SCR system [30]: 1—Engine control module; 2—AdBlue® pipe; 3—AdBlue® pump; 4—AdBlue® tank; 5—Solenoid valve; 6—Coolant line for preheating the AdBlue® tank; 7—Exhaust gas temperature sensor; 8—Coolant line for cooling the proportioner; 9—Dispenser; 10—SCR catalytic converter; 11—NOx sensor.

**Table 2.** Additional parameters of DEUTZ TCD 12.0 V6 diesel engine [30].

| Parameter | Value | Units |
|---|---|---|
| Number of cylinders | 6 | |
| Piston stroke | 145 | mm |
| Cylinder bore | 132 | mm |
| Displacement | 11.900 | cc |
| Specific fuel consumption | 194 | g/kWh |
| Euro standards | Euro 5 | |

The typical parameters of the NOx sensor are given in Table 3. The current regulation implies the minimum NOx mass measurement accuracy requirements of either ±20% or ±0.1 g/bhp-h [44]. However, the majority of NOx sensors are not able to meet these demands under transient loading due to different factors of noise like NOx sensor tolerance, exhaust flow rate, cross-sensitivity to ammonia (NH3), mass airflow (MAF), and sensor position. The noise of many sensors is about 10 ppm at a zero NOx concentration, which can be caused by residual NOx in the exhaust system. While at 100 ppm NOx concentration, the accuracy is approximate ±10% and achieves a better than ±10% accuracy at NOx values of 500 ppm or higher [45].

**Table 3.** Parameters of NOx sensor.

| Parameter | Value | Units |
|---|---|---|
| Measuring range (NOx) | 0–1500 | ppm |
| Accuracy | ±10 (20) | % |
| Operating temperature | −40–105 | °C |
| Exhaust gas temperature | <800 | °C |

In the data given for analysis taken from the onboard monitoring system, any of the above-mentioned combustion process model parameters were not available. Instead, the list of parameters stored on the server of the mining enterprise is given in Table 4.

**Table 4.** Parameters of the monitoring system taken for analysis.

| Nr | Parameter | Description | Units |
|---|---|---|---|
| | **ENGNOX** | **NOx Emissions** | **ppm** |
| 1 | ENGCOOLT | Coolant temperature | °C |
| 2 | ENGOILT | Oil Temperature | kPa |
| 3 | ENGRPM | Engine rotations | rpm |
| 4 | ENGTPS | Engine acceleration | % |
| 5 | FUELUS | Fuel consumption | L/h |
| 6 | GROILP | Gear oil pressure | kPa |
| 7 | GROILT | Gear oil temperature | C |
| 8 | HYDOILP | Hydraulic oil pressure | MPa |
| 9 | INTAKEP | Intake air pressure | kPa |
| 10 | INTAKET | Intake air temperature | °C |
| 11 | SPEED | Machine speed | km/h |

The sampling frequency of all parameters is 1 s except for INTAKEP, with 5 s. Results of preliminary data analysis of are represented in Figures 4 and 5. On the first graph, several cycles are shown of LHD loading and unloading, each lasting about 4 min (240–250 s). In every cycle, the largest oscillating values of NOx are observed at the beginning and during the completion of the cycle. On the other graphs, for the different engine parameters, the first cycle is shown along with moving average (10 points) curves. The most correlated with a NOx parameter is intake pressure (INTAKEP), although it has the longest sampling period (5 s). The other parameters, namely, rotation speed (ENGRPM). fuel use (FUELUSE) and engine acceleration (ENGTPS) are less statistically related to emission (ENGNOX) due to larger short-time deviations. The smoothed curves have fewer deviations and follow the ENGNOX curve more clearly. The other parameters from Table 4 react more slowly to NOx changes. The highest accuracy of the second-order polynomial regression function ($R^2$ = 0.5216 for INTAKEP) is enough low. Other types of regression functions with a single input parameter do not increase fitting accuracy. Hence, further improvement of prediction methodology is needed based on multivariate regression models.

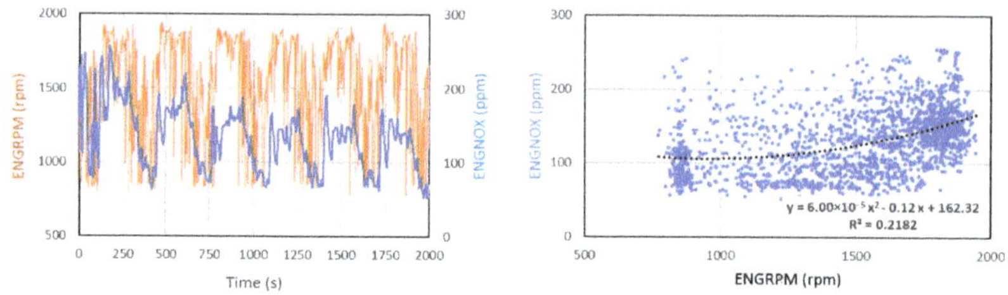


**Figure 4.** Relation of NOx emission (ENGNOX) with engine rotation speed (ENGRPM) over several working cycles with corresponding second-order polynomial regression.

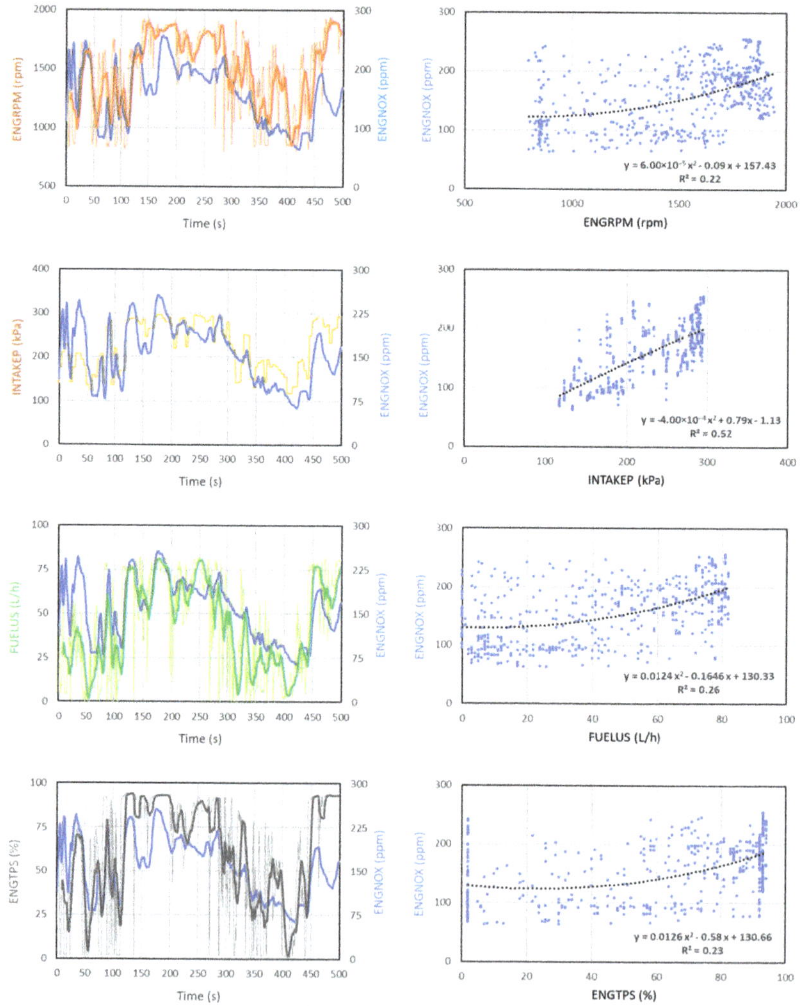


**Figure 5.** Relations of NOx emission (ENGNOX) with engine rotation speed (ENGRPM); engine intake pressure (INTAKEP); fuel use (FUELUS); and engine acceleration (ENGTPS) with the corresponding second-order polynomial regressions.

Variables for multivariate analysis were selected based on their impact on the fit of the predictive model to real data. The indicator referred to by the authors was primarily the coefficient of determination $R^2$. Additionally, information value and weight of evidence analysis have been conducted, and the results are consistent with the $R^2$ assessment.

## 4. Methodology

After the input variables selection, the algorithm of data processing assumed in this paper consists of the following steps:

1. Selection of recorded data segment not less than 10–20 working cycles.
2. Data is split into 90% training part and 10% test part.
3. A model is selected for the training data set (polynomial).
4. Model is tested on the test data.
5. From the original data and model output, $R^2$ and RMSE statistics are calculated, which allows evaluation of the prediction quality.
6. Moreover, the RMSE is calculated normalized to the segment length to be able to calculate the learning and test parts and compare them correctly.
7. The above steps have been done for two different model orders: 3 and 4, to see which model is good enough on the learning part, but not too good that there is no over-fitting on the test part, as seen in the higher order predictions.
8. The above-mentioned steps are fulfilled for two scenarios: first, for all 11 variables that are initially selected, and then for those 10 variables, for which the predicted and initial data are correlated most strongly.

### 4.1. Model Estimation

Multivariate polynomial (sometimes called "multinomial") fitting procedure solves for the coefficients of a polynomial regression model using traditional linear least squares technique [46]. It is implemented using QR factorization with pivoting to solve the system. This is more stable than the simple, unpivoted QR. We have also used automatic variable scaling to deal with a simple cause of ill-conditioning.

Once the model has been specified, the estimation procedure itself is rather simple. The problem becomes that of estimation of the vector $x$, given the linear system of equations:

$$A * x = y \tag{1}$$

For this estimation to have a unique solution, matrix A should be both non-singular and have more rows than columns. Problems with fewer rows than columns are called under-determined, and in such cases, it is strongly recommended to either obtain more data or reduce the order of the model. Assuming matrix $A^{n \times p}$ with $n > p$, we can solve this system via many different approaches, such as the pseudoinverse method, least-squares method, normal equations, i.e., $x = (A'A) \setminus (A'y)$, QR or pivoted QR, just to name a few. Of these methods, only those based on the QR factorization will also directly yield estimated variances for the parameters. A pivoted QR is also reasonably efficient, as well as numerically stable, which is why this approach has been chosen.

## 5. Data Analysis

For testing the methodology, the authors used data describing a single work shift in the mine. In the first step, a subset of usable variables have been selected from all of the channels since most of them contained only (or almost only) empty values. Out of those, the authors selected the ones that presented any meaningful behavior at all. For example, the variable describing if the engine is on or off is not helpful, since it is on during the entire shift. At this point, 11 variables remained, and those were used for the analysis (see Table 4). Raw signals of the selected variables are presented in Figure 6.

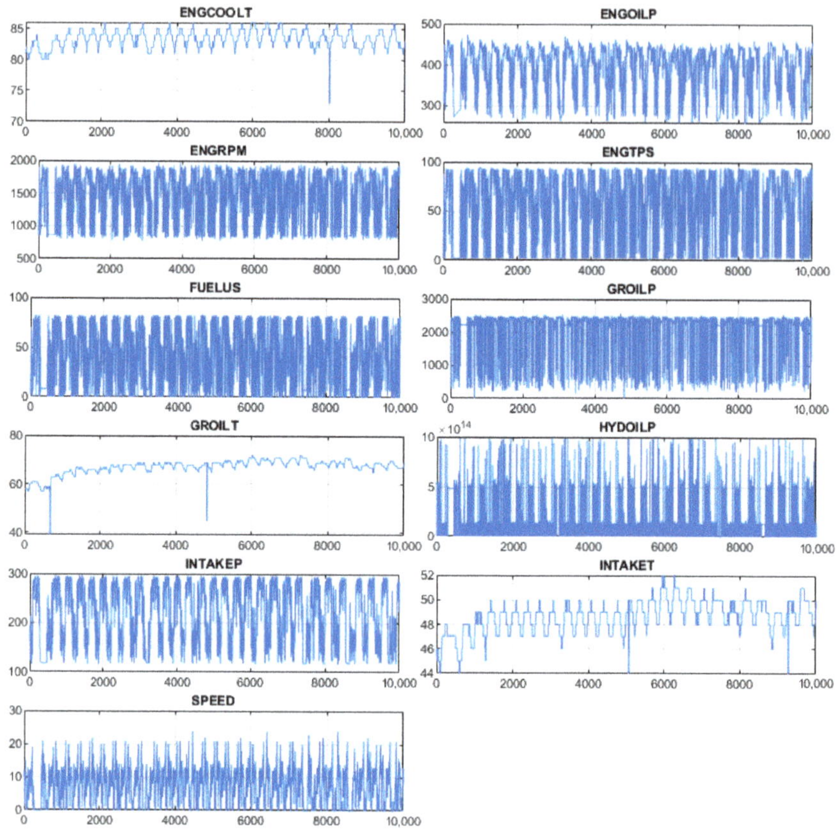


**Figure 6.** Full set of input variables.

The remaining data set has been divided into the training set and a testing set with a proportion of 90–10%, respectively. The input variables of the training set are then provided to the multivariate polynomial fitting procedure (see Section 4.1). When the models are fitted, they are evaluated in the testing segment. Models of orders 3 and 4 were tested. Those orders were chosen because orders lower than 3 presented poor quality of fit, and orders higher than 4 experienced too strong over-fitting errors. The results of fitting the models can be observed in Figure 7. RMSE values are summarised in Table 5.

**Table 5.** RMSE values for the different input variables taken in the model.

| | 10 Variables | 11 Variables |
|---|---|---|
| Model of order 3 | 23.231 | 21.179 |
| Model of order 4 | 17.342 | 14.131 |

Additionally, Table 6 presents the normalized RMSE (NRMSE) values for both orders with the distinction of training and testing segments. It is obtained by calculating the ordinary RMSE value and then dividing it by the number of samples in each respective segment. This way, the values can be compared. It is clearly visible that with higher order, the quality of fitting the model to the training segment increases, but it also causes increasing over-fitting problems in the testing segment. Moreover, the error value is always lower for the training segment in comparison to the testing segment, which is understandable.

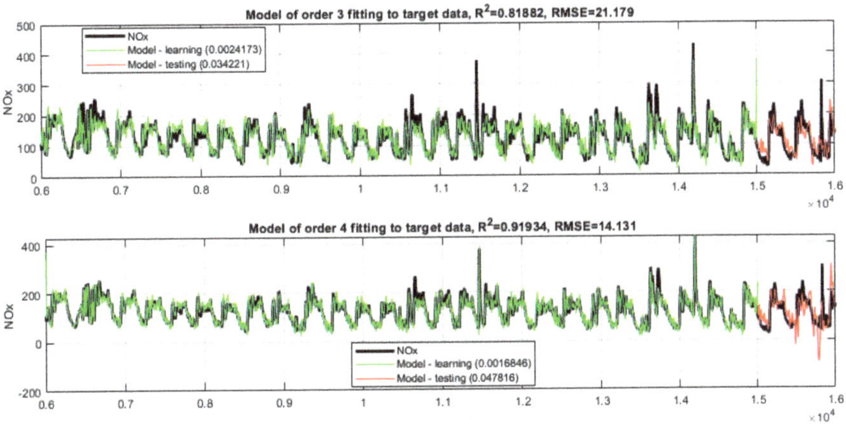


**Figure 7.** Orders 3 and 4, all 11 variables.

**Table 6.** NRMSE errors for respective model orders and 11 variables.

| | Training Segment | Testing Segment |
|---|---|---|
| Model of order 3 | 0.0024 | 0.0342 |
| Model of order 4 | 0.0016 | 0.0478 |

After that, the authors measured the cross-correlation between the obtained models and the individual input variables. One variable with the lowest correlation factor (ENGCOOLT—temperature of engine coolant) has been removed from the set of input variables. A reduced set of 10 input variables was used again to fit the models. The results are presented in Figure 8. Similarly, Table 7 presents the NRMSE values. In this case, one can also observe the effect where with increasing model order, the fit quality increases but also over-fitting values increase the error on the testing set.

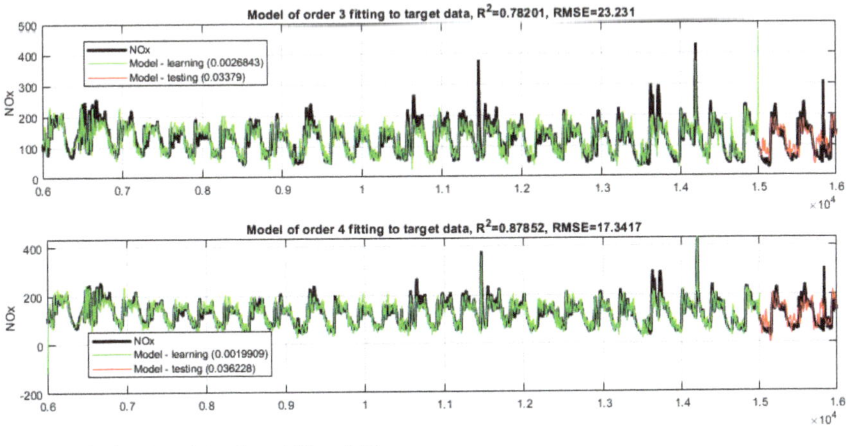


**Figure 8.** Orders 3 and 4, selected 10 variables.

**Table 7.** NRMSE errors for respective model orders and 10 variables.

| | Training Segment | Testing Segment |
|---|---|---|
| Model of order 3 | 0.0026 | 0.0337 |
| Model of order 4 | 0.0019 | 0.0362 |

It is also interesting to compare NRMSE tables. For 11 variables, errors for training segments are lower than for 10 variables. There is more input data to work on, so the fit is better. However, for 11 variables, errors for testing segments are higher than for 10 variables, because the over-fitting is more significant, and models with 10 input variables have better generalizing quality.

One can observe that NRMSE values for the training segment are lower for a full set of variables than for the reduced set. It comes from the fact that the model has more information to learn on. Similarly, NRMSE values are higher for a full set of variables than for the reduced set, because the model fitted on the full set is more specialized and has worse generalization properties, hence it tends to display over-fitting problems. The correlation of predicted and original data on NOx emission is shown in Figures 9 and 10.

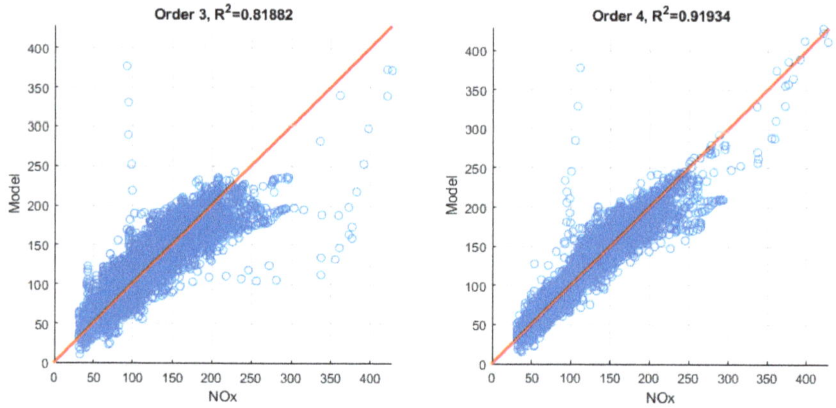


**Figure 9.** Correlation of predicted and original data for all 11 variables.

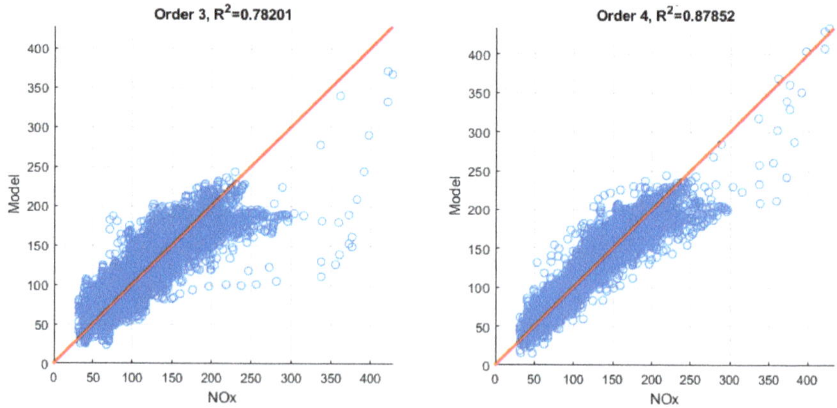


**Figure 10.** Correlation of predicted and original data for selected 10 variables.

## 6. Discussion

The research presented in this paper allows putting forth several points for the discussion:

- The decision of the model order selection has to be balanced between the quality of fit during the fitting phase and the amount of over-fitting exhibited while testing on a new portion of data.
- It is important to select the meaningful input variables. The authors made an attempt to use the full data set of 56 variables, no matter how irrelevant they were, because in theory, it would maximize the information. In practice, the results were meaningless and unusable. Moreover, even if one thinks that they selected the meaningful data set, it becomes visible that removing some variables can improve the generalization capability of the model, even if it slightly decreases the quality of fit on the training segment.
- For a larger data set (in terms of time), it is not obvious if the modeling with the same parameters will yield better or worse results. The authors attempted to use not one but two shifts of time as a working data set. It turned out that for the second shift, some other operator was driving the machine, and the data looked completely different. One could say that more data did not provide more information, and with the same parameters, the results were much worse.
- Some other parameters of LHD vehicle's operation, which are not explicitly reflected in the signals available via CAN bus, may have an influence on the NOx prediction, such as road waviness and its watering. However, this factor is reflected in several working parameters (FUELUS and partly ENGRPM). Temperature and humidity are also among the factors influencing the fuel combustion process. However, they are enough stable in a certain location of every underground mine and easily measured by the simple, low-cost digital sensors, which signals can easily be added to the onboard monitoring system.
- The achieved deviation of the values predicted by the model from the values measured by the permanently installed sensor is less than 15%, which is comparable with the accuracy of the NOx sensor itself (up to 20% under sharp loads application). Following Figure 9, the certain role in accuracy level plays the outliers, which are also visible in time series graphs as rare peaks during the transient periods of work. The duration of these separate periods and in total is very short; hence, they can be rejected to increase the accuracy of NOx prediction over the majority of remaining data samples.
- Subjectively, the authors assess that the result of the presented research was successful. However, it is obvious that further research needs to be conducted to try different modeling techniques. In this paper, the authors made the first attempt to solve the problem of the lack of NOx sensors on a large scale using the simplest possible technique, which is modeled using polynomial fitting. This way, it is possible to provide a practical solution for NOx emission assessment for the industry using simple software tools and not time-consuming procedures. The authors are well aware that there are a lot of different techniques of mathematical modeling, and the investigation of their potential for such use cases will be the subject of further work.

## 7. Conclusions

The conducted research resulted in the development of the statistical model for NOx emission values prediction.

1. **The model structure is optimized by order of polynomials and the number of input parameters.** During the research, it was decided to include 11 parameters measured with SYNAPSA systems (Table 4). In addition, models of orders 3 and 4 were compared. The choice is explained in Section 5—Data Analysis. The best achieved prediction accuracy by the 4th order polynomial model for 11 and 10 input variables is about 8% and 13%, respectively. It is comparable with the sensors' accuracy of 10% at a stable regime of loading and 20% in the transient periods of operation.

2. **The developed model can be considered as a "soft sensor" and used for NOx emissions monitoring and prediction in heavy-duty LHD vehicles with diesel engines** The solution presented in the article will contribute to a better understanding of the working atmosphere of underground crews at workplaces near LHDs. Since most machines are not equipped with sensors for measuring the concentrations of harmful gases, the proposed statistical methods can help improve working conditions by predicting the possible threat of atmospheric pollution by nitrogen oxides.
3. **These data of predicted NOx emissions can be utilized as input information for ventilation system power demand and capacity planning based on the production plan, required fleet of vehicles, the length of the transportation routes, and underground mining conditions.** With the statistical method proposed in the article for predicting the emission of NOx concentrations into the mine atmosphere, it will be possible to optimize the ventilation system for underground workings. Knowing the production plan and the demand for diesel-powered machinery will make it possible to estimate the value of nitrogen oxide concentrations at the workplace. This will enable ventilation services to manoeuvre the ventilation system accordingly, for example, by increasing the air volume flow to ventilate the workings faster.

**Author Contributions:** Conceptualization, A.B. and R.Z.; methodology, J.W.; software, J.W.; validation, P.Ś. and J.W.; formal analysis, P.K., P.Ś. and R.Z.; investigation, A.B.; resources, R.Z.; data curation, P.Ś., P.K.; writing—original draft preparation, A.B., J.W., P.K. and R.Z.; writing—review and editing, A.B. and P.K.; visualization, J.W.; supervision, R.Z.; project administration, A.B.; funding acquisition, R.Z. All authors have read and agreed to the published version of the manuscript.

**Funding:** This research received no external funding.

**Data Availability Statement:** Data is unavailable due to privacy restrictions.

**Acknowledgments:** This work is supported by the European Institute of Innovation and Technology (EIT), a body of the European Union, under the Horizon Europe, the EU Framework Programme for Research and Innovation. EIT RawMaterials GmbH under Framework Partnership Agreement No 21119 (VOT3D—Ventilation Optimizing Technology based on 3D-scanning).

**Conflicts of Interest:** The authors declare no conflict of interest.

## Abbreviations

The following abbreviations are used in this manuscript:

| | |
|---|---|
| LHD | Load-haul-dumping vehicle |
| NOx | Nitrogen Oxides |
| NMHC | Non-Methane Hydro Carbons |
| SCR | Selective Catalytic Reduction |
| LNT | Lean NOx Trap |
| DOC | Diesel Oxidation Catalysts |
| EAT | Exhaust Aftertreatment |
| EGR | Exhaust Gas Recirculating |
| DPF | Diesel Particulate Filter |
| ECU | Electronic Control Unit |
| CRDI | Common Rail Direct Injection |
| PLT | Progressive load test |
| VTT | Vehicle transient test |
| PM | Particulate matter |
| VOC | Volatile organic compounds |
| ANN | Artificial Neural Networks |
| MAF | Mass airflow |
| RMSE | Root mean square error |
| NRMSE | Normalized Root Mean Square Error |

## References


1. Jiskani, I.M.; Cai, Q.; Zhou, W.; Chang, Z.; Chalgri, S.R.; Manda, E.; Lu, X. Distinctive model of mine safety for sustainable mining in Pakistan. *Mining Metall. Explor.* **2020**, *37*, 1023–1037. [CrossRef]
2. Hebda-Sobkowicz, J.; Gola, S.; Zimroz, R.; Wyłomańska, A. Identification and statistical analysis of impulse-like patterns of carbon monoxide variation in deep underground mines associated with the blasting procedure. *Sensors* **2019**, *19*, 2757. [CrossRef] [PubMed]
3. Wodecki, J.; Stefaniak, P.; Michalak, A.; Wyłomańska, A.; Zimroz, R. Technical condition change detection using Anderson–Darling statistic approach for LHD machines—Engine overheating problem. *Int. J. Min. Reclam. Environ.* **2017**, *32*, 392–400. [CrossRef]
4. Wyłomańska, A.; Zimroz, R. The analysis of stochastic signal from LHD mining machine. In *Stochastic Models, Statistics and THEIR Applications*; Springer: Berlin/Heidelberg, Germany, 2015; pp. 469–478.
5. Michalak, A.; Śliwiński, P.; Kaniewski, T.; Wodecki, J.; Stefaniak, P.; Wyłomańska, A.; Zimroz, R. Condition Monitoring for LHD Machines Operating in Underground Mine—Analysis of Long-Term Diagnostic Data. In *Proceedings of the 27th International Symposium on Mine Planning and Equipment Selection-MPES 2018*; Springer: Cham, Switzerland, 2019; pp. 471–480.
6. Dong, L.; Tong, X.; Li, X.; Zhou, J.; Wang, S.; Liu, B. Some developments and new insights of environmental problems and deep mining strategy for cleaner production in mines. *J. Clean. Prod.* **2019**, *210*, 1562–1578. [CrossRef]
7. Ziętek, B.; Banasiewicz, A.; Zimroz, R.; Szrek, J.; Gola, S. A portable environmental data-monitoring system for air hazard evaluation in deep underground mines. *Energies* **2020**, *13*, 6331. [CrossRef]
8. Wallace, K.; Prosser, B.; Stinnette, J.D. The practice of mine ventilation engineering. *Int. J. Min. Sci. Technol.* **2015**, *25*, 165–169. [CrossRef]
9. Wróblewski, A.; Banasiewicz, A.; Gola, S. Heat Balance Determination Methods for Mining Areas in Underground Mines-A Review. In Proceedings of the IOP Conference Series: Earth and Environmental Science, Surakarta, Indonesia, 24–25 August 2021; IOP Publishing: Bristol, UK, 2021; Volume 942, p. 012011.
10. Struminski, A.; Madeja-Struminska, B. Mine ventilation practice in Polish copper mines. In *Mining in the New Millennium Challenges and Opportunities*; CRC Press: Boca Raton, FL, USA, 2020; pp. 173–179.
11. Slazak, N.; Obracaj, D.; Borowski, M. Methods for controlling temperature hazard in Polish coal mines. *Arch. Min. Sci.* **2008**, *53*, 497–510.
12. Yin, Z.; Linga, P. Methane hydrates: A future clean energy resource. *Chin. J. Chem. Eng.* **2019**, *27*, 2026–2036. [CrossRef]
13. Shaw, S.; Van Heyst, B. An Evaluation of Risk Ratios on Physical and Mental Health Correlations due to Increases in Ambient Nitrogen Oxide (NOx) Concentrations. *Atmosphere* **2022**, *13*, 967. [CrossRef]
14. Ghose, M.K.; Majee, S. Sources of air pollution due to coal mining and their impacts in Jharia coalfield. *Environ. Int.* **2000**, *26*, 81–85. [CrossRef]
15. Oluwoye, I.; Dlugogorski, B.Z.; Gore, J.; Oskierski, H.C.; Altarawneh, M. Atmospheric emission of NOx from mining explosives: A critical review. *Atmos. Environ.* **2017**, *167*, 81–96. [CrossRef]
16. Banasiewicz, A.; Janicka, A.; Michalak, A.; Włostowski, R. Photocatalysis as a method for reduction of ambient NOx in deep underground mines. *Measurement* **2022**, *200*, 111453. [CrossRef]
17. Kampa, M.; Castanas, E. Human health effects of air pollution. *Environ. Pollut.* **2008**, *151*, 362–367. [CrossRef] [PubMed]
18. Fukuto, J.M.; Cho, J.Y.; Switzer, C.H. The Chemical Properties of Nitric Oxide and Related Nitrogen Oxides. In *Nitric Oxide*; Elsevier: Amsterdam, The Netherlands, 2000; pp. 23–40. [CrossRef]
19. Abdelsalam, E.M.; Mohamed, Y.; Abdelkhalik, S.; El Nazer, H.A.; Attia, Y.A. Photocatalytic oxidation of nitrogen oxides (NOx) using Ag-and Pt-doped $TiO_2$ nanoparticles under visible light irradiation. *Environ. Sci. Pollut. Res.* **2020**, *27*, 35828–35836. [CrossRef] [PubMed]
20. Almetwally, A.A.; Bin-Jumah, M.; Allam, A.A. Ambient air pollution and its influence on human health and welfare: An overview. *Environ. Sci. Pollut. Res.* **2020**, *27*, 24815–24830. [CrossRef]
21. Galbreath, K.C.; Zygarlicke, C.J.; Tibbetts, J.E.; Schulz, R.L.; Dunham, G.E. Effects of NOx, $\alpha$-$Fe_2O_3$, $\gamma$-$Fe_2O_3$, and HCl on mercury transformations in a 7-kW coal combustion system. *Fuel Process. Technol.* **2005**, *86*, 429–448. [CrossRef]
22. Hori, M.; Matsunaga, N.; Malte, P.C.; Marinov, N.M. The effect of low-concentration fuels on the conversion of nitric oxide to nitrogen dioxide. In Proceedings of the Symposium (International) on Combustion, Sydney, Australia, 5–10 July 1992; Elsevier: Amsterdam, The Netherlands, 1992; Volume 24, pp. 909–916.
23. Özmen, İ.; Aksoy, E. Respiratory emergencies and management of mining accidents. *Turk. Thorac. J.* **2015**, *16*, S18. [CrossRef]
24. Shriwas, M.; Pritchard, C. Ventilation Monitoring and Control in Mines. *Min. Metall. Explor.* **2020**, *37*, 1015–1021. [CrossRef]
25. Iqbal, M.Y.; Wang, T.; Li, G.; Li, S.; Hu, G.; Yang, T.; Gu, F.; Al-Nehari, M. Development and Validation of a Vibration-Based Virtual Sensor for Real-Time Monitoring NOx Emissions of a Diesel Engine. *Machines* **2022**, *10*, 594. [CrossRef]
26. Söderena, P.; Laurikko, J.; Weber, C.; Tilli, A.; Kuikka, K.; Kousa, A.; Väkevä, O.; Venho, A.; Haaparanta, S.; Nuottimäki, J. Monitoring Euro 6 diesel passenger cars NOx emissions for one year in various ambient conditions with PEMS and NOx sensors. *Sci. Total Environ.* **2020**, *746*, 140971. [CrossRef]
27. Krot, P.; Sliwinski, P.; Zimroz, R.; Gomolla, N. The identification of operational cycles in the monitoring systems of underground vehicles. *Measurement* **2020**, *151*, 107111. [CrossRef]

28. Krot, P.; Zimroz, R.; Sliwinski, P.; Gomolla, N. Safe Operation of Underground Mining Vehicles Based on Cyclic Fatigue Monitoring of Powertrains. In *Structural Integrity and Fatigue Failure Analysis*; Lesiuk, G., Szata, M., Blazejewski, W., Jesus, A.M.D., Correia, J.A., Eds.; Springer International Publishing: Cham, Switzerland, 2022; pp. 283–292. [CrossRef]
29. KGHM ZANAM. LHD LD1701. Available online: https://www.kghmzanam.com/wp-content/uploads/2021/05/LKP_1701_EN.pdf (accessed on 16 January 2023).
30. DEUTZ TCD 12.0 and TCD 16.0 Diesel Engine. Available online: https://www.deutzsupport.com/product-details/tcd-12-0-and-tcd-16-0/ (accessed on 16 January 2023).
31. Stachulak, J.; Allen, C. Evaluation of the effects of diesel oxidation catalysts on $NO_2$ emissions from diesel-powered mining vehicles. *CIM J.* **2020**, *11*, 104–110. [CrossRef]
32. Demuynck, J.; Favre, C.; Bosteels, D.; Bunar, F.; Spitta, J.; Kuhrt, A. Diesel Vehicle with Ultra-Low NOx Emissions on the Road. In Proceedings of the 14th International Conference on Engines & Vehicles; SAE International, Napoli, Italy, 15–19 September 2019. [CrossRef]
33. Stobart, R.; Yang, Z. A control-oriented NOx emissions model for diesel engines. *Int. J. Powertrains* **2016**, *5*, 191–210. [CrossRef]
34. Rao, V.; Honnery, D. A comparison of two NOx prediction schemes for use in diesel engine thermodynamic modelling. *Fuel* **2013**, *107*, 662–670. [CrossRef]
35. d'Ambrosio, S.; Finesso, R.; Fu, L.; Mittica, A.; Spessa, E. A control-oriented real-time semi-empirical model for the prediction of NOx emissions in diesel engines. *Appl. Energy* **2014**, *130*, 265–279. [CrossRef]
36. Tschanz, F.; Amstutz, A.; Onder, C.H.; Guzzella, L. Control of diesel engines using NOx-emission feedback. *Int. J. Engine Res.* **2013**, *14*, 45–56. [CrossRef]
37. Yuan, Z.; Shi, X.; Jiang, D.; Liang, Y.; Mi, J.; Fan, H. Data-Based Engine Torque and NOx Raw Emission Prediction. *Energies* **2022**, *15*, 4346. [CrossRef]
38. Asad, U.; Kelly, C.; Wang, M.; Tjong, J. Effects of Intake Air Humidity on the NOX Emissions and Performance of a Light-Duty Diesel Engine. In Proceedings of the 2012 Internal Combustion Engine Division Fall Technical Conference, Vancouver, BC, Canada, 23–26 September 2012. [CrossRef]
39. Obodeh, O.; Ajuwa, C.I. Evaluation of Artificial Neural Network Performance in Predicting Diesel Engine NOx Emissions. *Eur. J. Sci. Res.* **2009**, *33*, 642–653.
40. Jassim, H.S.; Lu, W.; Olofsson, T. Determining the environmental impact of material hauling with wheel loaders during earthmoving operations. *J. Air Waste Manag. Assoc.* **2019**, *69*, 1195–1214. [CrossRef]
41. Krijnsen, H.C.; Bakker, R.; van Kooten, W.E.J.; Calis, H.P.A.; Verbeek, R.P.; van den Bleek, C.M. Evaluation of Fit Algorithms for NOx Emission Prediction for Efficient DeNOx Control of Transient Diesel Engine Exhaust Gas. *Ind. Eng. Chem. Res.* **2000**, *39*, 2992–2997. [CrossRef]
42. Ozgul, E.; Bedir, H. Fast NOx emission prediction methodology via one-dimensional engine performance tools in heavy-duty engines. *Adv. Mech. Eng.* **2019**, *11*, 1–16. [CrossRef]
43. Wen, H.T.; Lu, J.H.; Jhang, D.S. Features Importance Analysis of Diesel Vehicles NOx and $CO_2$ Emission Predictions in Real Road Driving Based on Gradient Boosting Regression Model. *Int. J. Environ. Res. Public Health* **2021**, *18*. [CrossRef] [PubMed]
44. Funk, S. Real world NOx sensor accuracy assessment and implications for REAL NOx tracking. In *SAE Technical Paper Series*; SAE International: Warrendale, PA, USA, 2021; Number 2021-01-0593.
45. Kawamoto, Y.; Todo, Y.; Shimokawa, H.; Aoki, K.; Kawai, M.; Ide, K. Development of High Accuracy NOx Sensor. In Proceedings of the WCX SAE World Congress Experience, Detroit, MI, USA, 9–11 April 2019; SAE International: Warrendale, PA, USA, 2019. [CrossRef]
46. Draper, N.R.; Smith, H. *Applied Regression Analysis*; John Wiley & Sons: Hoboken, NJ, USA, 1998; Volume 326.

energies

MDPI

Article

# Detection of Blockages of the Belt Conveyor Transfer Point Using an RGB Camera and CNN Autoencoder

Piotr Bortnowski [1,*], Horst Gondek [2], Robert Król [1], Daniela Marasova [3] and Maksymilian Ozdoba [1]

1 Department of Mining, Faculty of Geoengineering, Mining and Geology, Wrocław University of Science and Technology, Na Grobli 15, 50-421 Wrocław, Poland
2 VSB—Department of Machine and Industrial Design, Technical University of Ostrava, 17 Listopadu 2172/15, 708 00 Ostrava, Czech Republic
3 Institute of Logistics and Transport, Faculty BERG, Technical University of Košice, Park Komenského 14, 043 84 Košice, Slovakia
* Correspondence: piotr.bortnowski@pwr.edu.pl

**Abstract:** In the material transfer area, the belt is exposed to considerable damage, the energy of falling material is lost, and there is significant dust and noise. One of the most common causes of failure is transfer chute blockage, when the flow of material in the free fall or loading zone is disturbed by oversized rock parts or other objects, e.g., rock bolts. The failure of a single transfer point may cause the entire transport route to be excluded from work and associated with costly breakdowns. For this reason, those places require continuous monitoring and special surveillance measures. The number of methods for monitoring this type of blockage is limited. The article presents the research results on the possibility of visual monitoring of the transfer operating status on an object in an underground copper ore mine. A standard industrial RGB camera was used to obtain the video material from the transfer point area, and the recorded frames were processed by a detection algorithm based on a neural network. The CNN autoencoder was taught to reconstruct the image of regular transfer operating conditions. A data set with the recorded transfer blockage state was used for validation.

**Keywords:** belt conveyor; transfer point; chute monitoring; anomaly detection; image processing; blockages state



**Citation:** Bortnowski, P.; Gondek, H.; Król, R.; Marasova, D.; Ozdoba, M. Detection of Blockages of the Belt Conveyor Transfer Point Using an RGB Camera and CNN Autoencoder. *Energies* **2023**, *16*, 1666. https://doi.org/10.3390/en16041666

Academic Editor: Duarte Valério

Received: 14 January 2023
Revised: 3 February 2023
Accepted: 5 February 2023
Published: 7 February 2023



## 1. Introduction

A belt conveyor allows bulk materials to be moved over long distances [1,2] while providing optimal electricity consumption [3,4] and reliability [5] indicators. For this reason, belt conveyors are one of the most popular means of transport in industry and mining [6]. The length of the belt conveyor route is limited by the permissible belt stresses in the upper belt [7], resulting from the resistance to movement along the length of the route [8]. As a standard, it is assumed that the operating stresses should be 10% of the nominal belt strength [9]. Intermediate drives are a solution that extends the length of the route [10], but their use is not always possible or cost effective [11]. In such cases, it is necessary to use parallel transfer chutes, which allows the material to be reloaded from the feed conveyor to the receiving conveyor in the transport line [12]. In a situation where it is required to feed the material to the conveyor at a significant angle and change the flow direction, angular chutes are used [13].

Transfer chutes are crucial infrastructure elements that connect conveyors that are part of complex transport systems [14]. Proper operation of transfers determines the continuity of production. Blockage causes production shutdown and, consequently, losses related to downtime and removal of failures [15]. In addition, at the transfer points, there are significant energy losses [16] associated with additional resistance to accelerating the loaded material to the speed of the receiving belt [17] and friction against skirt boards [18]. The appropriate shape of the transfer chute structure allows for feeding the material steadily and at the right speed tangential to the receiving belt. Thanks to this, the resistance to movement is reduced, as well as the wear of the belt. Statistical studies and a review of available failure analyses show that transfer points are where the conveyor belt most often fails [19]. Due to the described issues, transfer chute projects are subject to increased supervision compared to other route elements. The design of modern transfer points is supported by numerical methods [20,21], thanks to which it is possible to optimize new [22] and already working structures [23,24]. There is also considerable dust [25,26] and noise [27] at the transfer points.

A well-designed transfer chute increases production capacity by controlling dust and noise [28], reducing belt wear [28] and mistracking [29], and eliminating potential blockages of the loading zone [30]. Blockages occurring is one of the most dangerous failures of chutes, the conveyor belt may be completely cut or broken, and significant damage to the steel structure of the route may occur [30,31]. Blockages are caused by the appearance of oversized or foreign bodies in the transported material or the feeding of too much material that the conveyor cannot receive [32]. At the moment of blockage, the loading space is gradually filled until it is filled. In this case, the receiving conveyor stops due to a significant increase in resistance or damage or a breakng of the blocked belt [33].

Most standard monitoring methods and non-contact blockage detection are adapted only to chutes with a gutter in the free fall zone and loading zone [34]. The most basic solution is proximity sensors, emitting a beam in the infrared band [35]. Information on the status of regular operation and blockage is sent with the detected movement of material. RF sensors are less and less used due to numerous operation problems in conditions of extreme dust and vibration, which generate false failure states. Over time, radiometric detectors have replaced RF sensors, which are considered one of the most effective solutions [36]. However, the sensor's construction requires using radioactive isotopes to detect the presence of a blockage in the hopper. The radiation source must be shielded and placed in the direction of the material flow in a chute, and the detected radiation level on the opposite side allows for blockage state determination [37]. The measurement method's most significant disadvantage is the cost, safety, and environmental impact. The use of radiometric sensors requires increased supervision of external services that allow the device to work and numerous training of service employees. The most popular solution is microwave sensors, consisting of a low-power microwave emitter and a receiver [38]. This device is relatively cheap compared to radiometric detectors, does not require additional training and safety regulations, and is not sensitive to vibration and dust. Among the contact methods, the most popular are limit switches signaling the status of the chute blockage at the moment of contact with the material at a safety level.

Many transfer points are structures without a chute in the zone of free fall of material; therefore, the measurement methods mentioned in the previous paragraph cannot be used. The article presents the research results on monitoring the transfer operation in an underground copper ore mine using images from an industrial RGB camera. The anomaly detection was based on the CNN autoencoder, which was taught the parameters of regular transfer operation. The detection effectiveness was validated based on collected data frames illustrating the actual lock states. The results of the neural network operation were illustrated based on the network learning loss function, the algorithm's effectiveness, and the error matrix was determined. Anomalous operating states were visualized using color maps superimposed on the anomalous test dataset.

## 2. Related Works

Vision methods in industry and mining are widely used in production quality supervision [39,40] and condition monitoring [41,42]. One of the most developed areas is detecting damage to idlers [43], pulleys, and conveyor belt drives based on anomalies in the image of the thermal imaging camera [44,45]. Increasingly, monitoring using thermal imaging methods is carried out using robots [46,47] and automated diagnostic platforms [48]. The thermal image can also be used to identify people [49]. The RGB image is also used for diagnostic purposes to determine rollers' anomalous operating states [50] and estimate the rotational speed [51]. In the case of conveyor belts, vision algorithms are used to detect surface damage [52–54]. Industrial cameras are used to detect the type of machines and vehicles [55] and to monitor the state of their operation. It is also possible to forecast performance and plan production in real-time based on parameters read from the image [56]. Another area of application is assessing grain size composition [57] based on images from drones [58], industrial cameras on conveyors [59], or cyclic transport vehicles [60]. Cameras installed on machines are used to monitor the wear status of components [61–63]. The issue most similar to the subject discussed in the article is the monitoring of crusher blockages based on the image from the chute [64,65]. Monitoring based on CNN is very often used to evaluate images when supervising production quality [66], monitoring the condition of structures [67], medicine [68] or nanotechnologies [69]. Many of the presented works are limited only to laboratory conditions. This article focuses on the practical use of industrial cameras and artificial intelligence in image processing to detect blockages under operating conditions.

## 3. Materials and Methods

### 3.1. Research Object

The research was carried out on a perpendicular, angular transfer (approx. 90°) in an underground copper ore mine (Figure 1). The nominal capacity of the feeding conveyor is 600 t/h, while the capacity of the receiving conveyor is 1700 t/h. The increased efficiency of the receiving conveyor is because several transfer chutes are located along its route. The receiving conveyor collects material from three feeding conveyors from several branches in the mine. The analyzed transfer is the first construction along the route of the receiving conveyor. The belt speed of the conveyor is 2 m/s, and the speed of the receiving conveyor is 2.7 m/s. The density of the transported material is approx. 1.7 t/m$^3$.

The transfer is a Rock-Box construction, where the material going into the box loses its kinetic energy and then slides onto the conveyor belt. At the moment of contact of the material with the belt, its acceleration occurs. The zone where the material is accelerated has been built with skirt boards, preventing material spillage and guiding material in the acceleration zone.

**Figure 1.** Research object—perpendicular, angular transfer point in an underground copper ore mine.

The selected transfer station is a construction with a tendency to blockades. Transported material moves the phase of preliminary grinding and separation on the grate [70], but this stage does not allow for maintaining a balanced grain size distribution of the ore stream. For this reason, many elongated lumps of rock in the material can cause blockage of the transfer point. In addition, there are often wastes in the stream that are the remnants of exploitation, such as rock bolts. The object also has a disadvantage in its design, as the falling stream of material hits the chute plate directly and does not slide freely down the side chute plate onto the receiving conveyor belt in the loading zone.

### 3.2. Measuring Equipment

Images were captured with an AXIS P1354-E standard industrial RGB camera (Figure 2). The camera is a regular part of the mine monitoring infrastructure, used to monitor the transfer space by the mine dispatcher. The article also tested the recorded image regarding the possibility of automatic detection of transfer blockages and unwanted foreign bodies in material drop and loading areas. The recorded images were in the mine's internal data acquisition system, while the processing was already performed in external software. The Ethernet protocol carried out power, communication, and data transmission. The camera was placed under the roof of the excavation at a distance of approx. 6 m from the transfer point. The camera is resistant to dust, vibrations, and climatic conditions, which is important due to the constantly deteriorating micro climate [71] and temperature above 30 °C. For this reason, no additional cooling of the apparatus was required. The camera has been able to record images with a frame rate of 10 FPS. The data rate was 13,197 kb/s. The video image is distributed into data frames measuring 1920 × 1080 px, 96 dpi resolutions in two directions, and 24 bits deep.

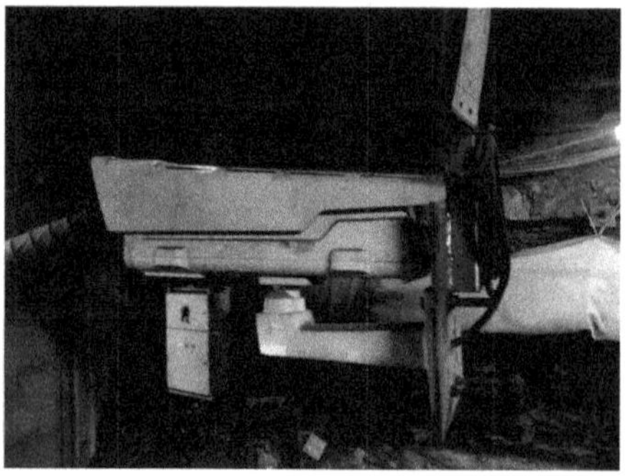

**Figure 2.** Measuring equipment—RGB camera to monitor the working conditions of the transfer point.

### 3.3. Methodology

Data acquisition was the first stage of the work. Data frames in video materials divided over individual frames were collected for different periods of transfer chute operation so that the training data set contained all the characteristic features of the regular operating state. The training set consisted of 1078 photos. To validate the model, 205 images were used, on which there were various types of blockages in the form of oversized lumps and foreign bodies blocking the free flow of material. Figure 3 shows sample images later used to create training and test collections.

The original images were processed in the second stage to isolate distinctive features and reduce interference. Through subsequent trials with different image processing methods and subsequent iterations of algorithm learning, the best results were obtained by degrading the complexity of the input images. The data frames were cropped and resized to represent the area of free fall and loading of material onto the receiving conveyor. Frequency domain filtering is used to reduce noise and image features while preserving all edges of objects. Then, using threshold operators and morphology, the elements of the transfer chute and the flowing material were separated from the background.

The last, third stage included neural network learning in order to obtain the best results of the reconstruction of the input image. A previously trained convolutional neural network VGG16 [72] was used, which was additionally modified for the purposes of the article. The network consists of six blocks. Each block contained convolutional layers and a pooling layer (MaxPooling). Some layers were preserved in a learned state, and the remaining layers were trained. After the section of subsequent blocks, one more layer of pooling (Global Average Pooling) was added. The last layer defined the class of the object being analyzed. Multi-class classification was used for two predicted states with Softmax activation. The autoencoder was validated against a test data set containing anomalous operating states. The loss of image reconstruction and the performance metrics of the prediction model were determined. The results were visualized using modified code [73] that allowed the density of outliers in images classified as abnormal to be determined using a color scale. Figure 4 presents the algorithm of conduct and subsequent tasks discussed in the article.

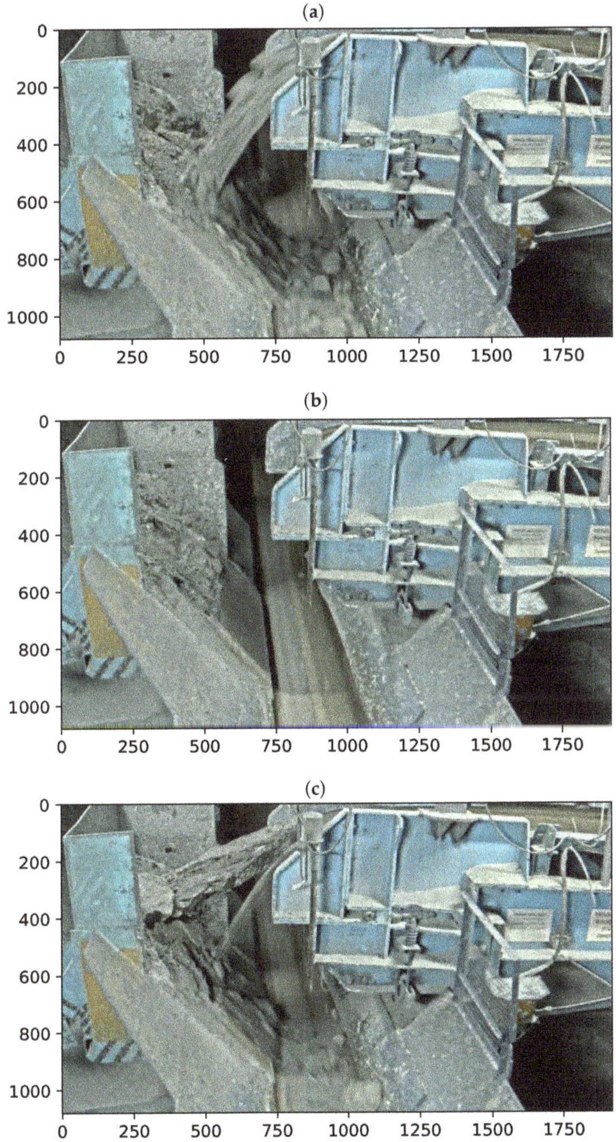


**Figure 3.** Recorded original data frames: (**a**) normal operation of the transfer point—free flow of bulk material; (**b**) normal operation of the transfer point—no flow; (**c**) anomalous operating state of the transfer point—blockage.

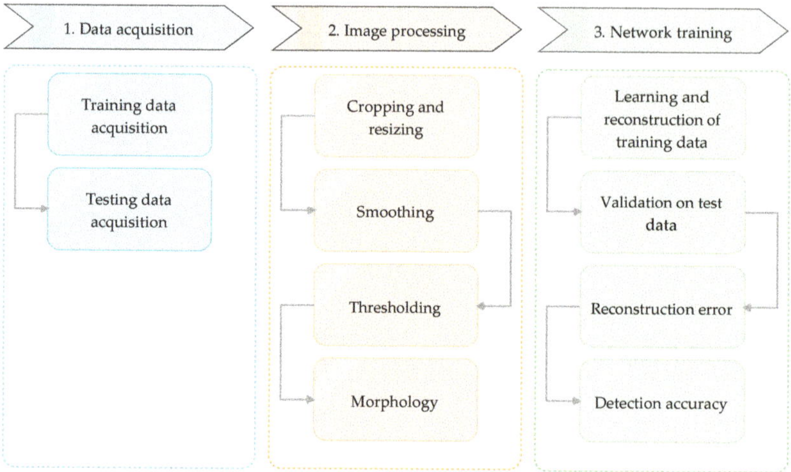


**Figure 4.** Procedure algorithm.

## 3.4. Image Processing

### 3.4.1. Crop

In the first stage of processing, the recorded data frames were cut only to the area where the transfer box was visible, and therefore the moment of free fall of the material and loading on the receiving belt was analyzed. Images from the original size were cropped to 1000 × 1000 for network training, and the input image size was further reduced to 225 × 225 without changing the aspect ratio. Figure 5 shows what an example image of regular (Figure 5a) and failure working state (Figure 5b) looks like after cropping and resizing.

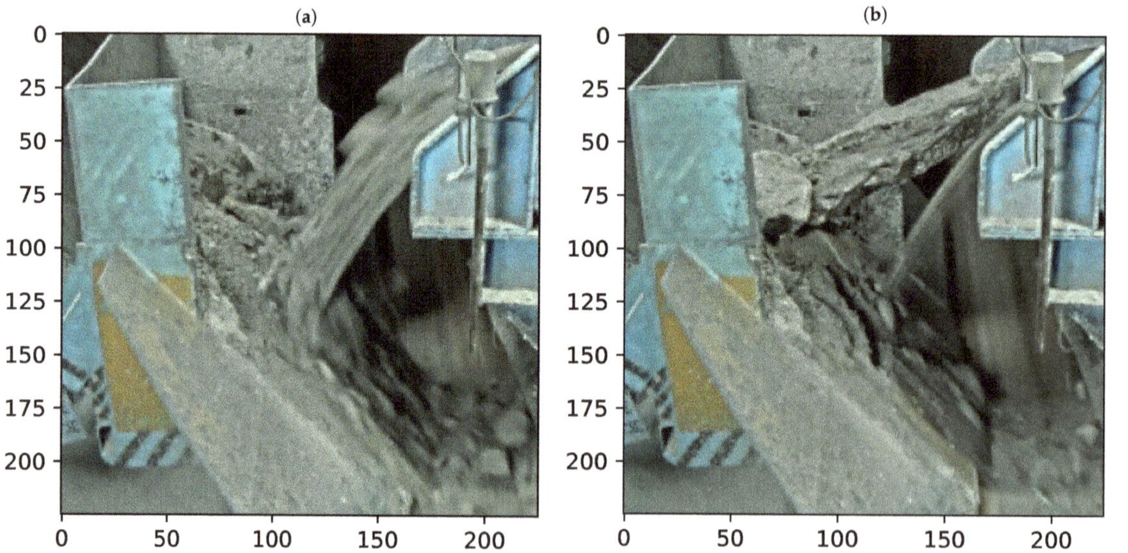


**Figure 5.** Original images after cropping and resizing: (**a**) normal working condition; (**b**) blockage state.

### 3.4.2. Smoothing

The Gaussian averaging operator was chosen as the most optimal way to smooth the image. Using Gaussian blur is the same as combining an image with Gauss. Thanks to this filter, it is possible to reduce components from high-frequency images (low-pass filter), which allows limiting interferences recorded in underground camera conditions. It is a type of image blur filter that uses the Gaussian function to determine the normal distribution in statistical issues to calculate transformations for each pixel of the image being processed. The Gaussian function $g$ for a one-dimensional system with a coordinate $x$, controlled by the variance of the $\sigma^2$, is [74]:

$$g(x,\sigma) = \frac{1}{\sqrt{2\pi\sigma^2}} e^{-\left(\frac{x^2}{2\sigma^2}\right)} \tag{1}$$

Image processing requires the use of the Gaussian function $g$ for a two-dimensional system with coordinates $x, y$, which is described by the equation:

$$g(x,y,\sigma) = \frac{1}{2\pi\sigma^2} e^{-\left(\frac{x^2+y^2}{2\sigma^2}\right)} \tag{2}$$

Equation (2) allows the calculation of the Gaussian template, which is then entangled with the image. For processing, use a 7 × 7 template that reduces many image features. Once entangled, each pixel has a new weighted average neighborhood value for that pixel. The original pixel has the highest weight value, and the weight of adjacent pixels is based on the distance from the original pixel. Thanks to this, the image obtains blurring and reduces many components and distortions, while maintaining the most important feature, which is the objects' edges and boundaries [74]. Figure 6 shows the picture of regular transfer operation and its spectrum before and after filtration.

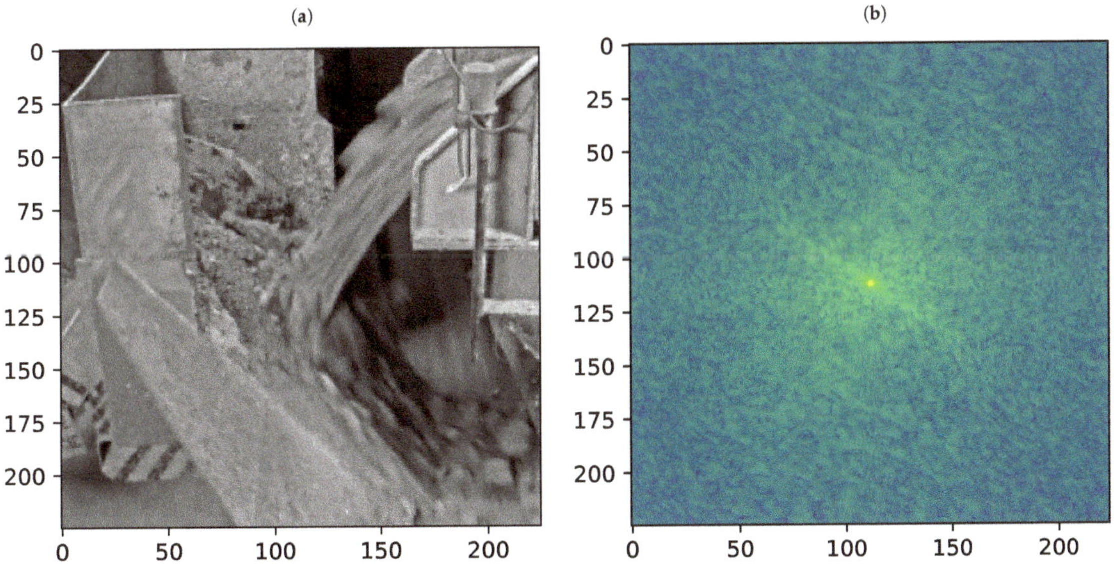


**Figure 6.** *Cont.*

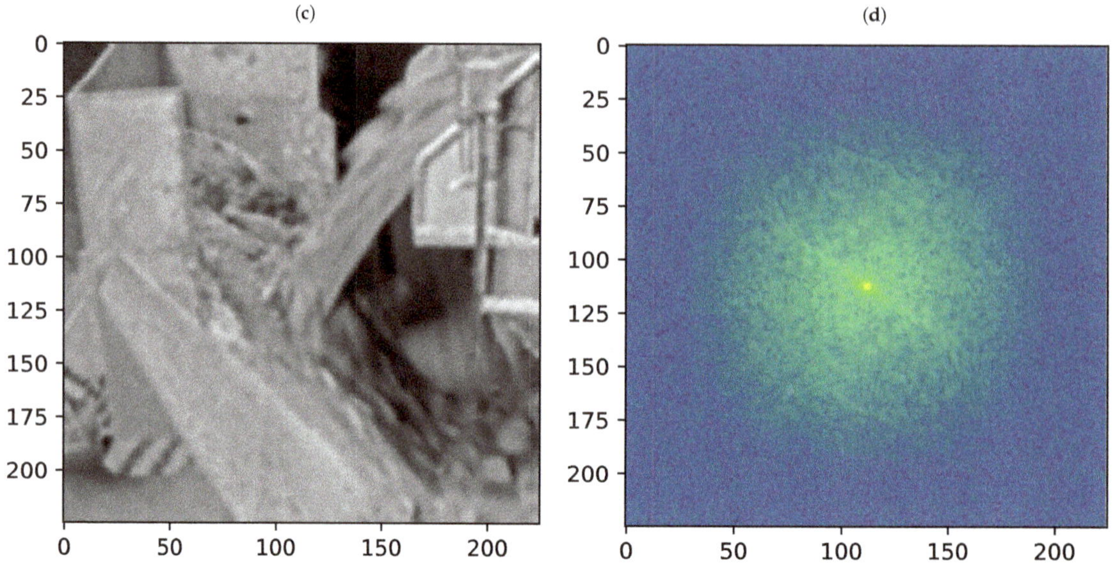


**Figure 6.** Filtering with the use of Gaussian blur of the image of the normal operation of the transfer point: (**a**) normal conditions; (**b**) normal conditions spectrum; (**c**) normal conditions filtered out; (**d**) normal conditions filtered out spectrum.

### 3.4.3. Thresholding

In the next stage, to distinguish the characteristic features of the images, a threshold was used, where the pixels above the designated level received white and below the black color [74]. Thanks to this procedure, it is possible to separate the background elements of the transfer from its construction and excavation, which isolates objects of interest from the monitoring point of view. In this case, binary thresholding combined with the Otsu method [75] and color inversion was used. The Otsu method allows the determination of the optimal value of the threshold of pixel separation and separation of the object from the background based on the difference in intensity. The research shows that the Otsu method maximizes the probability that the threshold will be chosen to divide the image between the object and its background [76–78]. This is achieved by selecting a threshold that provides the best class separation for all pixels in the image. This method uses a standardized histogram in which the number of each level point is divided by the total number of image points. Therefore, this represents the probability distribution for intensity levels as [74]:

$$p(l) = \frac{\mathbf{N}(l)}{N^2} \tag{3}$$

The formula is used to calculate the cumulative moments of the zero and first-order normalized histogram to the k-th level as:

$$\omega(k) = \sum_{l=1}^{k} p(l) \tag{4}$$

$$\mu(k) = \sum_{l=1}^{k} l \cdot p(l) \tag{5}$$

The total average image level is:

$$\mu T = \sum_{l=1}^{N_{mat}} l \cdot p(l) \tag{6}$$

The variance of class separation then has the form:

$$\sigma_B^2(k) = \frac{(\mu T \cdot \omega(k) - \mu(k))^2}{\omega(k)(1 - \omega(k))} \quad \forall k \in 1, N_{max} \tag{7}$$

The optimal threshold is the level at which the variance of class separability is maximum. That is, the optimal threshold is one where the variance is:

$$\sigma_B^2(T_{opt}) = \max_{1 \le k < N_{max}} \left(\sigma_B^2(k)\right) \tag{8}$$

The threshold results are shown in Figure 7 on the example of an image for a normal operating state and blockage.

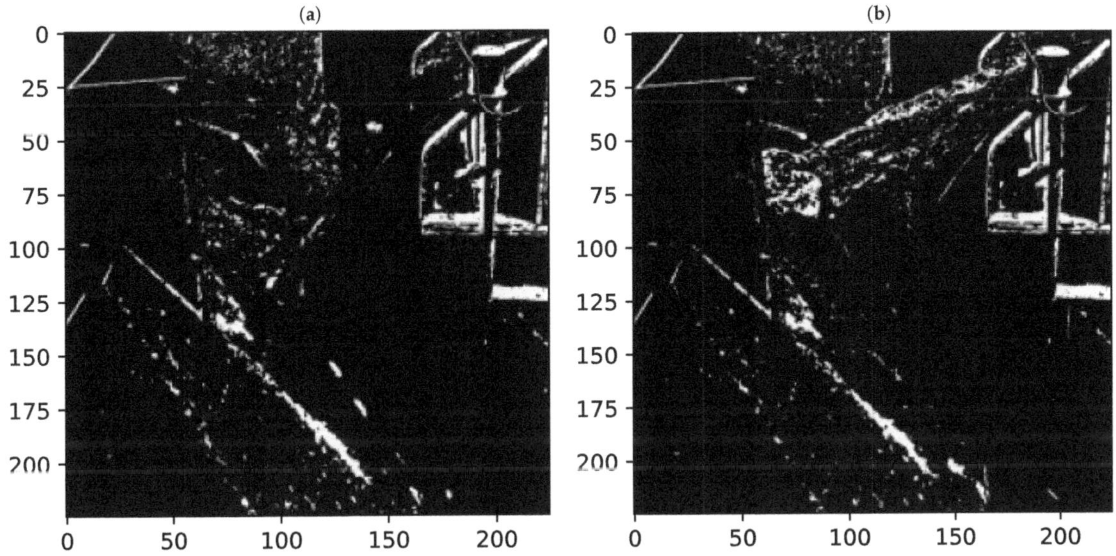


**Figure 7.** Thresholding results: (**a**) normal condition; (**b**) blockage state.

### 3.4.4. Morphology

The operators used in mathematical morphology were developed based on set [79]. Thanks to them, we process images according to shape, treating both of these elements as separate sets of points [80]. Local transformations defined that changing the value of the pixels represented as sets. Changing pixel values is strictly formalized by a miss-and-hit transformation. The object is represented as a $X$ set and tested by a structural element represented by the $B$ set. The transformation is defined as the operator point [74]:

$$X \otimes B = \left\{x \mid B_x^1 \subset X \cap B_x^2 \subset X^c\right\} \tag{9}$$

In Equation (9), $x$ represents one pixel in the image, which is one element from the set of $X$. $X^c$ is the complement of $X$, that is, the set of pixels of the image that are not in the set $X$. Dement structuring $B$ is represented by two parts $B^1$ and $B^2$, which apply to the set of $X$ or its complement $X^c$. The structural element is the shape, which is how mathematical morphological operations process images according to the properties of the shape. An operation $B^1$ on a set of $X$ is a hit transformation, and $B^2$ on a $X$ is a miss. The lower index indicates that the structuring element has moved to the position of the $x$ element. The $B$ element defines the window that is moved through the image [74]. In image processing, the simplest forms of morphological operators, i.e., erosion and extension, were used. Erosion otherwise known as reduction, occurs when $B^1$ is empty:

$$X \ominus B = \left\{x \mid B_x^1 \subset X\right\} \tag{10}$$

In the erosion operator, the hit or miss transformation determines whether a pixel $x$ belongs to the set. It is possible when each point of the element $B^1$ transferred to $x$ is in the set of $X$. Since all points in $B^1$ must be in $X$, this operator removes pixels on the boundaries of objects in the $X$ set. Thus, it erodes or shrinks the collection. The operator was used to remove noise from images after thresholding. The erosion operator sometimes causes the loss of valuable edges, so the image has been corrected using dilation, otherwise known as increase, which occurs when $B^2$ is empty [74]:

$$X \oplus B = \left\{x \mid B_x^2 \subset X^c\right\} \tag{11}$$

The results of image processing using morphological operators are shown in Figure 8

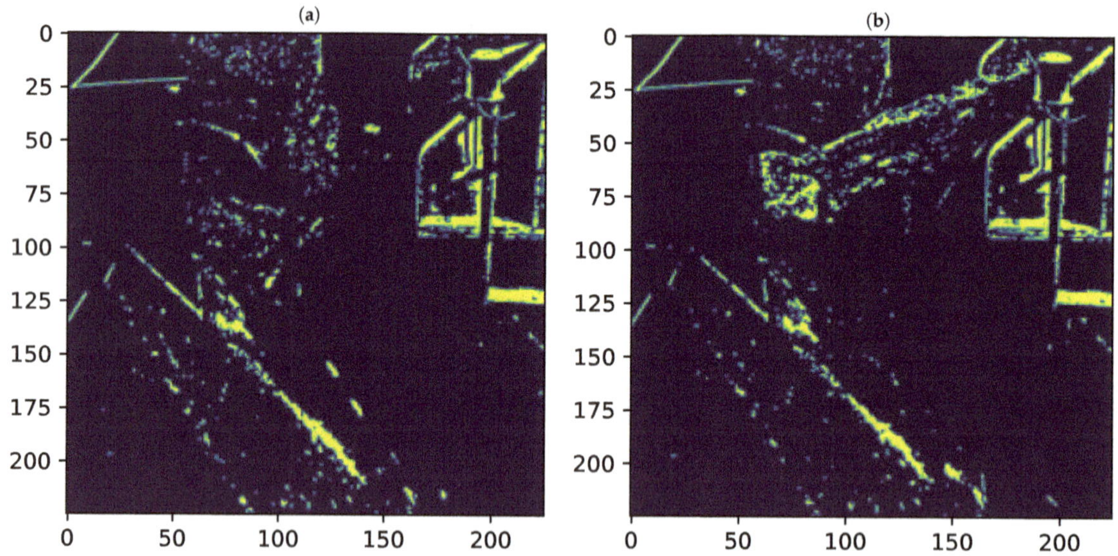


**Figure 8.** Morphology operators results: (**a**) normal condition; (**b**) blockage state.

### 3.5. CNN Autoencoder Structure

An autoencoder based on a convolutional neural network was used to detect anomalies on subsequent frames from the video material. This algorithm was taught to reconstruct the input data. Autoencoders are taught to encode data in an unsupervised manner, which is important in the case of binary classification of anomalous events at the transfer point. Preparing a dataset containing regular- and anomalous-state labels is difficult. Blockages and other anomalous events are rare, so it is difficult to prepare a proper database. In addition, anomalous events in the case of transfer are so diverse that detection and manual labeling of all possible cases are practically impossible [81].

Autoencoders consist of three main parts. The first part, an encoder, compresses the input into an encoded representation. The size of the encoded representation is usually several times smaller than the size of the input data. The second part, called the bottleneck, is the latent space where the encoded representation of the input in a compressed form is stored. The last, third part, called the decoder, is used to reconstruct compressed data. The reconstruction data are then compared with the input data, and a reconstruction error is determined [82]. Figure 9 shows a diagram of a typical autoencoder used to reconstruct an input image.

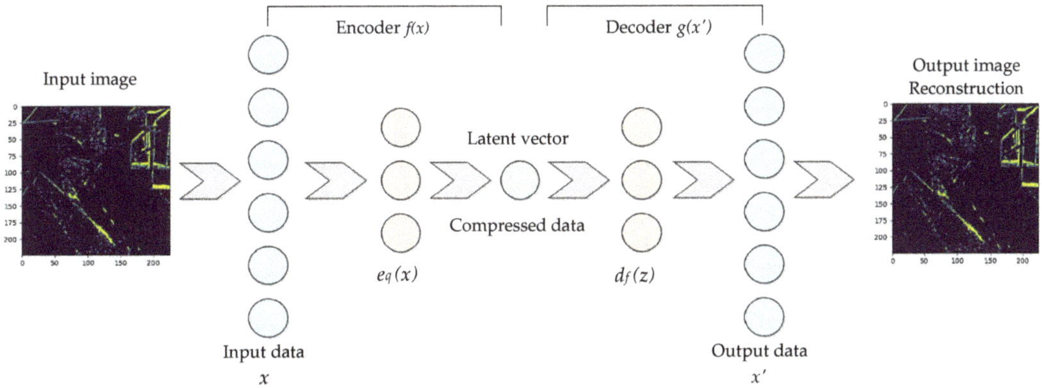


**Figure 9.** Autoencoder structure of a convolutional neural network.

Iterative neural network learning involves specifying a loss function that determines the error of the decoder reconstruction relative to the input image. The autoencoder loss function can be written as [83]:

$$Loss = \|x - d_{\phi}(z)\|^2 = \|x - d_{\phi}(e_{\theta}(x))\|^2 \quad (12)$$

The proposed neural network model determines the probability of anomalies in the studied image. For this reason, categorical cross-entropy was chosen as the parameter characterizing the autoencoder loss function. The categorical cross-entropy function can be used for two-class or multi-class classification. It allows for calculating the average difference between the actual value and the predicted probability distributions for all classes in the analyzed classification problem. In subsequent iterations, the result is minimized, and the ideal cross-entropy value is 0 [84]. The equation can describe the cross-entropy for a class problem:

$$Loss = -\sum_{i=1}^{N} t_i \cdot \log(p_i) \quad (13)$$

where $t_i$ is the truth label and $p_i$ is the Softmax probability for each class $i$ from the class set $N$.

## 4. Results

Figure 10 shows how the autoencoder loss function changed during subsequent iterations of neural network learning. For the first 10 iterations of the model for training data, the cross-entropy value varies quite a lot from 0.58 to 0.01. After this learning stage, the loss function's course is smoothed and stabilized at an average level of 0.002. The loss for validation data is less than 0.01 in a complete iterative cycle, and for the waveform stabilization range, it is only 0.001 on average. The model was trained until the cross-entropy value was approximately 0, which occurred in about 50 iterations.

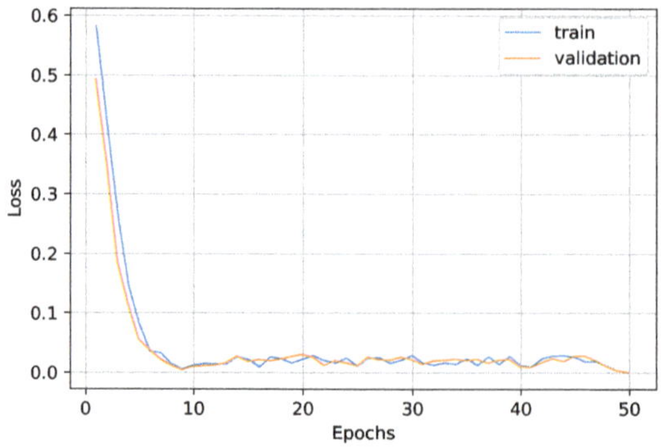


**Figure 10.** Autoencoder network learning loss—categorical crossentropy.

As an indicator of the quality of the prediction model, a parameter called accuracy was chosen, which allows the determination of how often the predictions were equal to the actual labels defining the class of the analyzed image. Figure 11 shows how the accuracy value changed during subsequent iterations of model learning. As with the loss function, the most significant changes were recorded in the first 10 iterations. After this stage, the mileage for test and training data stabilized. For iterations, when the autoencoder loss function was 0, the forecast accuracy was 1.

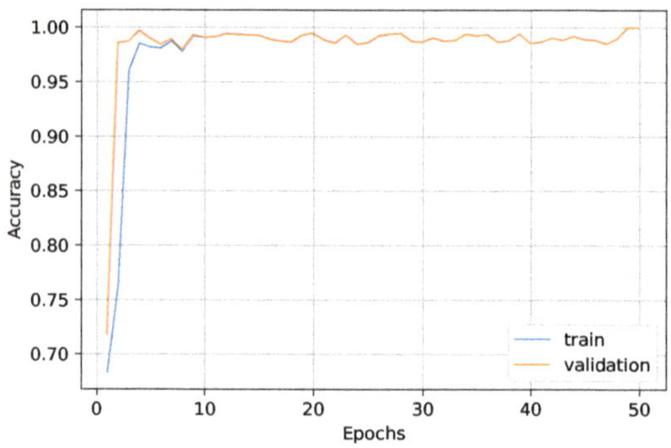


**Figure 11.** Autoencoder network training metric—accuracy.

The results of the prediction algorithm were visualized using a matrix of errors (Figure 12). The error matrix is used for assessing the quality of binary classification, where there were regular and anomalous transfer chute operation classes. Data labeled positive and negative are classified as either a predicted positive class or a predicted negative class. A given originally marked as positive may be mistakenly classified as negative. Due to the effectiveness of network detection equal to unity, there are only two classes in the error table: truly positive (TP) and truly negative (TN). The TP class accounted for 72.45% of the forecasted data, while the TN 27.55%.

The prediction algorithm allows the determination of the location of anomalous pixels on reconstructed images. Heatmaps were plotted based on their location and using the function of detecting maximum values on forecasted images, where colors show the density of outlier pixels. Figure 13 shows the detected anomaly caused by the blockage of an oversized rock block in the zone of free fall of material. Figure 14 shows the detected foreign body as a rock bolt stuck between the transfer box and the feed conveyor structure. The last example (Figure 15) shows the detected complete blockage. Each of the presented cases was positively classified as an anomaly, and the additional marking in the form of a field with a red border indicates only the area of the detected most significant number of outlier pixels.

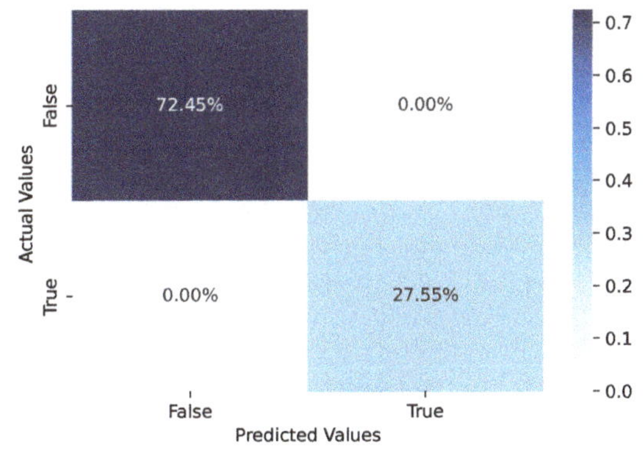


**Figure 12.** Confusion matrix.

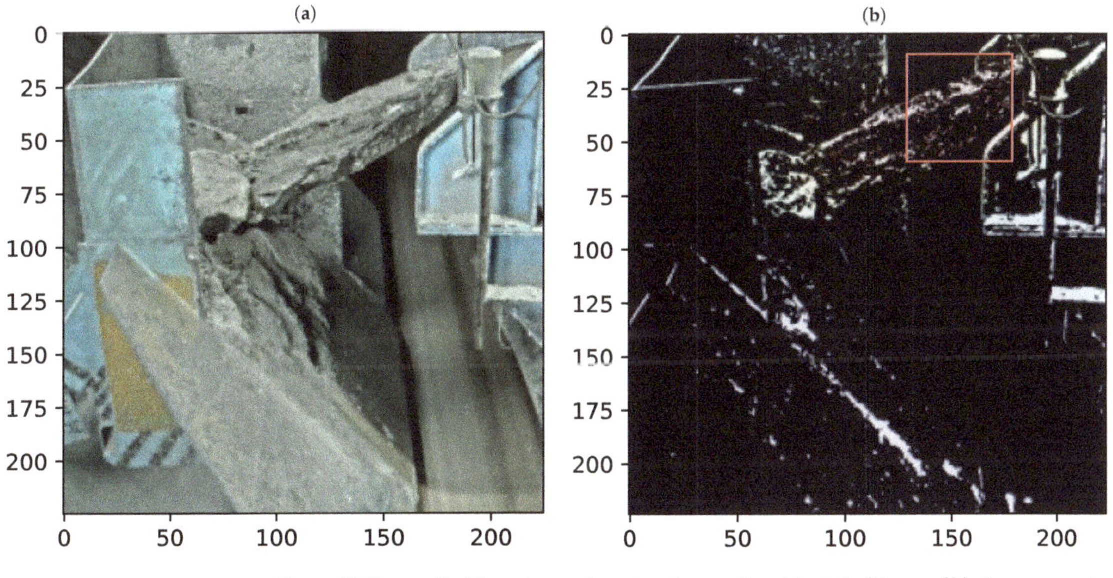


**Figure 13.** The result of the autoencoder network operation: (**a**) original image of blockage—oversized rock; (**b**) location of the anomaly most clearly.

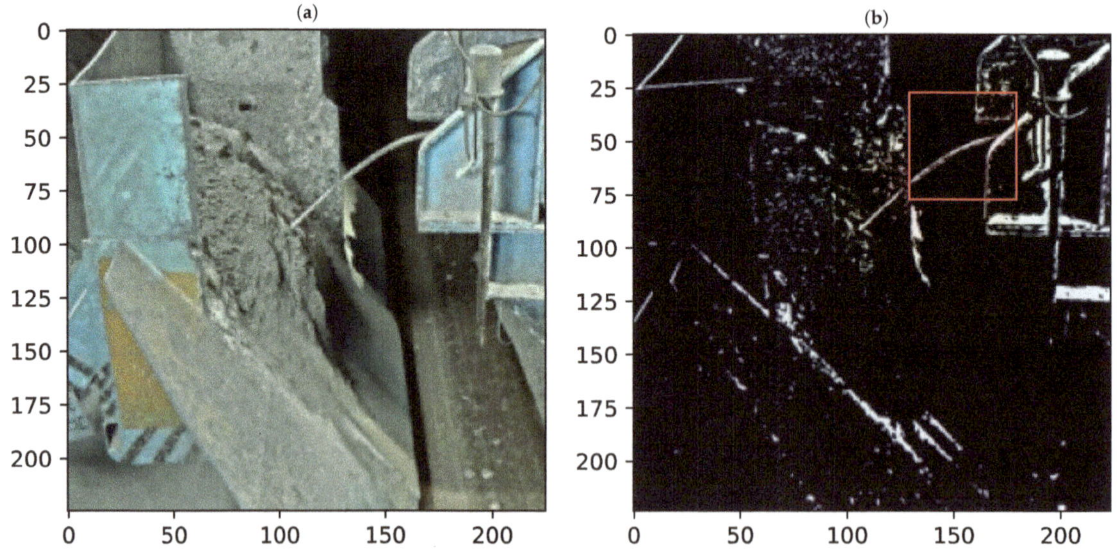


**Figure 14.** The result of the autoencoder network operation: (**a**) original image of blockage—foreign body mining anchor; (**b**) location of the anomaly most clearly.

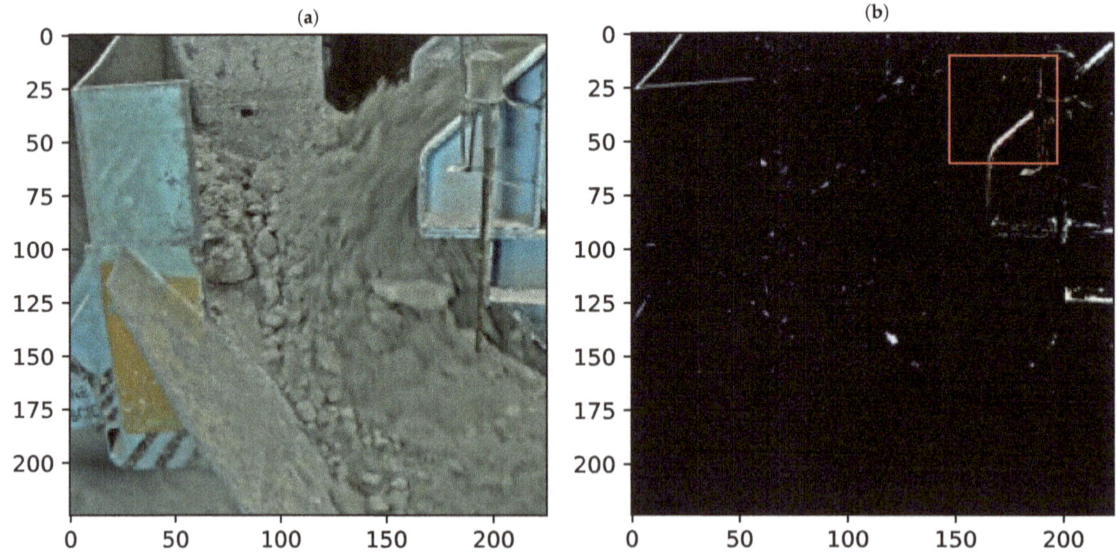


**Figure 15.** The result of the autoencoder network operation: (**a**) original image of blockage—full blockage; (**b**) location of the anomaly most clearly.

## 5. Conclusions

The article presents the possibility of detecting blockages of the conveyor transfer point using the proposed detection algorithm. The best results were obtained by filtering the images with a Gaussian filter and reducing complexity by thresholding and transforming using morphology operators. By pointing out, the image examples used for validation turned out to be a trivial problem for the trained autoencoder algorithm. The effectiveness and loss of learning validation data were more beneficial than training data because the anomalous images contained typical examples of blockages and the presence of solids in the zone of fall and loading of material. The real anomalous states vary widely, and it is impossible to record them all for model validation. The algorithm managed to detect the blockage of transfer through an oversized rock fragment and a rock bolt. Blocks caused by these objects are the most common reason for reduced capacity and can be considered the beginning of a potential blockage.

At this stage, the possibility of using the method proposed in industrial conditions should be considered. Since this method is in the initial stage of development, the detection algorithm should provide a security alert before the state of total blockage occurs. The tested model also detected total blockages when the receiving conveyor loading space was filled with excavated material, causing material spillage. The following stages of research will focus on validating the algorithm based on real-time image processing. In much more diverse operating conditions, there will undoubtedly be a decrease in detection efficiency and the occurrence of false alarm states. The presented algorithm will require modification and taking into account other parameters, e.g., information about the movement of the material.

**Author Contributions:** Conceptualization, P.B., H.G., R.K., D.M. and M.O.; methodology, P.B and M.O.; software, P.B. and M.O.; validation, H.G., R.K. and D.M.; formal analysis, R.K.; investigation, P.B., R.K. and M.O.; resources, R.K.; data curation, P.B.; writing—original draft preparation, P.B. and M.O.; writing—review and editing, H.G., R.K. and D.M.; visualization, P.B.; supervision, H.G., R.K. and D.M.; project administration, R.K.; funding acquisition, R.K. All authors have read and agreed to the published version of the manuscript.

**Funding:** The research work was funded with the research subsidy from the Polish Ministry of Science and Higher Education granted for 2023.

**Institutional Review Board Statement:** Not applicable.

**Informed Consent Statement:** Not applicable.

**Data Availability Statement:** Not applicable.

**Conflicts of Interest:** The authors declare no conflict of interest.

## References


1. Alspaugh, M. Latest developments in belt conveyor technology. In Proceedings of the MINExpo 2004, Las Vegas, NV, USA, 27 September 2004.
2. Woodcock, C.; Mason, J. *Bulk Solids Handling: An Introduction to the Practice and Technology*; Springer Science & Business Media: Berlin/Heidelberg, Germany, 1988.
3. Mathaba, T.; Xia, X. Optimal and energy efficient operation of conveyor belt systems with downhill conveyors. *Energy Effic.* **2017**, *10*, 405–417. [CrossRef]
4. Bajda, M.; Hardygóra, M. Analysis of the Influence of the Type of Belt on the Energy Consumption of Transport Processes in a Belt Conveyor. *Energies* **2021**, *14*, 6180. [CrossRef]
5. Mazurkiewicz, D. Computer-aided maintenance and reliability management systems for conveyor belts. *Eksploat. I Niezawodn.* **2014**, *16*, 377–382.
6. Bajda, M.; Hardygóra, M.; Marasová, D. Energy Efficiency of Conveyor Belts in Raw Materials Industry. *Energies* **2022**, *15*, 3080. [CrossRef]
7. Hou, Y.F.; Meng, Q.R. Dynamic characteristics of conveyor belts. *J. China Univ. Min. Technol.* **2008**, *18*, 629–633. [CrossRef]
8. Hager, M.; Hintz, A. The energy-saving design of belts for long conveyor systems. *Bulk Solids Handl.* **1993**, *13*, 749–758.
9. Woźniak, D.; Hardygóra, M. Aspects of Selecting Appropriate Conveyor Belt Strength. *Energies* **2021**, *14*, 6018. [CrossRef]
10. Bortnowski, P.; Gładysiewicz, A.; Gładysiewicz, L.; Król, R.; Ozdoba, M. Conveyor Intermediate TT Drive with Power Transmission at the Return Belt. *Energies* **2022**, *15*, 6062. [CrossRef]
11. Masaki, M.S.; Zhang, L.; Xia, X. A comparative study on the cost-effective belt conveyors for bulk material handling. *Energy Procedia* **2017**, *142*, 2754–2760. [CrossRef]
12. Wensrich, C.M. Evolutionary optimisation in chute design. *Powder Technol.* **2003**, *138*, 118–123. [CrossRef]
13. Ilic, D.; Roberts, A.; Wheeler, C.; Katterfeld, A. Modelling bulk solid flow interactions in transfer chutes: Shearing flow. *Powder Technol.* **2019**, *354*, 30–44. [CrossRef]
14. Grincova, A.; Andrejiova, M.; Marasova, D. Failure analysis of conveyor belt in terms of impact loading by means of the damping coefficient. *Eng. Fail. Anal.* **2016**, *68*, 210–221. [CrossRef]

15. Gerard, B.; O'Rourke, L. Optimisation of overland conveyor performance. *Aust. Bulk. Handl. Rev.* **2009**, 26–36. Available online: https://cemanet.org/wp-content/uploads/2019/09/Optimisation-of-overland-conveyor-performance.pdf (accessed on 13 January 2023).
16. Molnár, V.; Fedorko, G.; Husáková, N.; Král', J., Jr.; Ferdynus, M. Energy calculation model of an outgoing conveyor with application of a transfer chute with the damping plate. *Mech. Sci.* **2016**, *7*, 167–177. [CrossRef]
17. Roberts, A. Design and Application of Feeders for the Controlled Loading of Bulk Solids onto Conveyor Belts. Available online: https://www2.hcmuaf.edu.vn/data/dangnh/file/DESIGN%20AND%20APPLICATION%20OF%20FEEDERS%20FOR.pdf (accessed on 13 January 2023).
18. Zhang, S.; Xia, X. A new energy calculation model of belt conveyor. In Proceedings of the AFRICON 2009, Nairobi, Kenya, 23–25 September 2009; pp. 1–6.
19. Bortnowski, P.; Kawalec, W.; Król, R.; Ozdoba, M. Types and causes of damage to the conveyor belt-review, classification and mutual relations. *Eng. Fail. Anal.* **2022**, *140*, 106520. [CrossRef]
20. Rossow, J.; Coetzee, C. Discrete element modelling of a chevron patterned conveyor belt and a transfer chute. *Powder Technol.* **2021**, *391*, 77–96. [CrossRef]
21. Ilic, D.; Roberts, A.; Wheeler, C. Modelling bulk solid interactions in transfer chutes: Accelerated flow. *Chem. Eng. Sci.* **2019**, *209*, 115197. [CrossRef]
22. Doroszuk, B.; Krol, R. Analysis of conveyor belt wear caused by material acceleration in transfer stations. *Min. Sci.* **2019**, *26*, 189–201.
23. Hustrulid, A.I.; Mustoe, G.G. *Engineering Analysis of Transfer Points Using Discrete Element Analysis*; Geomechanics Research Center, Colorado School of Mines: Golden, CO, USA, 1996.
24. Doroszuk, B.; Król, R.; Wajs, J. Simple Design Solution for Harsh Operating Conditions: Redesign of Conveyor Transfer Station with Reverse Engineering and DEM Simulations. *Energies* **2021**, *14*, 4008. [CrossRef]
25. Seibel, R.J. *Dust Control at a Transfer Point Using Foam and Water Sprays*; US Department of the Interior, Bureau of Mines: Washington, DC, USA, 1976; Volume 97.
26. Gao, G.; Shen, J.; Liang, Q. Study of dust diffusion at transfer point of belt conveyor based on FLUENT. *J. Phys. Conf. Ser.* **2018**, *1064*, 012016. [CrossRef]
27. Mandal, B.B.; Bhattacharya, S.; Manwar, V.D.; Hussain, S.A. Health risk of exposure to noise in coal preparation and mineral processing plants. In *Innovative Exploration Methods for Minerals, Oil, Gas, and Groundwater for Sustainable Development*; Elsevier: Amsterdam, The Netherlands, 2022; pp.139–157.
28. Kesler, F.; Prener, M. DEM: Simulation of conveyor transfer chutes. *FME Trans.* **2009**, *37*, 185–192.
29. Scott, O.; Choules, P. The use of impact plates in conveyor transfers. *Tribol. Int.* **1993**, *26*, 353–359. [CrossRef]
30. Rudolf, L.; Durna, A.; Hapla, T. The issue of the transfer points on belt conveyors. *Int. Multidiscip. Sci. Geoconference SGEM* **2018**, *18*, 989–996.
31. Cleary, P.; Sinnott, M.; McBride, W. Prediction of particle flows and blockage problems in realistic 3D transfer chutes. In Proceedings of the World Congress on Chemical Engineering, Edinburgh, UK, 1–4 July 2005; p. 10.
32. Hastie, D.B.; Wypych, P. Experimental validation of particle flow through conveyor transfer hoods via continuum and discrete element methods. *Mech. Mater.* **2010**, *42*, 383–394. [CrossRef]
33. De-Song, B.; Xun-Sheng, Z.; Guang-Lei, X.; Zheng-Quan, P.; Xiao-Wei, T.; Kun-Quan, L. Critical phenomenon of granular flow on a conveyor belt. *Phys. Rev. E* **2003**, *67*, 062301. [CrossRef]
34. Owen, P. Condition monitoring for conveyors. In Proceedings of the 9th International Materials Handling Conference, Johannesburg, South Africa, 1997.
35. Norgia, M.; Svelto, C. RF-capacitive proximity sensor for safety applications. In Proceedings of the 2007 IEEE Instrumentation & Measurement Technology Conference IMTC 2007, Warsaw, Poland, 1–3 May 2007; pp. 1–4.
36. Jaafar, A. Nucleonic Gauging and Instrumentation. 2005. Available online: https://www.osti.gov/etdeweb/biblio/20756985 (accessed on 13 January 2023).
37. Byrne, B. Nucleonic Instrumentation. *Meas. Control* **1985**, *18*, 166. [CrossRef]
38. Kulp, M. Enhanced Technology for Blocked Chute Detection. *Eng. Min. J.* **2012**, *213*, 158.
39. Progorowicz, J.; Skoczylas, A.; Anufriiev, S.; Dudzik, M.; Stefaniak, P. Estimation of Final Product Concentration in Metalic Ores Using Convolutional Neural Networks. *Minerals* **2022**, *12*, 1480. [CrossRef]
40. Kuzba, B.; Pawlos, W.; Konieczny, A.; Krzeminska, M. Optimisation Platform for copper ore processing at the Division of Concentrator of KGHM Polska Miedz SA. *E3S Web Conf.* **2016**, *8*, 01037. [CrossRef]
41. Cao, W.; Zhang, H.; Wang, N.; Wang, H.W.; Peng, Z.X. The gearbox wears state monitoring and evaluation based on on-line wear debris features. *Wear* **2019**, *426*, 1719–1728. [CrossRef]
42. Castejón, M.; Alegre, E.; Barreiro, J.; Hernández, L. On-line tool wear monitoring using geometric descriptors from digital images. *Int. J. Mach. Tools Manuf.* **2007**, *47*, 1847–1853. [CrossRef]
43. Barszcz, T.; Siami, M.; Wodecki, J.; Zimroz, R. Automated IR Image Segmentation for Identification of Overheated Idlers in Belt Conveyor Systems Based on Outlier-Detection Method. *Sensors* **2022**, *22*, 10004. [CrossRef]
44. Szurgacz, D.; Zhironkin, S.; Vöth, S.; Pokorný, J.; Spearing, A.; Cehlár, M.; Stempniak, M.; Sobik, L. Thermal imaging study to determine the operational condition of a conveyor belt drive system structure. *Energies* **2021**, *14*, 3258. [CrossRef]

45. Jia, Z.; Liu, Z.; Vong, C.M.; Pecht, M. A rotating machinery fault diagnosis method based on feature learning of thermal images. *IEEE Access* **2019**, *7*, 12348–12359. [CrossRef]
46. Dabek, P.; Szrek, J.; Zimroz, R.; Wodecki, J. An Automatic Procedure for Overheated Idler Detection in Belt Conveyors Using Fusion of Infrared and RGB Images Acquired during UGV Robot Inspection. *Energies* **2022**, *15*, 601. [CrossRef]
47. Szrek, J.; Wodecki, J.; Błażej, R.; Zimroz, R. An inspection robot for belt conveyor maintenance in underground mine—Infrared thermography for overheated idlers detection. *Appl. Sci.* **2020**, *10*, 4984. [CrossRef]
48. Carvalho, R.; Nascimento, R.; D'Angelo, T.; Delabrida, S.; GC Bianchi, A.; Oliveira, R.A.; Azpúrua, H.; Uzeda Garcia, L.G. A UAV-based framework for semi-automated thermographic inspection of belt conveyors in the mining industry. *Sensors* **2020**, *20*, 2243. [CrossRef]
49. Uth, F.; Polnik, B.; Kurpiel, W.; Kriegsch, P.; Baltes, R.; Clausen, E. An innovative person detection system based on thermal imaging cameras dedicate for underground belt conveyors. *Min. Sci.* **2019**, *26*, 275–288. [CrossRef]
50. Hongmei, J.; Changbin, H.; Zhanli, L.; Pengfei, S. Video-based roller abnormality detection. In Proceedings of the 2021 International Conference on Computer Engineering and Application (ICCEA), Kunming, China, 25–27 June 2021; pp. 380–385.
51. Dąbek, P.; Krot, P.; Wodecki, J.; Zimroz, P.; Szrek, J.; Zimroz, R. Measurement of idlers rotation speed in belt conveyors based on image data analysis for diagnostic purposes. *Measurement* **2022**, *202*, 111869. [CrossRef]
52. Trybała, P.; Blachowski, J.; Błażej, R.; Zimroz, R. Damage detection based on 3d point cloud data processing from laser scanning of conveyor belt surface. *Remote Sens.* **2020**, *13*, 55. [CrossRef]
53. Zhang, M.; Shi, H.; Zhang, Y.; Yu, Y.; Zhou, M. Deep learning-based damage detection of mining conveyor belt. *Measurement* **2021**, *175*, 109130. [CrossRef]
54. Zhang, M.; Zhang, Y.; Zhou, M.; Jiang, K.; Shi, H.; Yu, Y.; Hao, N. Application of Lightweight Convolutional Neural Network for Damage Detection of Conveyor Belt. *Appl. Sci.* **2021**, *11*, 7282. [CrossRef]
55. Kim, J.; Chi, S. Multi-camera vision-based productivity monitoring of earthmoving operations. *Autom. Constr.* **2020**, *112*, 103121. [CrossRef]
56. Kim, J.; Hwang, J.; Chi, S.; Seo, J. Towards database-free vision-based monitoring on construction sites: A deep active learning approach. *Autom. Constr.* **2020**, *120*, 103376. [CrossRef]
57. Nefis, M.; Talhi, K. A model study to measure fragmentation by blasting. *Min. Sci.* **2016**, *23*, 91–104.
58. Bamford, T.; Esmaeili, K.; Schoellig, A.P. A real-time analysis of post-blast rock fragmentation using UAV technology. *Int. J. Min. Reclam. Environ.* **2017**, *31*, 439–456. [CrossRef]
59. Al-Thyabat, S.; Miles, N.; Koh, T. Estimation of the size distribution of particles moving on a conveyor belt. *Miner. Eng.* **2007**, *20*, 72–83. [CrossRef]
60. Maerz, N.H. Image sampling techniques and requirements for automated image analysis of rock fragmentation. In *Measurement of Blast Fragmentation*; Routledge: London, UK, 2018; pp. 115–120.
61. Shariati, H.; Yeraliyev, A.; Terai, B.; Tafazoli, S.; Ramezani, M. Towards autonomous mining via intelligent excavators. In Proceedings of the IEEE/CVF Conference on Computer Vision and Pattern Recognition Workshops, Long Beach, CA, USA, 16–17 June 2019; pp. 26–32.
62. Lim, S.N.; Zhou, N.; Soares, J.V.B. System and Method for Detecting Missing Tooth in Mining Shovel. US Patent 9,886,754, 6 February 2018.
63. Bilandi, S.T.; Ramezani, M.; Suzani, A.; Parnian, N.; Baumann, M.A.; Nouranian, S.; Hamzei, N.; Sameti, M.; Karimifard, S. Method and Apparatus for Locating a Wear Part in an Image of an Operating Implement. US Patent 10,339,667, 19 September 2019.
64. Kim, H.; Choi, Y. Lab Scale Model Experiment of Smart Hopper System to Remove Blockages Using Machine Vision and Collaborative Robot. *Appl. Sci.* **2022**, *12*, 579. [CrossRef]
65. Yao, J.; Wang, Z.; Liu, C.; Huang, G.; Yuan, Q.; Xu, K.; Zhang, W. Detection Method of Crushing Mouth Loose Material Blockage Based on SSD Algorithm. *Sustainability* **2022**, *14*, 14386. [CrossRef]
66. Staar, B.; Lütjen, M.; Freitag, M. Anomaly detection with convolutional neural networks for industrial surface inspection. *Procedia CIRP* **2019**, *79*, 484–489. [CrossRef]
67. Tang, Z.; Chen, Z.; Bao, Y.; Li, H. Convolutional neural network-based data anomaly detection method using multiple information for structural health monitoring. *Struct. Control Health Monit.* **2019**, *26*, e2296. [CrossRef]
68. Jain, S.; Seal, A.; Ojha, A.; Yazidi, A.; Bures, J.; Tacheci, I.; Krejcar, O. A deep CNN model for anomaly detection and localization in wireless capsule endoscopy images. *Comput. Biol. Med.* **2021**, *137*, 104789. [CrossRef]
69. Napoletano, P.; Piccoli, F.; Schettini, R. Anomaly detection in nanofibrous materials by CNN-based self-similarity. *Sensors* **2018**, *18*, 209. [CrossRef]
70. Krauze, K.; Rączka, W.; Sibielak, M.; Konieczny, J.; Kubiak, D.; Culer, H.; Bajus, D. Automated transfer point URB/ZS-3. *Min.–Inform. Autom. Electr. Eng.* **2017**, *2*, 80–91. [CrossRef]
71. Wróblewski, A.; Banasiewicz, A.; Gola, S. Heat Balance Determination Methods for Mining Areas in Underground Mines-A Review. *Iop Conf. Ser. Earth Environ. Sci.* **2021**, *942*, 012011. [CrossRef]
72. Chen, L.C.; Papandreou, G.; Kokkinos, I.; Murphy, K.; Yuille, A.L. Semantic image segmentation with deep convolutional nets and fully connected crfs. *arXiv* **2014**, arXiv:1412.7062.

73. Bhattiprolu, S. Localizing Anomalies in Images. 2022. Available online: https://github.com/bnsreenu/python_for_microscopists; https://www.youtube.com/watch?v=P9NdQG_vIvo (accessed on 13 January 2023).
74. Nixon, M.; Aguado, A. *Feature Extraction and Image Processing for Computer Vision*; Academic Press: Cambridge, MA, USA, 2019.
75. Otsu, N. A threshold selection method from gray-level histograms. *IEEE Trans. Syst. Man Cybern.* **1979**, *9*, 62–66. [CrossRef]
76. Sahoo, P.K.; Soltani, S.; Wong, A.K. A survey of thresholding techniques. *Comput. Vis. Graph. Image Process.* **1988**, *41*, 233–260. [CrossRef]
77. Lee, S.U.; Chung, S.Y.; Park, R.H. A comparative performance study of several global thresholding techniques for segmentation. *Comput. Vis. Graph. Image Process.* **1990**, *52*, 171–190. [CrossRef]
78. Glasbey, C.A. An analysis of histogram-based thresholding algorithms. *CVGIP Graph. Models Image Process.* **1993**, *55*, 532–537. [CrossRef]
79. Serra, J. Introduction to mathematical morphology. *Comput. Vis. Graph. Image Process.* **1986**, *35*, 283–305. [CrossRef]
80. Serra, J.; Soille, P. *Mathematical Morphology and Its Applications to Image Processing*; Springer Science & Business Media: Berlin/Heidelberg, Germany, 2012; Volume 2.
81. Mehrotra, K.G.; Mohan, C.K.; Huang, H. *Anomaly Detection Principles and Algorithms*; Springer: Berlin/Heidelberg, Germany, 2017; Volume 1.
82. Jana, D.; Patil, J.; Herkal, S.; Nagarajaiah, S.; Duenas-Osorio, L. CNN and Convolutional Autoencoder (CAE) based real-time sensor fault detection, localization, and correction. *Mech. Syst. Signal Process.* **2022**, *169*, 108723. [CrossRef]
83. Khare, N.; Thakur, P.S.; Khanna, P.; Ojha, A. Analysis of Loss Functions for Image Reconstruction Using Convolutional Autoencoder. In Proceedings of the International Conference on Computer Vision and Image Processing, Rupnagar, India, 3–5 December 2021; Springer: Cham, Switzerland, 2022; pp. 338–349.
84. Koidl, K. Loss Functions in Classification Tasks. School of Computer Science and Statistic Trinity College Dublin. 2013. Available online: https://www.scss.tcd.ie/~koidlk/cs4062/Loss-Functions.pdf (accessed on 13 January 2023).

energies

Article

# Minimizing Internal Leaks of a Powered Roof Support's Hydraulic Prop Based on Double Block with Charging

Dawid Szurgacz [1,2], Beata Borska [3], Ryszard Diederichs [1], Anthony J. S. Spearing [4] and Sergey Zhironkin [5,6,*]

1 Center of Hydraulics DOH Ltd., ul. Konstytucji 147, 41-906 Bytom, Poland
2 Polska Grupa Górnicza S.A., ul. Powstańców 30, 40-039 Katowice, Poland
3 KWK Ruda Ruch Halemba, ul. Halembska 160, 41-717 Ruda Śląska, Poland
4 School of Mines, China University of Mining and Technology, 1 Daxue Road, Tongshan District, Xuzhou 221116, China
5 Department of Trade and Marketing, Siberian Federal University, 79 Svobodny av., 660041 Krasnoyarsk, Russia
6 Department of Open Pit Mining, T.F. Gorbachev Kuzbass State Technical University, 28 Vesennya st., 650000 Kemerovo, Russia
* Correspondence: zhironkinsa@kuzstu.ru

**Abstract:** The hydraulic system of a powered roof support performs two functions. The first function is to control the powered roof support in the extraction wall. The second function is to protect against adverse overloads resulting from rock mass pressing directly on the powered roof support. This damaging phenomenon is prevented by the protection of the powered roof support, with a safety valve built into the hydraulic system or directly into the prop. However, the third function proposed by the authors based on the research results is to minimize leaks. These leaks usually develop in the props or in the hydraulic system. The authors propose implementing changes to the hydraulic system for this purpose. The change consists of replacing the existing support block with a double block with charging. Tests were carried out in real conditions, that is, a mining wall. Tests in the mining wall were carried out on the powered roof support's leaking prop. As a result of charging, the actual load capacity of the prop increased by about 10–50% in relation to the load capacity before charging. The use of a double block with charging ensured that the pressure in the under-piston space of the prop was maintained at a minimum of 250 bar. The results allowed us to determine the usefulness of the proposed solution and eliminate its disadvantages—the designated direction of research and development on the powered roof support allowed us to expand its functionality by minimizing leaks.

**Keywords:** internal leaks; hydraulic prop; powered roof support; tests under real conditions



**Citation:** Szurgacz, D.; Borska, B.; Diederichs, R.; Spearing, A.J.S.; Zhironkin, S. Minimizing Internal Leaks of a Powered Roof Support's Hydraulic Prop Based on Double Block with Charging. *Energies* **2023**, *16*, 1341. https://doi.org/10.3390/en16031341

Academic Editor: Paul Stewart

Received: 22 November 2022
Revised: 15 January 2023
Accepted: 25 January 2023
Published: 27 January 2023



## 1. Introduction

At the current level of economic development, mining companies are expected to reduce their production costs and thus increase their operational efficiency [1–4]. At the same time, lowering costs must not affect work safety [5–8]. This determines the need to increase the efficiency of machines [9–11] as well as their reliability [12,13], and to reduce their energy intensity [14–16]. In mining, the monitoring of machine work processes is gradually being introduced [17,18]. The aim of this monitoring is to reduce the cost of servicing the machines and to maintain the highest possible reliability. This can be achieved using the latest technologies [19–21].

Continuous development is also needed due to deteriorating mining and geological conditions. Operating at ever-greater depths generates new challenges in the field of exploitation [22,23] and regarding the requirements for machines and devices [24,25]. It is necessary to develop and optimize the scope of applied technologies [26–28], machine fleets [29–31], as well as ways to combat natural hazards [32–34]. For this purpose, bench testing [35–37] and in situ testing [38–41] are carried out. They are complemented by model testing [42,43] and computer simulations [44–46] based on mathematical algorithms [47–55].

In hard coal mining, the development of powered roof-support complexes is crucial [56–59]. A longwall complex is a set of machines constituting the essential equipment of a mining wall. It is used to mechanize the process of mining, loading, and transporting the product. The interdependence of construction and movement characterizes machines in the wall complex. The primary function of the wall complex is to achieve the required performance while maintaining safety. The powered wall complex includes a powered roof support, a scraper conveyor and a mining machine. In this work, the authors focus mainly on the development of the powered roof support. Recent research on the development of powered roof supports can be found in [60–63].

The powered roof support has two primary functions in the mining wall. It protects the roof of the excavation, ensuring the safety of operation. The second is to move the entire powered wall complex along the progress of the wall front. The operation of the powered roof support consists of repetitive cycles. Each cycle comprises the following stages: drawing off powered roof support, moving, spreading and securing the roof (Figure 1). Withdrawing the powered roof-support section consists in lowering its height—so that the canopy loses contact with the excavation roof. Then it is possible to move the powered roof support section towards the coal. After moving the powered roof support, it is expanded between floor and roof of the excavation—so as to get the canopy in contact with the excavation roof. To ensure proper operation of the powered roof support, it is necessary to maintain the required load-capacity value. Load-carrying capacity is the force with which the support acts on the roof [63,64].

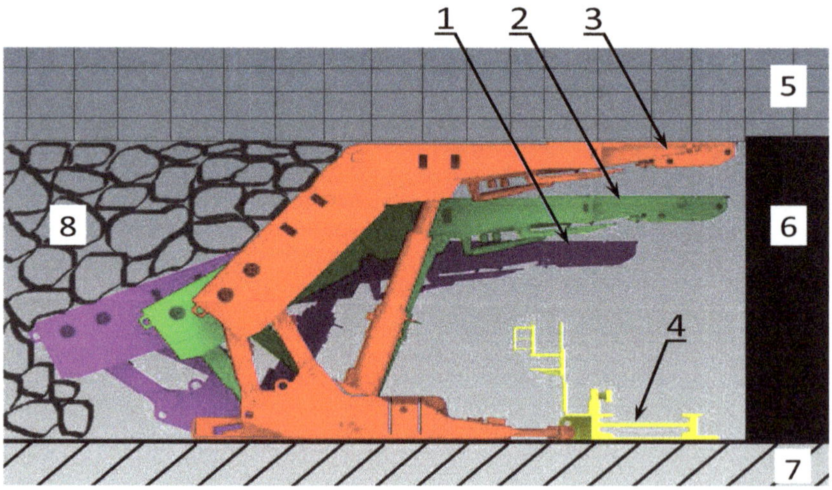


**Figure 1.** Work cycle for the powered roof-support section, where 1—withdrawing the powered roof-support section, 2—moving the powered roof-support section to a new location, 3—spragging the powered roof-support section, 4—longwall scraper conveyor, 5—roof, 6—coal, 7—floor, 8—goafs.

A powered roof support operates with three kinds of load capacity. Initial load capacity is obtained at the moment of expanding the powered roof support. After the powered roof support takes the pressure of roof rocks, it obtains working load capacity. However, at the maximum load-bearing value, the powered roof support reaches the nominal load capacity. Its value depends on the setting of the safety valves [65]. For the powered roof support to perform its tasks well, its reliable operation is necessary. Internal leaks represent one of the significant problems in the operation of the powered roof support.

The formation of internal leakage in the racks of the powered roof support significantly affects the loss of working load capacity. The simplest way to minimize them is to remove the cause of the leak. This is usually done by replacing the affected components, which generates high costs. The powered roof support usually consists of about 150–160 sections. The wall's extraction and the impact of the rock mass also affect the loss of the required load capacity. The proposed solution is to protect the powered roof support against loss of load capacity due to these events.

The presented test results were obtained in real conditions. The basis for the tests was the positive results of bench tests. The research is described in [65]. Researching the prototype solution in real-world conditions is a significant challenge. In conducting this research it was necessary to ensure that it did not influence the work safety or the safety of the powered roof support's operation. Another requirement was measurement to obtain test results. Satisfying those conditions allowed us to know the scope of the research and achieve our goal, described explicitly in this paper.

Section 2 describes our solution, which limits the consequences of leaks. We present the characteristics of a double valve with charging—the subject of this paper. This paper also concerns the measurement system and the excavation wall conditions, which was our research's location. The results of this research are described in detail in Section 3. The measurement results are presented in the form of graphs, and their analysis is provided afterwards. Section 4 discusses the effects of using the double valve with charging, while Section 5 summarizes the research and its results.

## 2. Materials and Methods

The research was conducted in an extraction wall (Figure 2), 166–245 m long and 970 m deep. The height of the wall was between 2.5 and 3.3 m. The longitudinal slope of the wall was up to 12°, while the transverse slope was up to 7°. The wall operation was carried out at 780–850 m. The operation was carried out with a longitudinal wall system with caving. On the selected deck's roof, there was shale with layers of coal and shale with coal. Above it, the shale was locally passing into sandy shale. Due to the presence of poorly compacted rocks on the direct deck roof, roof rocks were at risk of falling. The direct footwall contained shale. Below, there was a coal deck with a thickness of about 1.2–1.5 m. At the footwall of this deck laid shale locally, passing into sandy shale. The following hazards were present in the wall: methane hazard category III, first-degree water hazard, and dust hazard class B. The coal of the deck was classified as group III–IV self-ignition. The rock mass and rocks in the area of the wall were not prone to tremors. The wall had powered roof support, whose working range was from 2.4 to 4.4 m. The characteristics of the powered roof support's hydraulic prop are shown in Table 1.

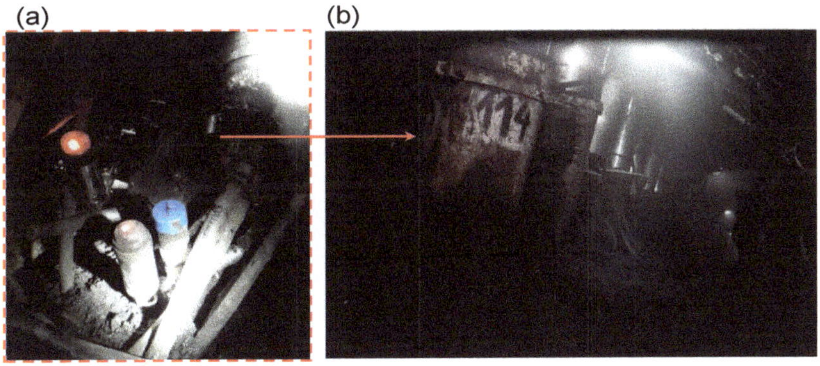


**Figure 2.** The extraction wall in which the research was carried out, where: (**a**) the measurement system with a prototype double block was installed on a prop; (**b**) view of the section on which the research was carried out.

**Table 1.** Technical characteristics of the hydraulic prop of the powered roof support.

| Operation Range | Work Unit |
|---|---|
| Working diameter | 300 mm/230 mm |
| Supply pressure | 25 ÷ 32 MPa |
| Nominal pressure | 43 MPa |
| Initial load capacity | 1767 ÷ 2262 kN |
| Working load capacity | 3039 kN |
| Hydraulic I stage stroke | 1212 mm |
| Hydraulic II stage stroke | 1129 mm |
| Min. length | 1897 mm |
| Max. length | 4238 mm |

A DOH DROPS-01 wireless pressure transducer was modified for testing purposes. Its measuring range was up to 60 MPa. Its modification consisted of changing the sampling value. The sampling frequency was 0.01 s (100 measurements per second). A DOH DROPS-01 are powered by batteries. A resistance strain gauge sensor is used as the measuring element. The sensors are equipped with a three-color LED. LEDs enable light signaling for specific pressure ranges. These pressure ranges are predeclared by the user.

The purpose of the tests was to minimize internal leaks of the powered roof support's prop, the parameters of which are summarized in Table 1. We propose the use of a prototype block with automatic charging in the hydraulic system to minimize these leaks. The operating parameters of the block are included in Table 2. A double block with charging is equipped with a threshold valve. This valve is set to 9 MPa. It means that below this value the charging is not carried out. Thanks to this, after withdrawing the powered roof support section, when the pressure in the over-piston space of the hydraulic prop is close to zero, the charging system is turned off. The prototype block is double—this means that the check valves used protect both the space above the piston and the space below the piston of the hydraulic prop against pressure loss. This is necessary in charging system. Charging is carried out by an additional connection of the double-charging block to the supply line. A shut-off valve is installed on the additional connection between the double-charging block and the supply line. This allows to turn off the charging any time.

**Table 2.** Technical characteristics of the prototype double block.

| Operation Range | Work Unit |
|---|---|
| Nominal pressure | 480 bar |
| Flow diameter | Ø 10 mm |
| Maximum flow | 400 l/min |
| Number of check valve cartridges | 3 |
| Work temperature | 40 °C ÷ 60 °C |

The purpose of the block is to minimize the effects of internal leaks and maintain the required value of working support. Working load capacity depends on the prop's diameter and the pressure in the under-piston space of the prop [66]:

$$F_W = \frac{\pi d^2}{4} \cdot p_w \quad (1)$$

where:

$F_W$—Working load capacity (N).

$d$—The prop's diameter (mm).

$p_w$—The pressure in the under-piston space of the prop (MPa).

When an internal leak occurs, there is a pressure drop in the under-piston space of the prop. Loss of pressure reduces the load capacity of the powered roof support. A double block charges the pressure, which was included in the formula to prevent this:

$$F_{RD} = \frac{\pi d^2}{4} \cdot \left(p_w - \sum \Delta p + \Delta p_d\right) \quad (2)$$

where:

$F_{RD}$—Actual load capacity after charging (N).
$d$—The prop's diameter (mm).
$p_w$—The pressure in the under-piston space of the prop (MPa).
$\sum\Delta p$—The sum of pressure losses in the under-piston space of the prop (MPa).
$\Delta p_d$—Increase in pressure as a result of charging (MPa).

## 3. Results

Bench tests differed from real-conditions tests in that they made it possible to determine the possible operation of the system. On the other hand, tests in real conditions showed the actual operation of the hydraulic system with the applied block with charging. Conducting this type of research is quite difficult in real conditions due to the lack of system operation characteristics with the use of a double block. The first days of testing allowed us to determine the features of the system's operation. The actual conditions adopted for the tests of the double block consisted of the mining wall, in which the powered roof support's prop had an internal leak.

Preparations for the actual tests began with the preparation of the system together with the measurement sensors at the surface station. These tests were aimed at verifying the sensors' correct operation and operation of the double block with charging—the site tests of the solution made it possible to verify the initial work before application in real conditions. Figure 3 shows the surface station on which the operation of the system and the measurement system were tested. In turn, Figure 4 shows the work of the block in which we observed the influence of charging time. The results of the bench tests allowed us to assess that the measurement system and the prototype block were ready for testing in real conditions. The bench test allowed us to look at the operation of the block and its functions and confirmed that the measurement system cooperated correctly with the double block.

Figure 5 shows the measurement obtained from the actual conditions of the mining wall. This figure shows a section of measurements from about 12 research days. The average pressure under the piston of the hydraulic prop was up to 30 MPa, and was mainly related to the pressure of rock layers on the powered roof support. To fully illustrate the measurement for the operation of the hydraulic system with a double block, Figure 5 has been divided into individually selected fragments of the operation of the double block with charging. The measurements presented in Figure 5 were analyzed, consisting of the selection of several work areas of the block charging. The mining and geological conditions as well as the pressure of the layers of roof rocks were omitted. The focus was solely on the analysis of the pressure charging function.

Figure 6 shows the analysis of the charging time, where four boosts of pressure were obtained. The first charge (a) was 1.9 min. The second charging started after 11.7 min and reached time (b), which was 2.4 min. For the subsequent charging (c), the maintenance time of the working load capacity of the powered roof support was 15.9 min. In the graph (Figure 6), the third charging (d) reached a time of 5.0 min. The working load capacity between the third and fourth charging (e) was 95 min. However, the fourth charge in graph (f) was captured within 1.2 min.

The graph in Figure 7 considers the charging time together with the relocation of the section of the powered roof support. The powered roof-support section's moving time was 6.9 min. The first charging (b) took 40 s, and the time load capacity of the powered roof support between the first and the second charging (c) was 48 min. The second charging time happened in (d) 2.6 min. The maintenance of the load capacity resulting from Figure 7 before the subsequent charging (e) was within 180.9 min. The following charging time (f) was about 2.4 min. After this time, the working load capacity (g) lasted 40.6 min, after which, the subsequent charging occurred in a time (h) of 1.9 min.

**Figure 3.** View of the test bench for verification of the double block charging function, where: 1—pressure sensor for the supra-block space, 2—pressure charging, 3—double block with pressure charging, 4—safety valve, 5—pressure sensor for the sub-block space, 6—hydraulic prop, 7—frame of test site.

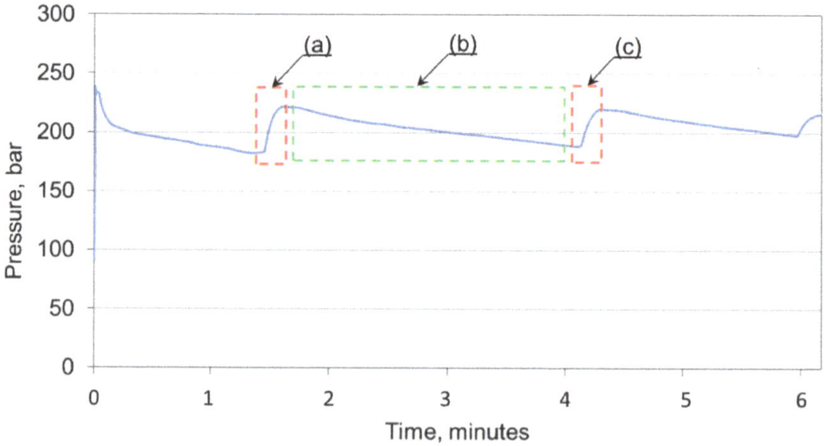


**Figure 4.** Stationary test of the double block charging function, where: (a) the charging time is 15.1 s; (b) the time of loss of the prop's load capacity is 141.8 s; (c) the charging time is 9.6 s.

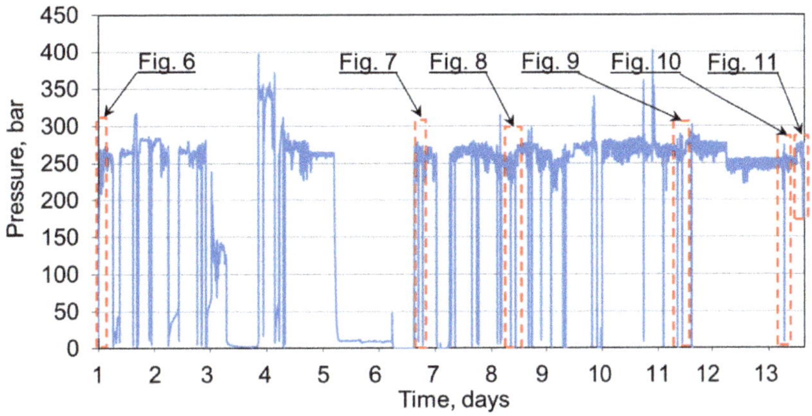


**Figure 5.** Testing the functionality of a double block in real conditions.

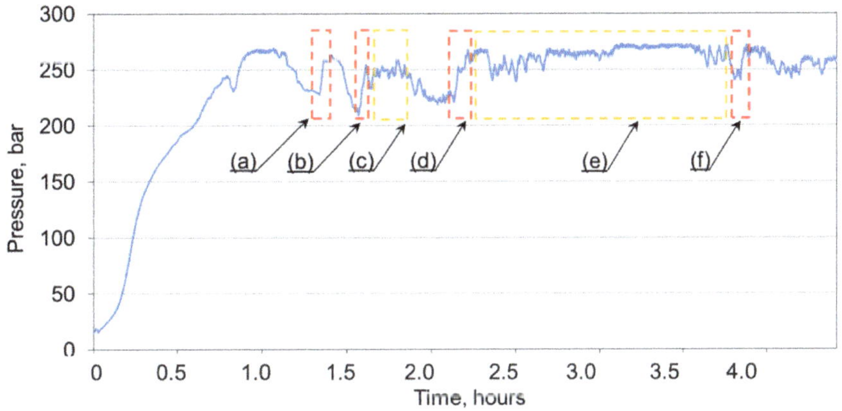


**Figure 6.** The four times ranges of pressure charging obtained to maintain the required working load capacity.

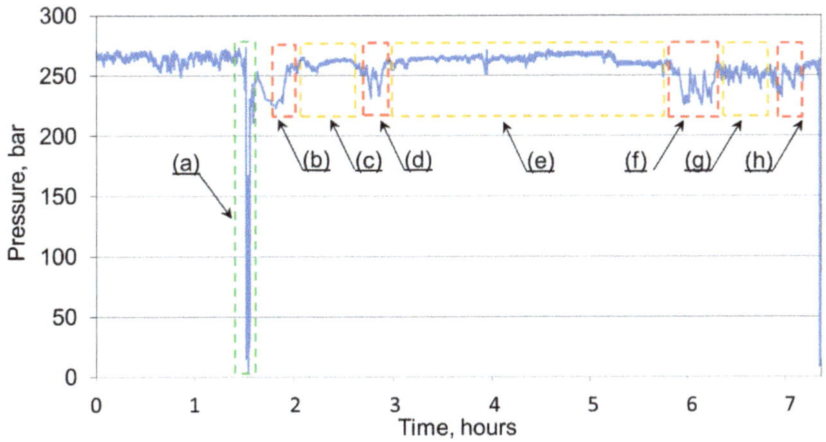


**Figure 7.** Analysis of the charging function in real conditions.

Figure 8 shows a graph of the obtained measurement for the charging function. The analysis considers the time of moving the section, which was (a) 2.2 min. After which, the first pressure charging time was recorded (b = 54 s). Maintaining the working load capacity (c) lasted about 24.4 min. On the other hand, its decrease caused a pressure charging which lasted about (d) 1.4 min. The working load capacity between the second and third charging was maintained within (e) 20.7 min. The third pressure boost was about (f) 50 s. On the other hand, the working load capacity was maintained for about (g) 23 min. The fourth charging took about (h) 22.5 s, after which, the working load capacity was maintained for the time of (i) 28.2 min. The analysis of the charging function was completed by moving the section to a new position, which took (j) 1.2 min.

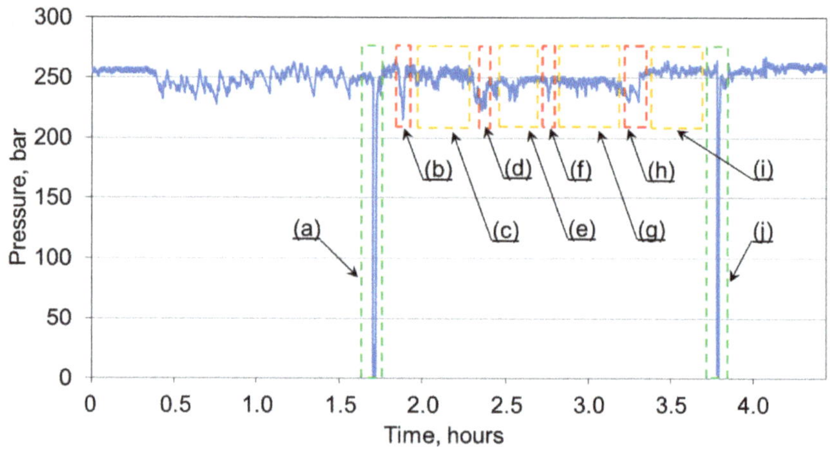


**Figure 8.** Analysis of the pressure charging function with the time of moving the section to a new position.

The graph in Figure 9 shows the charging time (c), which was 13 s. However, the time of the first change of the powered roof-support section to the new position (a) was 3.2 min. Load capacity time between the first switching of the powered roof support (b) reached 72.9 min. The second movement of the section to a new location (e) happened in 8.6 min, and the load capacity time between this operation and the charging time (d) was 284.4 min. Figure 10 shows the pressure charging time (a), which was about 1.2 min. On the other hand, the working load-capacity time (b) lasted about 37.3 min. Then, the section of the powered roof support was moved to a new place (c = 6.9 min). The measurement of the first charging time (Figure 11) was about (a) 18.3 s. On the other hand, the maintenance of the load capacity before the second charging (b) was about 2.8 h. The research analysis in Figure 11 ends with the time of the second charging (c) time, which was 23 s.

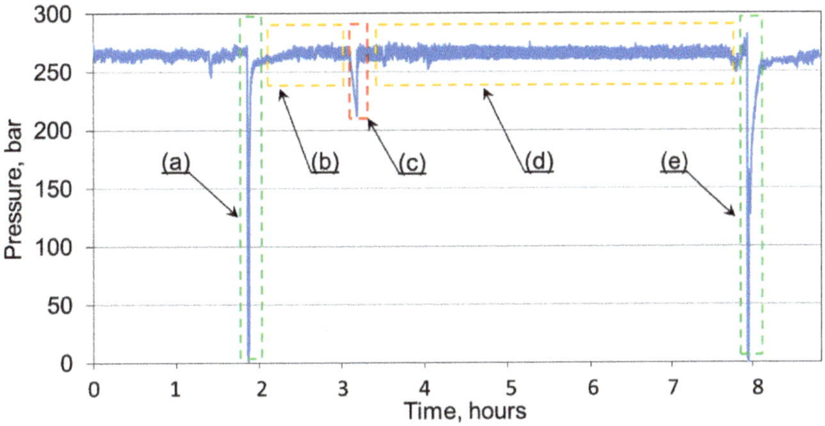


**Figure 9.** Analysis of the charging function during the tests.

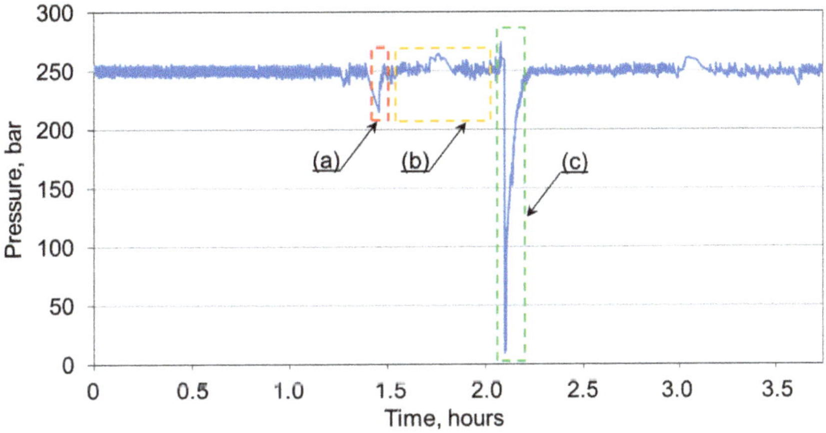


**Figure 10.** Testing the charging function along with the section relocation.

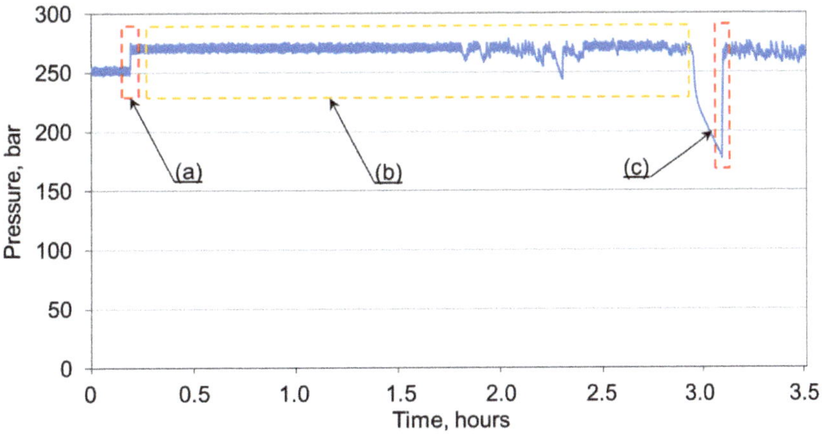


**Figure 11.** Analysis of the actual load capacity, including the charging function.

## 4. Discussion

The study considered pressure loss as an internal leak. It was formed on the inner seal, which caused a pressure drop in the prop. To obtain the best-possible operating parameters of the double block, we used a technically defective hydraulic prop employed in powered roof support in an extraction wall. The collected testing material presented in the graphs displayed in Figures 5–11 determined the actual charging time. Table 3 shows the aggregated results of the charging time ($T_D$) results for the recorded pressure drops ($\sum\Delta p$).

**Table 3.** The values obtained based on the research carried out.

| Supply Pressure $C_z$ (bar) | Sum of Pressure Losses $\sum\Delta p$ (bar) | Actual Load Capacity before Charging $F_R$ (kN) | Time of Charging $T_D$ (s) | Actual Load Capacity after Charging $F_{RD}$ (kN) | Increase in Load Capacity as a Result of Charging $\Delta F$ (%) |
|---|---|---|---|---|---|
| 260 | 42 | 1611 | 113 | 1837 | 14 |
| 255 | 50 | 1484 | 143 | 1795 | 21 |
| 270 | 37 | 1561 | 298 | 1893 | 21 |
| 270 | 30 | 1710 | 74 | 1900 | 11 |
| 255 | 23 | 1632 | 40 | 1795 | 10 |
| 265 | 24 | 1646 | 157 | 1844 | 12 |
| 255 | 30 | 1604 | 144 | 1780 | 11 |
| 260 | 22 | 1639 | 113 | 1837 | 12 |
| 260 | 49 | 1526 | 54 | 1837 | 20 |
| 250 | 35 | 1590 | 83 | 1766 | 11 |
| 255 | 21 | 1632 | 50 | 1773 | 9 |
| 255 | 20 | 1625 | 23 | 1802 | 11 |
| 270 | 56 | 1498 | 13 | 1900 | 27 |
| 260 | 43 | 1512 | 70 | 1809 | 20 |
| 270 | 96 | 1251 | 23 | 1879 | 50 |

The pressure drops resulting from the internal leakage of the prop ranged from 20 to 96 bar. This directly affected the significant reduction of the actual load capacity of the prop. The charging time ranged from 13 to 298 s. Cases that recorded longer charging times were probably due to insufficient pressure in the main power line. The charging function was interrupted at pressures which were too low in the main power line. Only after the pressure in the main power line increased did the block continue to charge. This had a direct effect on the increase in the charging time.

Based on Formula (1), the load capacity values for the tested prop were calculated. The temporary load capacity values were calculated for when the pressure drop occurred ($F_R$), and for after the pressure charging ($F_{RD}$). The results are recorded in Table 3 and shown in Figures 12 and 13. The research results suggest that a double block with a charging function ensured that the required working load capacity was maintained at a minimum level of 1766 kN. As a result of charging, the actual load capacity of the prop increased by about 10–20% in relation to the load capacity before charging. At higher pressure losses, the increase in load capacity after charging reached up to 50%. After charging, the prop's load capacity was directly proportional to the temporary pressure in the main power line. The results of real tests confirmed the results obtained in bench tests [65].

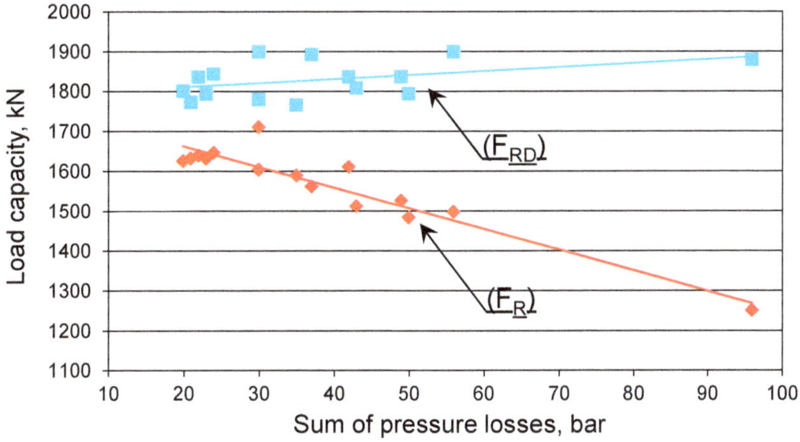


**Figure 12.** Analysis of the prop's actual load capacity before charging $F_R$ and after charging $F_{RD}$.

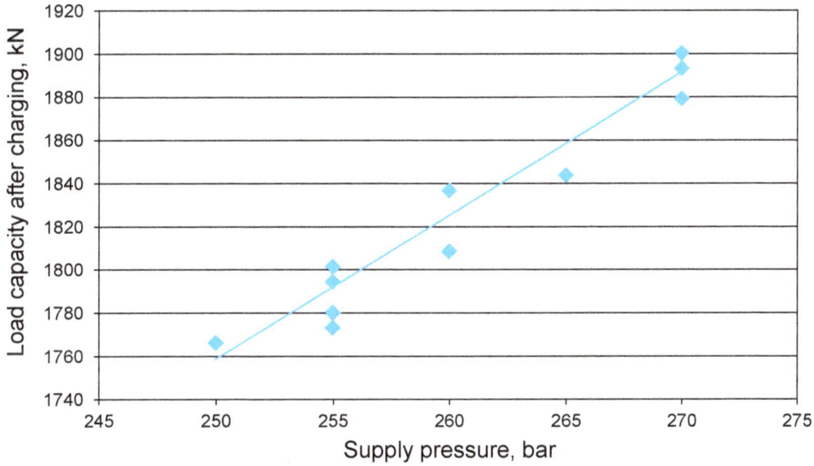


**Figure 13.** Analysis of the prop's actual load capacity after charging depending on the pressure in the main power line.

## 5. Conclusions

This work presents the implementation of the double-charging block in real condidions. The basis for the preparation for these tests were the previously conducted bench tests. Bench tests confirmed the correct operation of the double-charging block. Positive results of bench tests allowed to start testing in real conditions. The actual tests made it possible to determine the usefulness of the proposed solution in the event of an internal leak in the hydraulic prop of the powered roof support. Minimizing internal leaks is important to improving the reliability of powered roof support. This is important for improving safety in the longwall and maintaining the continuity of exploitation.

On the basis of the present research, it can be concluded that a double block with charging performs its function correctly in real conditions. The tests were carried out in a mining wall. For the research, a prop with an evolving internal leak was selected. As demonstrated by our measurements, after each pressure drop in the under-piston space of the prop, the block charged automatically. The charging time ranged from several seconds to several minutes. This time depended mainly on the instantaneous pressure value in the main power line.

This research proves the validity of the adopted theory. The solution minimizes the effects of internal leaks. Using a double block with charging ensured that the pressure in the under-piston space of the prop was maintained at a minimum of 250 bar. Thus, the prototype block maintained the required working load capacity in the prop, despite its internal leakage. It is essential to operate powered roof support and ensure safety adequately: the proposed solution allowed us to avoid costly and time-consuming prop replacement. Thus, using a block with charging may increase the efficiency of operation.

In connection with the test results, it is proposed that the powered roof support performs three functions. These functions already include two which are well-established, i.e., moving machines and devices using a control system as well as protecting against adverse mining and geological conditions. The third function proposed by the authors in this paper is eliminating any leaks that may affect the system's adverse mining and geological conditions and the control functionality.

**Author Contributions:** Conceptualization, D.S. and B.B.; methodology, D.S. and B.B.; software, B.B.; validation, D.S., B.B. and S.Z.; formal analysis, A.J.S.S.; investigation, D.S.; resources, R.D.; data curation, S.Z.; writing—original draft preparation, D.S., B.B. and A.J.S.S.; writing—review and editing, D.S. and B.B.; visualization, D.S.; supervision, D.S.; project administration, D.S., B.B. and S.Z.; funding acquisition, D.S. and S.Z. All authors have read and agreed to the published version of the manuscript.

**Funding:** This research received no external funding.

**Institutional Review Board Statement:** The study was conducted according to the guidelines of the Declaration of Helsinki and approved by the Mining Institute Review Board of T.F. Gorbachev Kuzbass State Technical University (Protocol Number 1101 from 29 November 2021).

**Informed Consent Statement:** Not applicable.

**Data Availability Statement:** Not applicable.

**Conflicts of Interest:** The authors declare no conflict of interest.

## References


1. Bazaluk, O.; Slabyi, O.; Vekeryk, V.; Velychkovych, A.; Ropyak, L.; Lozynskyi, V. A Technology of Hydrocarbon Fluid Production Intensification by Productive Stratum Drainage Zone Reaming. *Energies* **2021**, *14*, 3514. [CrossRef]
2. Patyk, M.; Bodziony, P.; Krysa, Z. A Multiple Criteria Decision Making Method to Weight the Sustainability Criteria of Equipment Selection for Surface Mining. *Energies* **2021**, *14*, 3066. [CrossRef]
3. Kumar, R.; Singh, A.K.; Mishra, A.K.; Singh, R. Underground mining of thick coal seams. *Int. J. Min. Sci. Tech.* **2015**, *25*, 885–896. [CrossRef]
4. Gil, J.; Kołodziej, M.; Szurgacz, D.; Stoiński, K. Introduction of standardization of powered roof supports to increase production efficiency of Polska Grupa Górnicza, S.A. *Min. Inform. Autom. Electr. Eng.* **2019**, *56*, 33–38. [CrossRef]
5. Pokorny, J.; Dlouhá, D.; Kucera, P. Study of the necessity of use virtual origin in assessment of selected fire plume characteristics. *MM Sci. J.* **2016**, *5*, 1424–1428. [CrossRef]
6. Pokorny, J.; Mozer, V.; Malerova, L.; Dlouhá, D.; Wilkinson, P. A simplified method for establishing safe available evacuation time based on a descending smoke layer. *Commun. Sci. Lett. Univ. Zilina* **2018**, *20*, 28–34. [CrossRef]
7. Uth, F.; Polnik, B.; Kurpiel, W.; Baltes, R.; Kriegsch, P.; Clause, E. An innovate person detection system based on thermal imaging cameras dedicate for underground belt conveyors. *Min. Sci.* **2019**, *26*, 263–276. [CrossRef]
8. Szurgacz, D.; Zhironkin, S.; Pokorný, J.; Spearing, A.J.S.; Vöth, S.; Cehlár, M.; Kowalewska, I. Development of an Active Training Method for Belt Conveyor. *Int. J. Environ. Res. Public Health* **2022**, *19*, 437. [CrossRef]
9. Bazaluk, O.; Velychkovych, A.; Ropyak, L.; Pashechko, M.; Pryhorovska, T.; Lozynskyi, V. Influence of Heavy Weight Drill Pipe Material and Drill Bit Manufacturing Errors on Stress State of Steel Blades. *Energies* **2021**, *14*, 4198. [CrossRef]
10. Góralczyk, M.; Krot, P.; Zimroz, R.; Ogonowski, S. Increasing Energy Efficiency and Productivity of the Comminution Process in Tumbling Mills by Indirect Measurements of Internal Dynamics—An Overview. *Energies* **2020**, *13*, 6735. [CrossRef]
11. Doroszuk, B.; Król, R. Conveyor belt wear caused by material acceleration in transfer stations. *Min. Sci.* **2019**, *26*, 189–201. [CrossRef]
12. Grzesiek, A.; Zimroz, R.; Śliwiński, P.; Gomolla, N.; Wyłomańska, A. A Method for Structure Breaking Point Detection in Engine Oil Pressure Data. *Energies* **2021**, *14*, 5496. [CrossRef]

13. Buyalich, G.; Buyalich, K.; Byakov, M. Factors Determining the Size of Sealing Clearance in Hydraulic Legs of Powered Supports. *E3S Web Conf.* **2017**, *21*, 3018. [CrossRef]
14. Bortnowski, P.; Gładysiewicz, L.; Król, R.; Ozdoba, M. Energy Efficiency Analysis of Copper Ore Ball Mill Drive Systems. *Energies* **2021**, *14*, 1786. [CrossRef]
15. Borkowski, P.J. Comminution of Copper Ores with the Use of a High-Pressure Water Jet. *Energies* **2020**, *13*, 6274. [CrossRef]
16. Kawalec, W.; Suchorab, N.; Konieczna-Fuławka, M.; Król, R. Specific energy consumption of a belt conveyor system in a continuous surface mine. *Energies* **2020**, *13*, 5214. [CrossRef]
17. Wodecki, J.; Góralczyk, M.; Krot, P.; Ziętek, B.; Szrek, J.; Worsa-Kozak, M.; Zimroz, R.; Śliwiński, P.; Czajkowski, A. Process Monitoring in Heavy Duty Drilling Rigs—Data Acquisition System and Cycle Identification Algorithms. *Energies* **2020**, *13*, 6748. [CrossRef]
18. Szurgacz, D.; Zhironkin, S.; Vöth, S.; Pokorný, J.; Spearing, A.J.S.; Cehlár, M.; Stempniak, M.; Sobik, L. Thermal Imaging Study to Determine the Operational Condition of a Conveyor Belt Drive System Structure. *Energies* **2021**, *14*, 3258. [CrossRef]
19. Kawalec, W.; Błażej, R.; Konieczna, M.; Król, R. Laboratory Tests on e-pellets effectiveness for ore tracking. *Min. Sci.* **2018**, *25*, 7–18. [CrossRef]
20. Zimroz, P.; Trybała, P.; Wróblewski, A.; Góralczyk, M.; Szrek, J.; Wójcik, A.; Zimroz, R. Application of UAV in Search and Rescue Actions in Underground Mine—A Specific Sound Detection in Noisy Acoustic Signal. *Energies* **2021**, *14*, 3725. [CrossRef]
21. Bardzinski, P.; Jurdziak, L.; Kawalec, W.; Król, R. Copper ore quality tracking in a belt conveyor system using simulation tools. *Nat. Resour. Res.* **2020**, *29*, 1031–1040. [CrossRef]
22. Wang, J.; Wang, Z. Systematic principles of surrounding rock control in longwall mining within thick coal seams. *Int. J. Min. Sci. Tech.* **2019**, *29*, 591–598. [CrossRef]
23. Jixiong, Z.; Spearing, A.J.S.; Xiexing, M.; Shuai, G.; Qiang, S. Green coal mining technique integrating mining-dressing-gas draining-backfilling-mining. *Int. J. Min. Sci. Tech.* **2017**, *27*, 17–27.
24. Krauze, K.; Mucha, K.; Wydro, T.; Pieczora, E. Functional and Operational Requirements to Be Fulfilled by Conical Picks Regarding Their Wear Rate and Investment Costs. *Energies* **2021**, *14*, 3696. [CrossRef]
25. Kotwica, K.; Stopka, G.; Kalita, M.; Bałaga, D.; Siegmund, M. Impact of Geometry of Toothed Segments of the Innovative KOMTRACK Longwall Shearer Haulage System on Load and Slip during the Travel of a Track Wheel. *Energies* **2021**, *14*, 2720. [CrossRef]
26. Ji, Y.; Zhang, Y.; Huang, Z.; Shao, Z.; Gao, Y. Theoretical analysis of support stability in large dip angle coal seam mined with fully-mechanized top coal caving. *Min. Sci.* **2020**, *27*, 73–87.
27. Xiaozhen, W.; Jialin, X.; Weibing, Z.; Yingchun, L. Roof pre-blasting to prevent support crushing and water inrush accidents. *Int. J. Min. Sci. Tech.* **2012**, *22*, 379–384.
28. Klishin, V.I.; Klishin, S.V. Coal Extraction from Thick Flat and Steep Beds. *J. Min. Sci.* **2010**, *46*, 149–159. [CrossRef]
29. Mo, S.; Tutuk, K.; Saydam, S. Management of floor heave at Bulga Underground Operations—A case study. *Int. J. Min. Sci. Tech.* **2019**, *29*, 73–78. [CrossRef]
30. Frith, R.C. A holistic examination of the load rating design of longwall shields after more than half a century of mechanised longwall mining. *Int. J. Min. Sci. Tech.* **2015**, *26*, 199–208. [CrossRef]
31. Król, R.; Kisielewski, W. Research of loading carrying idlers used in belt conveyor-practical applications. *Diagnostyka* **2014**, *15*, 67–74.
32. Ziętek, B.; Banasiewicz, A.; Zimroz, R.; Szrek, J.; Gola, S. A Portable Environmental Data-Monitoring System for Air Hazard Evaluation in Deep Underground Mines. *Energies* **2020**, *13*, 6331. [CrossRef]
33. Ji, Y.; Ren, T.; Wynne, P.; Wan, Z.; Zhaoyang, M.; Wang, Z. A comparative study of dust control practices in Chinese and Australian longwall coal mines. *Int. J. Min. Sci. Tech.* **2016**, *25*, 687–706. [CrossRef]
34. Prostański, D. Empirical Models of Zones Protecting Against Coal Dust Explosion. *Arch. of Min. Sci.* **2017**, *62*, 611–619. [CrossRef]
35. Bajda, M.; Hardygóra, M. Analysis of Reasons for Reduced Strength of Multiply Conveyor Belt Splices. *Energies* **2021**, *14*, 1512. [CrossRef]
36. Woźniak, D.; Hardygóra, M. Method for laboratory testing rubber penetration of steel cords in conveyor belts. *Min. Sci.* **2020**, *27*, 105–117. [CrossRef]
37. Bajda, M.; Błażej, R.; Hardygóra, M. Optimizing splice geometry in multiply conveyor belts with respect to stress in adhesive bonds. *Min. Sci.* **2018**, *25*, 195–206. [CrossRef]
38. Adach-Pawelus, K.; Pawelus, D. Influence of Driving Direction on the Stability of a Group of Headings Located in a Field of High Horizontal Stresses in the Polish Underground Copper Mines. *Energies* **2021**, *14*, 5955. [CrossRef]
39. Gładysiewicz, L.; Król, R.; Kisielewski, W.; Kaszuba, D. Experimental determination of belt conveyors artificial friction coefficient. *Acta Montan. Slovaca* **2017**, *22*, 206–214.
40. Huang, P.; Spearing, S.; Ju, F.; Jessu, K.V.; Wang, Z.; Ning, P. Control Effects of Five Common Solid Waste Backfilling Materials on In Situ Strata of Gob. *Energies* **2019**, *12*, 154. [CrossRef]
41. Wajs, J.; Trybała, P.; Górniak-Zimroz, J.; Krupa-Kurzynowska, J.; Kasza, D. Modern Solution for Fast and Accurate Inventorization of Open-Pit Mines by the Active Remote Sensing Technique—Case Study of Mikoszów Granite Mine (Lower Silesia, SW Poland). *Energies* **2021**, *14*, 6853. [CrossRef]
42. Rajwa, S.; Tomasz Janoszek, T.; Stanisław Prusek, S. Influence of canopy ratio of powered roof support on longwall working stability—A case study. *Int. J. Min. Sci. Tech.* **2019**, *29*, 591–598. [CrossRef]

43. Rajwa, S.; Janoszek, T.; Prusek, S. Model tests of the effect of active roof support on the working stability of a longwall. *Comput. Geotech.* **2020**, *118*, 103302. [CrossRef]
44. Juganda, A.; Strebinger, C.; Brune, J.F.; Bogin, G.E. Discrete modeling of a longwall coal mine gob for CFD simulation. *Int. J. Min. Sci. Tech.* **2020**, *30*, 463–469. [CrossRef]
45. Janus, J.; Krawczyk, J. Measurement and Simulation of Flow in a Section of a Mine Gallery. *Energies* **2021**, *14*, 4894. [CrossRef]
46. Świątek, J.; Janoszek, T.; Cichy, T.; Stoiński, K. Computational Fluid Dynamics Simulations for Investigation of the Damage Causes in Safety Elements of Powered Roof Supports—A Case Study. *Energies* **2021**, *14*, 1027. [CrossRef]
47. Dlouhá, D.; Pokorný, J.; Dlouhá, K. Necessity of knowledge about math in safety engineering. In Proceedings of the 14th Conference E-Learning: Unlocking the Gate to Education around the Globe, Prague, Czech Republic, 20–21 June 2019; pp. 380–386.
48. Baiul, K.; Khudyakov, A.; Vashchenko, S.; Krot, P.V.; Solodka, N. The experimental study of compaction parameters and elastic after-effect of fine fraction raw materials. *Min. Sci.* **2020**, *27*, 7–18. [CrossRef]
49. Dlouhá, D.; Dubovský, V. The improvement of the lake Most evaporation estimates. *Inż. Miner.* **2019**, *21*, 159–164.
50. Dlouhá, D.; Dubovský, V.; Pospíšil, L. Optimal calibration of evaporation models against Penman-Monteith Equatíon. *Water* **2021**, *13*, 1484. [CrossRef]
51. Dubovský, V.; Dlouhá, D.; Pospíšil, L. The calibration of evaporation models against the Penman-Monteith equation on lake Most. *Sustainability* **2021**, *13*, 313. [CrossRef]
52. Dlouhá, D.; Hamříková, R. Interactive distance materials of mathematics for VŠB-TU Ostrava. In Proceedings of the 13th Conference Overcoming the Challenges and the Barriers in Open Education, Prague, Czech Republic, 25–26 June 2018; pp. 67–72. Available online: https://www.fast.vsb.cz/230/cs/Veda-a-vyzkum/Publikace/2018/ (accessed on 1 November 2021).
53. Dlouhá, D.; Kozlová, K. Knowledge assessment of student's high school mathematics. In Proceedings of the 17th Conference on Applied Mathematics (APLIMAT 2019), Bratislava, Slovak Republic, 5–7 February 2019; Volume 1, pp. 243–252.
54. Hamříková, R.; Dlouhá, D. Video tutorials for students of the master's program. In Proceedings of the 12th Conference Open Education as a Way to a Knowledge Society, Prague, Czech Republic, 26–27 June 2017; pp. 446–451.
55. Dlouhá, D.; Hamříková, R. Our experience with the involvement of students in the creation of study materials. In Proceedings of the 17th Conference on Applied Mathematics (APLIMAT 2019), Bratislava, Slovak Republic, 5–7 February 2019; Volume 1, pp. 301–308.
56. Peng, S.S.; Feng, D.; Cheng, J.; Yang, L. Automation in U.S. longwall coal mining: A state-of-the-art review. *Int. J. Min. Sci. Tech.* **2019**, *29*, 151–159. [CrossRef]
57. Ralston, J.C.; Hargrave, C.O.; Dunn, M.T. Longwall automation: Trends, challenges and opportunities. *Int. J. Min. Sci. Tech.* **2017**, *27*, 733–739. [CrossRef]
58. Ralston, J.C.; Reid, D.C.; Dunn, M.T.; Hainsworth, D.W. Longwall automation: Delivering enabling technology to achieve safer and more productive underground mining. *Int. J. Min. Sci. Tech.* **2015**, *25*, 865–876. [CrossRef]
59. Hu, S.; Ma, L.; Guo, J.; Yang, P. Support-surrounding rock relationship and top-coal movement laws in large dip angle fully-mechanized caving face. *Int. J. Min. Sci. Tech.* **2018**, *28*, 533–539.
60. Buyalich, G.; Byakov, M.; Buyalich, K. Factors Determining Operation of Lip Seal in the Sealed Gap of the Hydraulic Props of Powered Supports. *E3S Web Conf.* **2017**, *41*, 1045. [CrossRef]
61. Szurgacz, D.; Zhironkin, S.; Cehlár, M.; Vöth, S.; Spearing, S.; Liqiang, M. A Step-by-Step Procedure for Tests and Assessment of the Automatic Operation of a Powered Roof Support. *Energies* **2021**, *14*, 697. [CrossRef]
62. Buyalich, G.; Byakov, M.; Buyalich, K.; Shtenin, E. Development of Powered Support Hydraulic Legs with Improved Performance. *E3S Web Conf.* **2019**, *105*, 3025. [CrossRef]
63. Szurgacz, D. Dynamic Analysis for the Hydraulic Leg Power of a Powered Roof Support. *Energies* **2021**, *14*, 5715. [CrossRef]
64. Stoiński, K.; Mika, M. Dynamics of Hydraulic Leg of Powered Longwall Support. *J. Min. Sci.* **2003**, *39*, 72–77. [CrossRef]
65. Szurgacz, D.; Borska, B.; Zhironkin, S.; Diederichs, R.; Spearing, A.J.S. Optimization of the Load Capacity System of Powered Roof Support: A Review. *Energies* **2022**, *15*, 6061. [CrossRef]
66. Irresberger, H.; Grawe, F.; Migenda, P. *Zmechanizowane Obudowy Ścianowe (Podręcznik dla Praktyków)*; Wyd. Tiefenbach Polska Sp. z o.o.: Piekary Śląskie, Poland, 2003.

energies

*Review*

# Review of Linear Electric Motor Hammers—An Energy-Saving and Eco-Friendly Solution in Industry

**Adam Wróblewski [1], Pavlo Krot [1,*], Radosław Zimroz [1], Timo Mayer [2] and Jyri Peltola [2]**

[1] Faculty of Geoengineering, Mining and Geology, Wroclaw University of Science and Technology, Na Grobli 15, 50-421 Wroclaw, Poland
[2] Lekatech Oy, 47400 Kausala, Finland
* Correspondence: pavlo.krot@pwr.edu.pl

**Abstract:** Standard hydraulic breaking hammers are widely used for crushing oversized blasted materials and concrete structures demolition in industry. These hammers, installed in on-surface working excavators or stationary manipulators at the dumping points of underground conveyors, provide the required limited sizes of bulk materials and enable the safe operation of other equipment (screens, crushers). In parallel, hydraulic hammers have an alternative—fully electric hammers. This paper aims to review existing linear electric motor (LEM) hammers as an energy-saving and eco-friendly solution in industry. Global market analysis is presented with potential branches of LEM hammers. Several aspects for implementation—design optimization, dynamics simulation, machine control, and performance estimation—are considered. Different case studies for LEM-hammer application are given. The preliminary measurements are demonstrated on the electric hammer of Lekatech Company, which is intended for the mining industry and construction demolition. Experiments showed that depending on the impact frequency, type of rock, and shape of the crushing tool, the time to fracture varies significantly. Optimal parameters exist for every case, for which adjusting requires online hammer control.

**Keywords:** linear inductive motor (LIM); linear electric motor hammer; mining; raw materials; crushing



**Citation:** Wróblewski, A.; Krot, P.; Zimroz, R.; Mayer, T.; Peltola, J. Review of Linear Electric Motor Hammers—An Energy-Saving and Eco-Friendly Solution in Industry. *Energies* **2023**, *16*, 959. https://doi.org/10.3390/en16020959

Academic Editors: Sergey Zhironkin and Dawid Szurgacz

Received: 13 December 2022
Revised: 4 January 2023
Accepted: 11 January 2023
Published: 14 January 2023



## 1. Introduction

Both open-pit and underground mining operations face the issue of oversized chunks of rock material that remain after blasting. To fit the rock material to further crushing stages and conveying systems, rock breakers are used. Currently, these operations are performed by hydraulic breaker hammers, mostly operated by a human in place. In the era of electric and digital transformation, especially within the mining industry, the efficiency and safety of this process can be significantly increased. The new standards for sustainable mining demand the mine of the future to be carbon-dioxide-free, digitized, and expectant of the replacement of inefficient, dirty, energy-driven, and manual processes through the implementation of electric, remote-controlled, and/or autonomous machines.

The fully electrical impact hammer is a promising alternative to the hydraulic hammers currently in use. It could be a functional machinery subsystem, which is environmentally friendly (without $CO_2$ emissions), energy-saving, with increased performance, and not requiring high pressure and hazardous oils. Lower noise and tremor levels, which characterize an electric device, in contrast to hydraulic hammers, are other significant advantages in the case of quarrying and civil-engineering operations that are performed near residential areas. Although many technical solutions in electric-hammer design have been patented in the world, there is no evidence in the market of products ready to be used in place of hydraulic or pneumatic hammers in heavy industries.

A recently started ECHO project aims to introduce an electrical, programmable hammer (LEH) to provide full control and digitization in these sectors of industry. This novel patented technological solution of Lekatech Company (Finland) is capable of boosting the rock-fragmentation process. Compared to hydraulic hammers, LEH eliminates $CO_2$ emissions, saves energy and hydraulic oil, increases safety, and lowers noise level and life-cycle cost. The increased performance is achieved by the fully adjustable force and frequency of impacts, depending on the hardness of the treated material, which are more limited in hydraulic hammers.

The research work in this project is related to measurements and the process of data collection on operations during LEH testing in several mining companies in Poland, Finland, and Spain. The aim is to confirm LEH reliability and to determine the optimal controllable parameters of vibration for different materials and operating conditions. The 3-year ECHO project funded by EIT Raw Materials is led by Lekatech in a consortium with Iberian Sustainable Mining Cluster (Spain), MNLT Innovations IKE (Greece), KGHM Polska Miedz S.A. and Wroclaw University of Science and Technology (Poland). This project will result in a fully new approach to the hammering process in mining and provide a cost-efficient, digitized, safe, environmentally friendly, and user-friendly solution, ready to be commercially distributed and used in the industry.

### 1.1. Market Analysis and Global Players

Following the results of the global market research in [1,2], rock-breaker sales were a little over 1 billion USD in 2017 and are projected to grow by more than 11% during 2018–2023, to nearly 2 billion USD by 2023. This sector of machines manufacturing is spread out because there are five top manufacturers of hydraulic hammers: Sandvik AB, Atlas-Copco (Krupp), Montabert (Joy Global), Furukawa, and Eddie, but their total production takes a total of only 24.29%; hence, there are many local manufacturers in different countries. Europe and China took 56.97% of the global production market of hydraulic hammers as of 2016.

For the application market (see Figure 1), the construction industry has always had largest share (33.58% in 2016). While the sector of municipal engineering and mining industries take 27.98% and 18.03%, respectively.

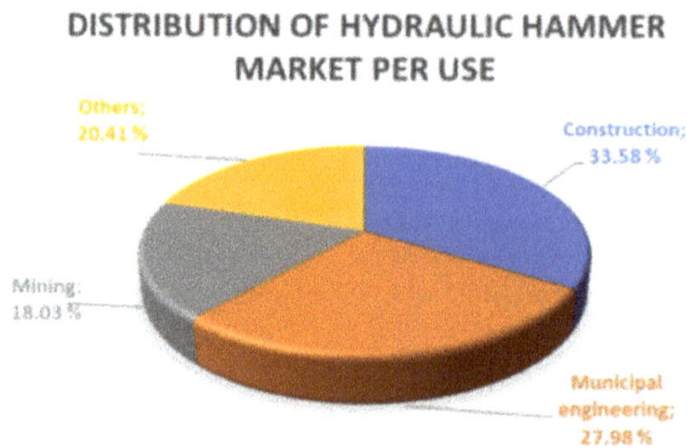


**Figure 1.** Market share of hydraulic hammering applications [2].

### 1.2. Energy Consumption

In the mining industry, although the alternative technologies of rock breaking are under consideration [3,4] to replace blasting, the conventional method of drilling and blasting has many advantages from the viewpoint of productivity. On the other hand, there are certain implications due to non-homogeneous geological structures, which resulted in oversized pieces of rocks influencing the rock-breaking hammers' performance and the fuel consumption of machines.

Opportunities for energy efficiency in the stone and asphalt industry are considered in [5]. The typical breakdown of energy costs in an average-size mining enterprise producing aggregate and the energy consumption at the different stages of crushing operations are represented in Figure 2. Besides electric energy, the biggest part of costs is spent on explosives. Therefore, blasting optimization and planning has significant potential for the economy. Just using the electronic detonators in place of non-electric initiation has shown such benefits as:

- A 32% decrease in the mean size of rock in the post-blast pile;
- A 37% increase in the amount of rock of less than 8-inch size;
- A 25% reduction in digging time to excavate the pile;
- A 6–10% savings in primary crushing costs measured by power consumption.

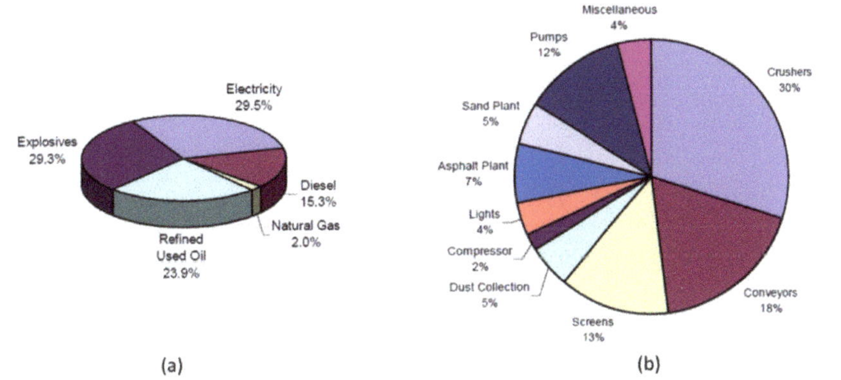


**Figure 2.** The typical breakdown of energy costs in an average-size mining enterprise (**a**); and electric-energy consumption share in the crushing facility of asphalt plant (**b**) [5].

The diagram in Figure 3 reveals the energy balance of mining and construction machines with hydraulic and electric hammers. Approximately 47% of the total energy is supplied by the machine combustion engine to the ancillary equipment. According to their average efficiency, a hydraulic pump spends 11.7%, hydraulic piping takes about 12.9% due to losses by throttling, and only 3.7% reaches the breaker. Meanwhile, impact energy constitutes only 8.5% [6].

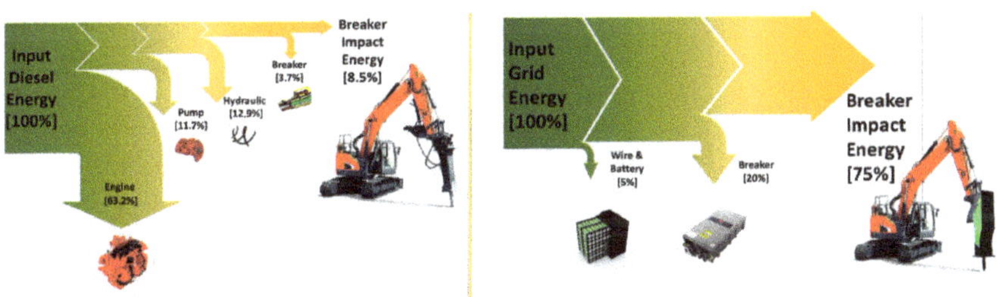


**Figure 3.** Sankey plot of energy flow through an excavator during operation with a hydraulic hammer adopted from [6] and fully electric hammer.

## 2. Problem Formulation

Pneumatic and hydraulic actuators have similar dimensions to tubular electric actuators; thus, they constitute an alternative choice for the designers of industrial machines. The force is created by applying pressure to a working gaseous or liquid medium in a cylinder and the piston converts the pressure to a force. Ficarella et al. [7] showed that the key parameter, which is responsible for better hammer performance, is the increased working pressure. This solution has several disadvantages. The most frequently occurring problems in hydraulic hammers during operation and maintenance are as follows:

- Sealing damage and oil leakage due to wear and surface contamination;
- Cylinder body deformations due to overloading by high pressure;
- Possible piston rod deformation under load;
- Hydraulic oil consumption and utilization;
- Impossibility of changing hammering parameters online within wide ranges.

Traditional actuators usually use open-loop force control but many applications often require controlled stroke and force, which requires the installation of additional sensors and special wiring with environmental and mechanical protection as well as signal-acquisition hardware. Instead, all these issues can be easily resolved with electric hammers.

An electric hammer based on a tubular linear inductive motor has better accuracy, readiness, and control reliability; it is maintenance-free, has easier energy-supply availability, and does not create any troubles with hydraulic fluid.

The increased costs of hydraulic-hammer operation and maintenance during the full life cycle make their replacement with electric hammering solutions profitable. However, certain considerations have to be taken into account for the successful implementation of innovative drives.

This paper is focused on the analysis of possibilities for the application of linear electric-motor hammering solutions to the rock-breaking process. The analysis of other similar domains is given in the paper to understand better the pros and cons of linear electric motors and potentially new domains for their application.

## 3. State of the Art

### 3.1. What Is a Linear Electric Motor

Gieras [8] and Budig [9] classified linear electric motors as a group of special electrical machines that convert electrical energy directly into the mechanical energy of axial motion.

Linear electric motors can drive a linear-motion load without intermediate gears, screws, or crankshafts. Linear electric motors can be classified as follows:

- DC motors;
- Induction motors;
- Synchronous motors, including reluctance and stepping motors;
- Oscillating motors;
- Hybrid motors.

The application of DC linear motors is less widespread, save for special automotive applications where a DC supply is provided by the onboard generator. The direct-drive electromagnetic active suspension system is developed by Shen et al. in [10], which consists of a double-stator air-core tubular permanent magnet linear motor (DATPLM) acting in parallel with a coil spring. The proposed DATPLM can also operate in generator mode, which provides an additional energy supply. This suspension system can simultaneously eliminate road disturbances and apply active roll and pitch control. Unlike the existing slotted topologies of magnetic poles (radial, axial), the proposed device has a quasi-Halbach topology with improved dynamic response and very-low-force ripple. Some other studies in this domain are presented by Gysen et al. in [11]. The dynamic model of the car suspension system based on a linear electric motor subjected to sharp loading impacts is proposed by van Casteren et al. in [12]. This model includes non-linear bump stops, actuator saturation, and a combination of Coulomb and viscous friction. The issues of using accelerometers are discussed. The use of the skyhook controller showed improvements in a frequency region of 0.7 to 8 Hz. It is shown that the non-linear model more accurately predicts the acceleration of the car mass.

The most popular are permanent magnet (PM) linear synchronous motors (LSMs) and linear induction motors (LIMs), which are manufactured commercially for many applications. The most known application of LIMs is in transportation based on electric traction systems, where the primary is mounted on the vehicle and the secondary is installed along the track. Linear induction motors are used in cranes for material handling. They are also used for liquid-metal pumping, actuators for door movement, high-voltage circuit breakers, and accelerators for rigs for testing vehicle performance under impacts. In recent years, significant attention has also been paid to LIM applications in kinetic weapons (rail guns). In hammering applications, similar to rail guns, tubular LIMs are used, whose structure is shown in Figure 4.

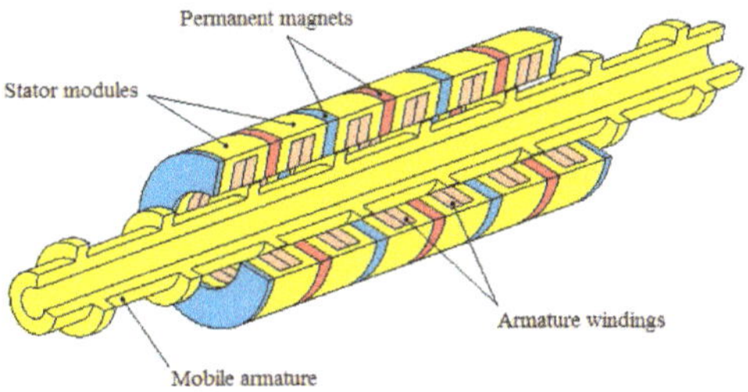


**Figure 4.** Typical structure of tubular LIM with permanent magnets (Lo et al. [13]).

In the general sense, the stator and the rotor of LIM are known as primary and secondary, respectively. If the primary of the linear induction motor is connected to a three-phase voltage supply, a traveling flux wave is produced, which travels along the primary body. Due to the relative motion between the traveling flux wave and the secondary, a current is induced in the conductor, which also produces a magnetic flux-wave interaction with the traveling flux wave of the primary, resulting in a linear force. Depending on what body is fixed, primary or secondary, the electromagnetic force will move the opposite body in the direction of the traveling flux wave.

The linear synchronous speed $v_s$ of the traveling flux wave produced by the primary of the linear induction motor is given by the formula:

$$v_s = 2fP, \tag{1}$$

where $f$ is the frequency of the supply voltage (Hz); and $P$—pole pitch.

If the LIM is an asynchronous or induction motor, the speed of the secondary is less than the synchronous speed and the difference between the two speeds is known as the slip. The slip of the linear induction motor is given by,

$$s = (v_s - v_r)/v_s, \tag{2}$$

The speed of the secondary $v_r$ in the linear induction motor is given by the formula:

$$v_r = (1 - s)v_s, \tag{3}$$

The linear force or thrust of the linear induction motor is given by the formula:

$$F = P_{gap}/v_s, \tag{4}$$

where $P_{gap}$—power at air gap.

Govindpure et al. [14] noted that during the LIM design process, basic equations of rotary induction motors (RIM) are commonly used, but they do not reflect the true merit of LIM. The end effect and edge effects are usually neglected. A double-sided LIM of 12 m in length is designed by the 3D Maxwell software and analyzed at no-load conditions.

The literature analysis shows that modern mining enterprises worldwide are aimed at digitizing and automating mining processes with autonomous working or robotized machines. Although the different versions of the design (Grinchenko [15]) of linear electric motors have been known for a long time (Brittain and Laithwaite [16]), they are still not widely represented in heavy industries such as mining, oil and gas, construction and metallurgy (Kurilin et al. [17]).

The actuators based on linear electric motors are mainly applied in precise manufacturing processes, robots, materials machining centers, additive manufacturing, and laser cutting for working tools positioning and parts moving (Gieras [18], Korendyi et al. [19,20]). The best linear electric motors for such applications are permanent magnet (PM) linear synchronous motors (LSM).

Combined linear induction motors (CLIMs) for moving robotic trolleys in sealed radiation chambers are considered by Tiunov [21]. These motors consist of induction driving units and braking units based on permanent magnets. Acting together, these units make it possible to obtain a low-speed trolley without a complicated control system or frequency converters.

One of LIM's applications, which can be considered for underground mines, is their combination with hydraulic cylinders in the powered roof supports of a long-wall shear complex. The main reason is to decrease the response time of hydraulic actuators to sharp dynamic loading occurring from the rock (Rudzki et al. [22]).

### 3.2. Design Optimisation

In most industrial applications, LIM requires a larger air gap as compared to a conventional rotary induction motor; hence, the magnetizing current of LIM is larger than that of a rotary induction motor of the same power. Therefore, the efficiency and the power factor of LIM are lower than that of a conventional induction motor of the same rating. Nasar et al. [23] estimated primary core and solid secondary eddy-current losses in the tubular linear induction motor (TLIM), considering the nonlinear B-H characteristic, hysteresis, and skin effect. Techniques for the parameter determination of the equivalent circuit and performance predictions are developed, which are in good agreement with calculated values.

The optimization of the electrical and mechanical parts of LIMs plays an important role in the reduction in losses and improvement of performance, which increases overall competitiveness in high-power applications.

A tubular linear induction motor (6 kW, 3-phase, 415 V, 50 Hz) is designed and analyzed by Kumar et al. [24] for continuous hammering application. Design analysis was carried out using ANSYS MAXWELL software, which allowed for the simulation of the thrust force, linear speed, and other parameters of the motor. In paper [25], the authors presented the design optimization of an LIM. The objective function of the optimization problem includes power efficiency, output thrust, and the total weight of the device. Various design parameters were used as constraints. Optimization resulted in significant improvements in machine efficiency and output power compared to the known solutions. Machine design and its optimization were carried out with RMxprt software.

In low-power applications of both hydraulic (pneumatic) or electrical drives, e.g., in manual hammers, vibrations and impacts can cause health-related issues for workers. Meanwhile, in large-scale mechanisms, impacts of high amplitudes cause cyclic fatigue and failures of structural elements and impact surrounding buildings, e.g., in pile-driving machines. Therefore, the development of an innovative vibration-damping solutions is of great importance for the safety and durability of hammering machines.

The development of an electric hammer using self-synchronization phenomena is conducted by Bonkobara et al. [26] to address the problem of hand–arm vibration syndrome. The authors showed, based on a dynamic model, that two oscillators acting together can reduce harmful vibration. Some other types of damping principles and technical solutions were also proposed by Harada [27].

Goman et al. [28] describes a mathematical model of interconnected electromechanical and thermal processes in an LIM. The thermal model consisted of eight control volumes on each tooth pitch of the LIM. Model verification was performed using the finite element method and using experimental data. The authors determined the limits of safe operation by considering the unevenness of heating along the length in two cases: natural cooling and forced cooling. For forced cooling, the required values of airflow were determined. For the arc-induction motor of the screw press, the influence of various factors (i.e., of the stroke, the use of a soft start, and the use of forced cooling) on heating was evaluated.

The recent developments patented by Peltola et al. [29–32] disclose the design of LIM-based devices for different hammering applications. The main efforts of these and many other known inventions are aimed at structural strength capacity, and impact-force increasing for the minimal power supply.

### 3.3. Material and Topology of Permanent Magnets

The performance of linear electric motors and hammers directly depends on the magnetic properties of permanent magnets. The two possible options—Ferrite and Neodymium-based—are given in Table 1. Although the price of Neodymium magnets is higher, they have superior properties (specific energy, coercive field) in comparison with ferrite magnets (Milanesi [33]). Ferrites have low remanence, are low-energy and have a strong temperature de-rating.

**Table 1.** Comparison among main parameters of ferrite and Nd-Fe-B magnets (Milanesi [33]).

| Parameter | Ferrite | Nd-Fe-B |
| --- | --- | --- |
| Remanence Br [T] | 0.2 ÷ 0.5 | 1.1 ÷ 1.3 |
| Coercive field Hc [kA/m] | 150 ÷ 295 | 700 ÷ 1000 |
| Relative permeability | 1.1 | 1.08 |
| Temp. coeff. of Br [% °C] | −0.11/−0.12 | −0.54/−0.60 |
| Temp. coeff. of Hc [% °C] | −0.2 | +0.3 |

In tubular electric motors, the topology of permanent magnets can be surface-mounted, radially magnetized or axially magnetized. The two main topologies of permanent magnets in linear electric motors—(a) surface-mounted; and (b) buried magnet arrangements—are shown in Figure 5. Both topologies produce similar thrust force under certain geometrical parameters of the electrical machine (van Zyl [34]).

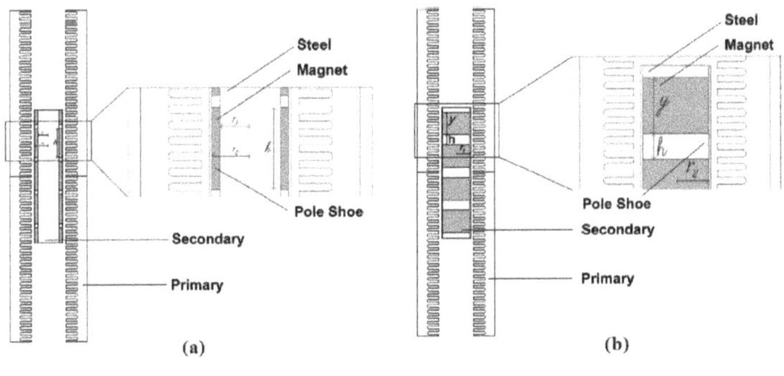


**Figure 5.** Two main topologies of permanent magnets in linear electric motors: (**a**) surface-mounted; and (**b**) buried magnet arrangements (van Zyl [34]).

Experimental tests and simulation results of buried magnet topology are shown in Figure 6. We can conclude that the advantage of LIM is in the wide range of force regulation without a significant increase in the required current in the coils. Increasing from about 5 to 10 A results in a greate force by 1000 N. In addition, the DC test gives remarkably higher thrust force than the AC test.

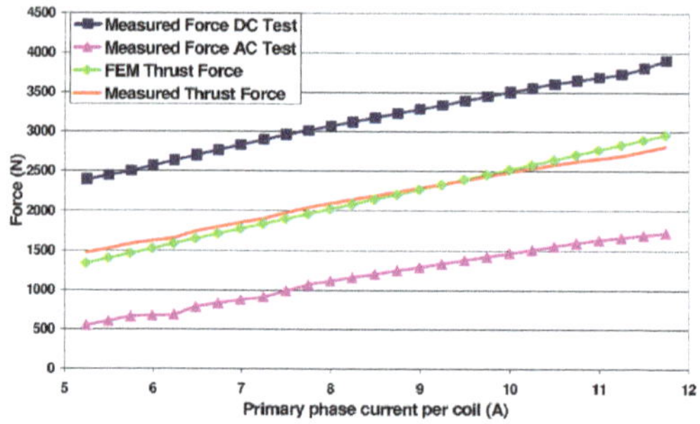


**Figure 6.** Measured and predicted forces for the buried magnet arrangement (van Zyl [34]).

Cogging forces appear as the result of the interaction between the permanent magnets and the steel teeth of the primary section. This cogging force results in a slightly jerky motion of the internal part, which is most noticeable at low speeds and creates a problem in the precise positioning of actuators. Certain methods are developed for cogging-force reduction by Lo et al. [13].

A comparative analysis of two variants of magnet topology was carried out by Wang et al. [35]. The authors concluded that the axially magnetized option has a higher force density, but is more material-consuming. If the same volume of the permanent magnet is supposed, the two topologies provide the same force density. Nevertheless, axially anisotropic rare-earth magnets are usually less expensive and widely used.

### 3.4. Dynamical Model of Electric Hammer

Although LIM has many potential applications with periodic motion (air compressors, hydraulic pumps) because of their good reliability, high power density, and convenient maintenance, most studies rarely concentrate on the dynamic behaviour of the linear oscillating actuator such as as a hammer under the action of external loads (Jiao et al. [36], Giuffrida et al. [37]). The main efforts are directed at the simulation of hydraulic pressure distributors and working-cycle optimization (Gorodilov [38]). The multi-body mathematical models of the technological vibratory and impacting units with electromagnetic excitation are developed by Neiman et al. [39,40]. Vibro-impact system dynamics under a Hertzian contact force and the influence of asymmetric electromagnetic actuators are analyzed by Herisanu et al. [41] near the primary resonance.

When modelling the mechanisms containing the hammers, it is important to account for the unavoidable clearances in the joints. Chen et al. [42] developed a planar hydraulic rock-breaker model with multiple joint clearances by combining the hydraulic cylinder model and the clearance in joints. The dynamic simulations showed that multiple clearances can reduce the dynamic responses of a rock breaker producing vibrations and slow movements. In addition, the friction could reduce the rapid vibrations significantly.

Song et al. [43] estimated the impact loads delivered to the housing of a hydraulic breaker quantitatively. Vibrations in the equipment housing, which were experimentally measured, and the impact loads in the chisel were derived from strain gauge measurements of the striking energy.

The proposed authors' generalized dynamical model of the electric hammer is depicted in Figure 7. This is a non-linear mechanical system with piece-wise linear characteristics of stiffness parameters due to internal clearance in contact of the moving part with the chisel ($\delta_{12}$). Although the operator intends to press the chisel to the rock before and during crushing, certain external clearance ($\delta_{23}$) can appear during hammer work. This calculation scheme is similar and applicable to both hydraulic and electrical hammers. The only difference is that excitation force ($F_e$) is created either by hydraulic pressure or the electric current in the coils. In addition, both types of hammers have a gas or hydraulic damper ($c_{01}$) on the top of the housing ($m_0$), to accumulate the kinetic energy of the moving part

($m_1$) and increase its impact on the working tool ($m_2$). The experimental investigation of linear permanent-magnet actuators with gas springs is conducted by Ummaneni et al. [44].

**Figure 7.** Hydraulic and electric hammers with different available shapes of chisels for rock-piece crushing (**a**); and the generalized multi-body dynamical model (own study) (**b**).

It is important to note that the energy portion transferred to the rock ($m_3$) by the moving part ($m_1$) via the working tool ($m_2$) depends not only on the contact conditions between colliding bodies but also on the technical condition of the supporting structure; in particular, the excavator or stationary manipulator. The wear of bearings (radial clearances) in the couplings ($\delta_0$) reduces the overall stiffness of the structure and creates additional shocks on its elements, which are damped ($c_0$) by the hydraulic cylinders of actuators, but still cause further degradation and looseness of joints in manipulators.

During hammer operation, the manipulator stiffness ($k_0$) is continuously changing, due to the arbitrary regulating of hammer position by a machine operator. When setting the frequency of hammering impacts within a wide range (1–20 Hz), it could coincide with certain natural modes of supporting-structure oscillations. In turn, those modes can also change their frequencies depending on the spatial position of the hammer. Those dynamical processes are similar both in hydraulic and electric hammering, but the supporting structure is less susceptible to higher frequency oscillations in the latter case.

Depending on the shape of the treated material and crushing conditions, i.e., flat surface and stable position or uneven shape on an unstable basement, different chisel ends can be used: cone, blunt, or wedge. Worth noting is that during the operation of the hammers, the shape of the chisel changes significantly but their replacement usually occurs after full failure (rod bending or cracking).

### *3.5. The Diagnostics and Control of Linear Electric Motors*

Since the hammers are subjected to severe impacts, they are susceptible to frequent failures, and diagnostics is of great importance during a crushing machine's operation. Different kinds of internal faults in linear electric motors as a subclass of electrical machines can be detected by the electrical signature analysis (ESA), motor voltage signature analysis (MVSA), or motor current signature analysis (MCSA). These methods allow the detection of damages not only in electrical parts but in mechanical elements too. Instead, hydraulic actuators including hammers require the installation of additional sensors to measure such parameters as pressure, oil flow, displacement, or acceleration.

The combined modeling and identification of the parameters of a tubular LIM (TLIM) are represented by Agnello et al. [45]. In addition, the TLIM speed control for this specific application is proposed. A fault diagnosis method is proposed by Wang et al. [46] to detect current sensor faults for the primary permanent-magnet linear motor (PPMLM) in the traction system. Only two current sensors are required in this method, and the gain fault and zero-offset fault can be distinguished.

Hydraulic hammers always apply maximum pre-installed impact energy, while the electric hammer can automatically or manually control the impact. To eliminate this shortcoming, a hybrid system of a hydraulic breaker system is suggested by Yoon et al. [47] with optimized impact forces and active control to improve energy efficiency. The characteristics of rock properties are obtained by a proximity sensor which can determine the depth, at which the piston stroke will reach the object. Moreover, a cascade control system for multiple levels of impact points is included and monitoring modules are developed with wireless communication. The field test showed the feasibility of the suggested breaking system. However, the rebuilding of existing hammering systems makes this approach difficult to implement, particularly, in mining machines working in harsh environmental conditions.

The original application of a three-phase LIM as a sensor of the damage in the continuously rolled steel sheet is proposed by Szewczyk and Walasek [48]. A data-based method to monitor linear electro-mechanical actuators is developed by Ruiz-Carcel and Starr [49]. The proposed algorithm uses the electric current and position data, which are typically available from the controller, to detect and diagnose mechanical damage in the actuator. The main parameter from the viewpoint of hammer performance is the impact energy, whose reduction can be a sign of internal damage either in electrical or mechanical parts. There are many methods of impact energy measurements but mainly in laboratory studies on the test rigs (Ficarella et al. [7,50,51], Park et al. [52]).

## 4. Case Studies of Hammers Applications

### *4.1. Deep Drilling in Oil And Gas*

Percussive-rotary drilling is an effective method for hard-rock drilling. Since hydraulic and pneumatic hammers have many problems in deep core drilling, a new electromagnetic hammer driven by a tube linear motor is introduced by Wu et al. [53]. The problem of the relatively high working temperatures in a deep down hole is considered and its influence on the electromagnetic thrust of the linear motor is analyzed. A model of the linear motor hammer is developed, allowing the authors to analyze the impact power and frequency under different stroke lengths and temperature conditions (see Figure 8). Impact frequency and impact power have an inverse relationship, but the first parameter is non-linearly, and the second is linearly, dependent. The impact stroke has an influence mainly below 50 mm.

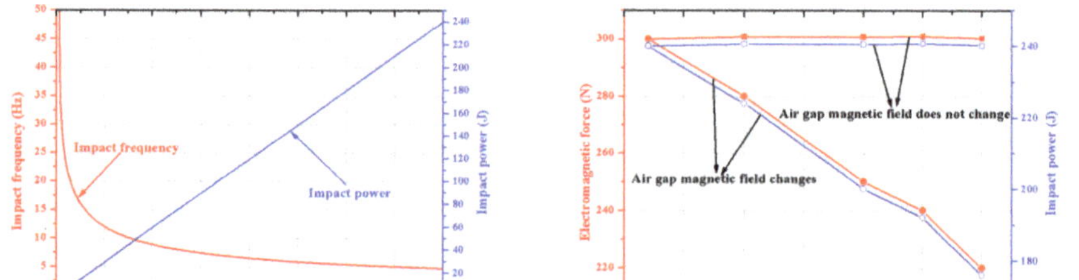


**Figure 8.** The impact frequency and impact power with different impact strokes (**a**); average electromagnetic thrust and impact power curves under different temperatures (**b**) (Wu et al. [53]).

The prototype test showed that the linear motor hammer can reach and even exceed the hydraulic hammer of the same diameter, especially in the low-frequency range. Wu et al. [54], using the developed model, estimated the loss of permanent magnet, coil resistance variation, and coil thermal expansion within the temperature range from 20 °C to 300 °C. The obtained results of model simulations showed the by less than 31% trust reduced by copper loss; the thrust reduced by 18% by the permanent magnet loss. On the other hand, the effect of the air-gap change by thermal expansion can be practically neglected.

A tubular linear permanent magnet motor used for drilling is described by Zhang et al. [55]. The advantage of this motor is in the gas springs, which allow for increasing the electric power transmitted to rock without additional elements. A mathematical model of the hammering process is composed and the dynamics of oscillatory motions are analyzed. The dynamic response of no-load mode and load mode is compared. Simulation results show the advantages of the additional spring in comparison to a simple forced motion of the hammer.

In the rigs for blasting holes, and drilling in the underground mines and open pit mines, there are many hydraulic cylinders (Wodecki [56]). They can be naturally substituted with LIMs of appropriate scale because usually the electric power is supplied to these drilling machines by additional connected cables, which can be used for electric units too. In this case, the percussion force, thrust force and rotation torque can be combined in one device. An example of an electric-hammer application for the drilling of blast holes is given by Petit [57]. In the case of electric percussive drilling machines, the diagnostics of bit wear become easier (Zhang et al. [58], Antonucci et al. [59]), due to simple hardware and software tools for operational-parameter recording and electrical-signal interpretation.

A comparative study of pneumatic and electric drilling devices regarding productivity, noise, vibration, and created dust is conducted by Rempel et al. [60]. The experiments showed that noise, vibration, and respirable dust levels were higher (by 13 dB, 5×, and 40×) with the pneumatic drill, while there were no differences in drilling productivity (ROP—rate of penetration) between these two types of units of similar mass.

### 4.2. Metallurgical Forging Press

Probably, the most spectacular example of a linear electric motor application is the metallurgical forging press shown in Figure 9. Due to the implementation of electrical driving technology instead of traditional hydraulic drive, many technological benefits have become available (Schuler [61]).

**Figure 9.** New linear electric forging hammer (Schuler [61]).

The embedded electronic control system can automatically adjust the energy of impact and the number of necessary forging blows in accordance with the actual state after each cycle, until the final part thickness is achieved. The influence of working-tool wear on process accuracy can, therefore, be compensated by regulating the energy input. This approach and design of an electrically driven press improves product quality and makes it possible to optimize the technological process.

### 4.3. Sieving Screens

Traditionally, hydraulic crushing hammers are used at the first stage of sieving screens (see Figure 10), to prevent the passing of oversized rocks for further sieving and other technological processes (Bembenek et al. [62], Korendiy et al. [63]).

**Figure 10.** New crushing hammer on the first in the technological chain sieving screen (METSO [64]).

One more industrial application of linear motors is for actuators of vibration in sieving machines. Such types of drives allow the manufacturing of equipment with very low energy consumption and material costs (Makarov et al. [65]).

The main problem, which should be solved in the actuators of sieving screens and other vibrating machines for bulk-material processing, is the tuning of frequency, amplitude, and trajectory (orbit) in the course of the operation when the properties of the sieved media change (Gursky et al. [66]) and these changes are quite difficult to predict or track (Bardzinski et al. [67–69]). Screen-parameter tuning is difficult to realize in the standard design with unbalanced vibrators and requires different technical solutions (Gurski et al. [70,71]). Using LIM as actuators can replace special spring supports requiring complicated methods of diagnostics (Krot et al. [72]) and, thus, reduce maintenance actions and cost.

The linear electric motors used as the actuators of vibrating machines can easily change all the above-mentioned technological parameters by the signals from accelerometers, realizing efficient feedback loop control. Aipov et al. [73] developed a sieve mill for a grain cleaning machine driven by a linear induction motor. It is shown that the advantage of a linear induction motor compared to standard drive designs is significantly less metal consumption for drive shafts, transmission elements, connecting shafts, bearings, and structure. Hence, energy consumption is also reduced due to the pulse drive, which makes it possible to perform technological-parameter regulation within a wide range for various crops with various physical and mechanical parameters.

### 4.4. Crushing Hammers in Mining

By different estimates, depending on variable geological conditions and mineral production technologies, about 15% of blasted rocks are oversized for further processing, including transportation by heavy vehicles and belt conveyors (Doroszuk et al. [74], Król et al. [75]), screening or comminution, which can cause machines failures or additional energy consumption, e.g., in tumbling mills (Góralczyk et al. [76], Bortnowski [77]).

The influence of the particle-size distribution of excavated material on hydraulic excavator productivity was experimentally investigated by Kujundžić et al. [78] by photogrammetric method and the excavator working cycle was measured by analysis of video recordings. It was discovered that a larger number of fine particles in granular materials with a higher uniformity increases the volume of the bucket load. In underground mines, the oversize pieces of the blasted ore generate additional loads on the structural elements and transmission of load–haul–dump (LHD) vehicles (Krot et al. [79]).

The concept of "active bucket" is proposed by Gorodilov et al. [80,81] for open-pit mining excavators, which supposes the dynamic impacts (vibrational or low-frequency) of buckets on the rock mass of high strength. Any bucket element (front wall, teeth) can be equipped with a dynamic actuator (electromagnetic, hydromechanical, pneumatic or hydraulic).

Unlike other types of crushers (jaw, cone, gyratory) designed for fine-fraction production, impact hammers, both stationary and mobile, are intended to be used at the first stage of blasted-rock-pieces treatment.

The energy consumption using the laboratory vibratory jaw crusher is estimated by Mazur [82] for limestone and diabase. In this study, electrical energy is measured as a function of changing parameters of the vibratory crusher: jaw stroke, the outlet gap section, and the frequency of jaw vibration. The author compared the expected crushing energy by the Bond hypothesis with the measured values and observed large differences between them. The crushing work index is determined in [82] on the basis of Bond's formula [83]:

$$WI_v = \frac{E_{jv}}{10(1/\sqrt{D_{80}} - 1/\sqrt{d_{80}})}, \tag{5}$$

where $WI_v$—crushing work index (kWh/Mg); $E_{jv}$—specific energy of crushing (kWh/Mg); $D_{80}$, $d_{80}$—control grain size for the feed and the crushing product, respectively (μm).

By analogy with the comminution process, the smaller the pieces after the material crushing, the more energy is required for the hammering process. Hence, the component $(1/\sqrt{D_{80}} - 1/\sqrt{d_{80}})$ could be a measure of crushing-process intensity. However, this parameter is well-suited for the small and fine fractions of materials with more similar structures, while its application for crushing large oversized pieces of rock may have a significant bias from theoretical values due to the action of many factors, e.g., piece position; chisel position on the rock (middle or end), bit shape, edge wear; impact direction (angle of inclination), rock stratification).

### *4.5. Application of Electric Hammers in Mining*

The electric hammers based on a tubular linear asynchronous motor have an armature, which produces an oscillating movement. To increase the power indices, the design can include a pneumatic power accumulator, whose characteristics are matched with the parameters of the linear asynchronous motor. In this way, the one-sided movement of the linear asynchronous motor armature under the action of the electromagnetic forces is provided, while the reversing operation happens due to a pneumatic spring. The pneumatic shock absorber also protects the base unit from shock impacts.

The energy to the oversize piece of rock is transmitted from the armature to a separate bit (chisel with a hard cone or wedge) via a certain clearance where the armature rod is accelerated to gain the appropriate kinetic energy. A required working voltage (DC or AC) is supplied to the linear electric motor and due to the traveling electromagnetic field in the stator, the armature goes up and down.

In the case of helical steel springs or gas accumulators, the moving armature compresses a spring or gas and when the forces acting on the armature body reach equilibrium at the upper position, the linear motor stator can be periodic with a certain periodicity disconnected from the line by a signal from the power controller or special sensor. Then, the armature moves down and applies an impact to the tool. The energy of a single impact may be controlled by changing the volume of the gas accumulator (stiffness of the spring) and the function of the supplied voltage (time that the motor for moving the armature is turned on).

The linear electric motor hammers do not require secondary energy converters (compressors, oil pumps, hydraulic lines), which increase the energy efficiency of the hammer since they are powered directly by the main electric lines. The mechanical impact energy of the electric hammer achieved by Kabachkov et al. [84] reached 18,500 J and the crushed volume of oversize magnesite up to 4.5 $m^3$. The hammer power supply voltage is 380 V; the total mass of the hammer is 3000 kg; impacting part mass is 110 kg; the mass of the tool is 250 kg and the diameter is 0.2 m. The pieces were treated at the frequency of 60–100 $min^{-1}$ (1–2 Hz). The conducted tests showed that after a series of impacts by the wedge (lance) end of the tool, a pulverized layer appears on the surface, which absorbs the energy of the impacts. The total number of impacts for the largest pieces was up to 12–14 and even 1–2 impacts were enough for small-piece (0.5–1.5 $m^3$) crushing with the impact of maximum energy.

In the construction industry, hammers are used for hard-concrete-element demolition (walls, foundations). The use of electric hammers to break up concrete foundations in the process of the overhaul and replacement of production equipment is described by Minaev et al. [85]. The cooling system of the electric hammer is designed for a duty factor of 15%. In the experiments, the authors suspended a hammer from an overhead crane in the workshop.

The operators of both hydraulic and electric hammers should put the impact tool on the oversize piece at a straight angle to prevent its sliding over the surface and possible bending. The oversized piece of rock should be crushed starting from the edge and, step by step, continued to the center. To prevent tool-bit heating, which promotes its strength reduction and excessive wear, a maximum of only 10–20 strikes should be produced with a subsequent stopping. The maintenance staff should provide scheduled lubrication of the tool and bit sharpening.

The current challenges for underground and open-pit mining include the improvement of safety as well as the optimization of production processes, reducing bottlenecks and downtime while increasing the operational efficiency of the machines. To achieve these goals, the hydraulic hammer can be remotely controlled, allowing for a more comfortable and safe work environment.

A specific stage in the production chain of an underground copper mine is a place where blasted bulk material is transported by the LHD vehicles and heavy trucks to the conveying system. Such a place is called a dumping point (see Figure 11), which is equipped with a screen classifying the material into coarse and fine fractions. The process of bulk material delivery has a cyclic schedule (Krot et al. [86]) and the performance of a hydraulic hammer breaking oversized rocks to prevent damage to the conveyor belt affects the overall productivity of the mine. Therefore, to make the hammering process continuous, a prototype test rig has been proposed for debugging the algorithm for the automatic detection of oversized rocks, crushing them along with sweeping of bulk material (Stefanik et al. [87]). The whole view of the hydraulic manipulator with a hammer is depicted in Figure 12.

**Figure 11.** The dumping point with a screen and hydraulic hammer on the manipulator in the underground mine (Stefaniak et al. [87]).

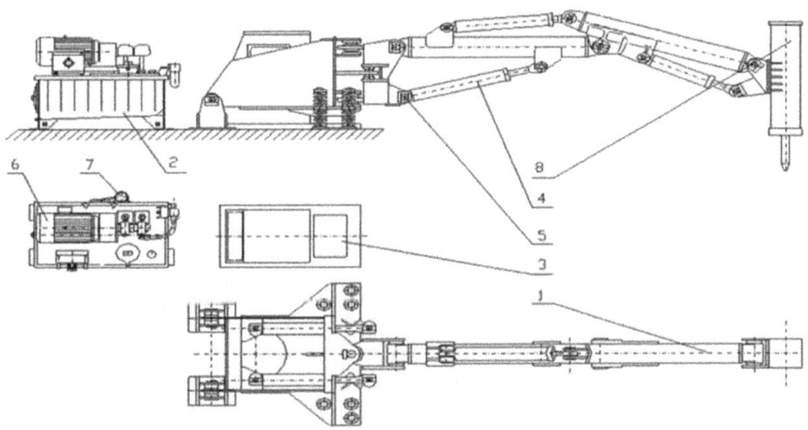


**Figure 12.** Typical design of hydraulic manipulator with the hydraulic hammer for crushing of oversized pieces of blasted material (KGHM ZANAM [88]) in the underground mine: 1—boom; 2—hydraulic unit; 3—air-conditioned cabin; 4—hydraulic system; 5—central control system; 6—electrical unit; 7—automatic fire extinguisher; 8—hydraulic hammer.

To improve the efficiency of the hydraulic system at the dumping point, an approach is proposed by Siwulski et al. [89] based on the analytical model of a stationary rock-breaking machine (see parameters in Table 2). The results of the calculations showed benefits from a modified hydraulic system in energy consumption.

**Table 2.** Parameters of the hydraulic manipulator and hammer (KGHM ZANAM [88]).

| Parameter | Value | Units |
|---|---|---|
| Height min | 5.0 | $m^2$ |
| Width min | 6.0 | $m^2$ |
| Length min | 10.0 | $m^2$ |
| Boom reach | 4700 | mm |
| Boom turning angle | ±40 | grad |
| Total weight | 7000 | kg |
| Impact energy | 400 ÷ 1900 | J |
| Working pressure | 10 ÷ 17 | MPa |
| Hydraulic oil flow | 20 ÷ 120 | L/min |

The concept of process monitoring and automation with visual sensors for autonomous operation and remote hammer control in such applications is represented by Krauze et al. [90]. The development of a telerobotic rock breaker is also described by Duff et al. [91].

A similar system for the automatic control of impact hammers used in underground mines is described by Cardenas et al. [92]. The position and angle are determined at which the impact hammer must strike a rock to break it. The automatic system uses sensor data composed of point clouds and images. The rock segmentation subsystem receives identifiers of the rock pieces above the grizzly screen and transfers information to a pose generation subsystem, which produces a list of rock-breaking targets. Then, it selects the best candidates among them. The conducted experiments showed the efficiency of the developed system.

Poor visibility in the underground environment may result in a collision between the hammer and the grizzly screen covered by fine fractions of bulk material. To prevent machine damage, Correa et al. [93] developed the haptic tele-operation system based on a 3D LIDAR scanning and point-cloud model of the environment, which can estimate repulsion forces transferred to the operator (see Figure 13).

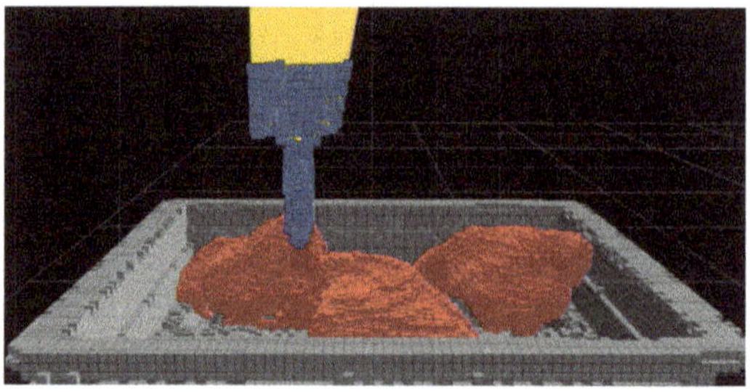

**Figure 13.** The 3D LIDAR scanning at the dumping point of the underground mine for tele-operation of hydraulic hammer (Correa et al. [93]).

Regarding the autonomous operation of impact hammers, although research started more than twenty years ago (Takahashi et al. [94]), the reported results are still not satisfactory for the application of this technology in a real mining environment. Lampinen et al. [95] obtained an average success rate of only 34% in the task of the remote control of rock breaking.

All such automated and semi-autonomous systems for hydraulic hammer positioning have certain functional limitations related to the absence of additional control of impact force and frequency, which can be realized in the electric hammers.

### *4.6. Prediction of the Impact-Hammer Performance*

While performing crushing operations, the constant setting of breaking force causes energy dissipation due to the various strengths of the material. In cases when the overall situation is not monitored, this may damage the mechanical and hydraulic components of the system. Frequent maintenance and parts replacement can result in costly losses and increased downtime. Therefore, a novel approach to rock breaking is needed that is able to predict the treated material properties in order to set an optimal impact force and frequency of hammer impacts.

There are many kinds of rock mass classifications and ratings (Bieniawski [96], Aksoy [97]). The typical classification of rock materials by the unconfined compressive strength (UCS), which is most applicable for separated rock pieces, is given in Table 3.

**Table 3.** The unconfined compressive strength (UCS) of rock materials (Bieniawski [96]).

| Material Grade | UCS (MPa) |
|---|---|
| Very soft | 30 |
| Soft | 70 |
| Medium hard | 100 |
| Hard | 130 |
| Extremely hard | 160 |

The main dynamical parameters in the rock-crushing process by hammers are as follows: strike length $L$; impact energy $E$; impact frequency $f$; and impact efficiency $\eta$. For hard rocks, longer strokes are better, while shorter and higher frequency impacts are used for softer rocks to reduce unnecessary energy losses, which improves the performance and durability of machine parts (Yoon et al. [47]).

Mezentsev [98] determined the crushing efficiency of a hydraulic hammer as follows:

$$\eta = \frac{A}{Q_c \Delta p}, \tag{6}$$

where $A$—crushing energy; $Q_c$—volumetric consumption in a cycle; and $\Delta p$—difference in inlet/outlet pressure of the hammer. The relation of hammer efficiency with power is shown in Figure 14. It was concluded in [98] that efficiency rises with power but, at the same time, greater power corresponds to higher frequencies (6.0–11.7 Hz) for almost equal-impact energy values. The author concluded that the most influencing factor of efficiency is the hydraulic distributor and the most advanced devices can provide a maximum of 60–70% efficiency.

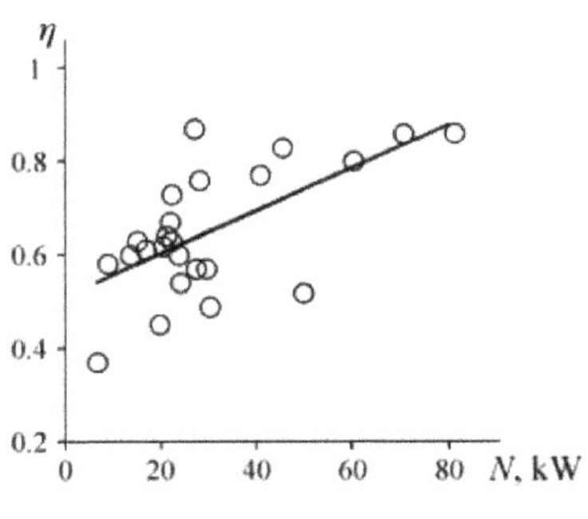


**Figure 14.** The efficiency of hydraulic hammering vs. power [98].

Crushing energy delivered to a working tool is determined (but not equal to) by the kinetic energy of the moving part of the hammer:

$$E = \frac{mv^2}{2}, \tag{7}$$

where $m$—mass of moving part; and $v$—velocity at the moment of impact.

However, not all of the kinetic energy generated by the moving part is transferred to the rock mass. It depends on the rebound rate in the contacts with the working tool (back end of the chisel) and with the rock surface (front end of the chisel). In mechanics, for collision with a stationary surface, this dynamic process is described by the coefficient of restitution ($C_R$):

$$C_R = \frac{u}{v}, \quad (8)$$

where $u$—speed after impact; and $v$—speed before impact. Its values correspond to two marginal cases of collision: 0—absolutely inelastic impact; and 1—absolutely elastic impact. The latter case is preferable in both contacts from the viewpoint of rock-crushing efficiency by the hammer. This coefficient does not account for the masses of colliding bodies; however, such influence exists.

The hardening of the working-tool material made of alloy steel at the top end can change the $C_R$ after a long series of impacts. Therefore, material properties (steel grades) of an electric core should be correctly chosen (or bottom-end hardened) to provide durability.

The other relations describing the hammering process depend on design parameters: The acceleration of the moving part:

$$a = \frac{F_e}{m}, \quad (9)$$

The maximum velocity before impact:

$$V_{max} = at, \quad (10)$$

The time of one-way motion (half-period of vibration):

$$t = \sqrt{\frac{2S}{a}}, \quad (11)$$

The required frequency of hammer:

$$f = \frac{1}{2t}, \quad (12)$$

where $m$—is the mass of the moving part; $F_e$—thrust force acting on the moving part from the electric system; and $S$—the distance between the working tool's upper end and the moving part's lower end in its upper position (see $\delta_{12}$ in Figure 7b).

Certain efforts have been undertaken to predict the properties of crushed materials to increase the efficiency of breaking hammers (Aksoy et al. [99], Ismael et al. [100], Kucuk et al. [101], Tumac et al. [102]). Research on the resonance characteristics of rock material under harmonic excitation is represented by Li et al. [103], where the influence of excitation frequency, the rock properties and dimensions are analyzed using FEM. The mass and shape of the rock are the main factors affecting its resonance frequency, besides the material properties. A 3D FEM model of impact is developed and experimentally verified by Chiang et al. [104], which permits simulation of the energy transmission to the rock via the bit–rock interaction, and to analyze the process of rock fragmentation. The majority of studies in this domain are devoted to the small-size bodies' (projectiles) penetration into different concrete walls with comparatively low energy. Anyway, results obtained by Higan et al. [105] can be useful in hammering analysis. For example, more dense materials produce less fractured masses and have larger dominant fragment sizes. The tip geometry influenced the penetration depth, which was in quadratic relation with the initial kinetic energy (Kumano et al. [106]). The dominant rock properties affecting the penetration rate in percussive drilling are determined by Kahraman et al. [107] and they are uniaxial compressive strength, the Brazilian tensile strength, the point load strength and the Schmidt hammer value. However, the rate of penetration has weak correlation with both elastic modulus and density, and the absence of correlation was noted with P-wave velocity.

The results of experimental and numerical studies of the impact breakage of granite with high ejection velocities are represented by Zhang et al. [108]. The dependencies of the high-speed collision parameters granite damage ratio ($\alpha_d$) and energy dissipation ($E_{nd}$) on impact velocity are shown in Figure 15. Although the usual range of hammer impact velocities is less than shown in these graphs, the general dependencies (almost linear) will also be valid for low-frequency hammering applications.

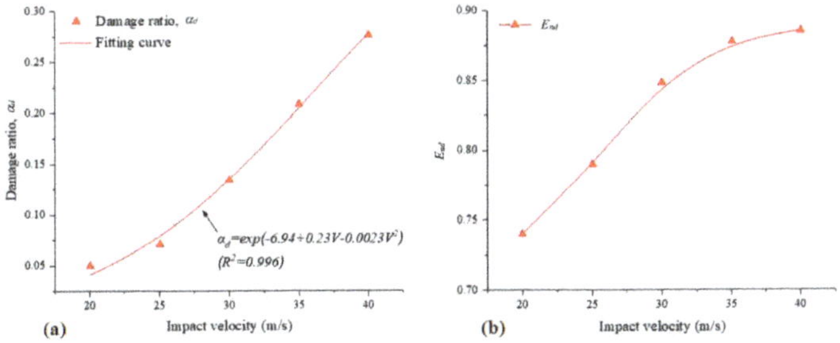


**Figure 15.** Dependencies of hammering parameters on impact velocity: (**a**) granite damage ratio ($\alpha_d$); (**b**) energy dissipation ($E_{nd}$) (Zhang et al. [108]).

Results of the experimental investigation are reported by Gorodilov and Efimov [109,110] of shock pulses in the piston-bit system in interaction with rock mass. The rock mass was simulated by a marble block of 0.95 × 0.95 × 1.5 m$^3$ in size and 400 kg in weight. The test cylindrical pistons (3.1, 5 and 16 kg and 565 mm) were impacted by a pendulum hammer into the bits (5 and 9.6 kg and 160 mm) with a tapered end (30 mm length and 60° angle), which was tightly contacted with a rock. The pre-blow velocities of the pistons (5.01, 4.01, 2.25 m/s) were set to obtain similar impact energies for the different weights. The highest efficiency of rock fracture is observed in the piston–bit pair of 5.5–5 kg weight, while the lightest piston (3.1 kg) has the highest recoil and the weakest fracture efficiency. The other two pistons have small recoil, which corresponds to the higher energy transfer coefficient to the bit. The shock duration (up to negative velocity moment) is 0.1 ms for the 5 kg piston and 0.6 ms for the heavier piston. The motion patterns of the 3.1 and 5.5 kg pistons differ from the 16 kg piston; namely, lighter pistons exhibit oscillations with a period close to the double wave travel time in these pistons. For the heavier piston, such oscillations were almost invisible. The logical continuation of these results is the development of a hammer design methodology and the creation of a control system adaptable to material properties (Gorodilov et al. [111]).

The Leeb hardness test (LHT) with a test value of LD for rock materials is investigated by Corkum et al. [112]. An updated methodology is developed and the correlation of Leeb hardness LD with UCS units is shown. Instead of manual measurements with special tools according to ASTM ASTM D5873-14 [113], this methodology can be efficiently implemented for automatic online monitoring and process control in electric hammers with additional sensors, e.g., proximity meters (Yoon et al. [47]).

## 5. Experiments on the Electric Hammer

The previous studies of the electric hammer of Lekatech Oy were conducted by Mayer [114] and Korhonen [115] concerning technical issues and to ensure the electric hammer complies with requirements set for the European Economic Area and product safety for North American markets (Ukkola [116]).

The current research project being carried out by the authors is related to:

- Performance assessment of the electric hammer in the various geological conditions of mining companies in Poland, Finland, and Spain;
- Measurement and collection of operational process data;

- Checking the reliability of electric-hammer elements in the conditions of underground and opencast mines;
- Determination of optimal controlled vibration parameters for various materials and grinding tasks.

Accessories required for the electric-hammer operation are as follows: cooling system (not always necessary), inverter, rectifier, fuses, relays, and chisel lubrication unit, which were provided for the hammer installed on the excavator in a separate box (see Figure 16a). In the pilot LEH design, the user interface is organized via touch screen PLC CPX-Terminal-Vemcon 7" (see Figure 16b).

(**a**) (**b**)

**Figure 16.** The Lekatech electric hammer (LEH) based on a linear electric motor installed on the excavator (**a**) and the user interface (**b**) (Lekatech [117]).

The LEH's main power is 500 V AC. This is converted to DC by the passive rectifier, after which its DC voltage may change a little. The inverter then changes the current back to AC, according to which mode the hammer is controlled. This AC is fully controlled by PLC and allows the LEH to produce impacts with many different frequencies and energies (see Table 4). Once the hammer's mode (frequency and impact energy) has been selected from the terminal, the joystick button (foot pedal) is used to operate the hammer, if the tool is pressed down, and the tool limit switch is, hence, on. Oil and water cooling happens periodically or if the respective temperatures become too high.

**Table 4.** Technical parameters of Lekatech electric hammer (LEH) [118].

| Parameter | Value | Units |
|---|---|---|
| Hammer weight | 454 | kg |
| Minimum working weight | 516 | kg |
| Impact frequency (adjustable) | 60–900 | $min^{-1}$ |
| Impact energy (adjustable) | 500–1500 | J |
| Working tool diameter | 90 | mm |
| Voltage (main) AC | 400 | V |
| | 50 | Hz |
| | $3 \times 63$ | A |
| Voltage (converter) DC | 700 | V |

The series of experiments were conducted with different frequencies of impacts and chisel end shapes (Mayer [114]). The graphs of impacts and corresponding kinetic energy of the electric hammer are shown in Figure 17 for the frequency of 18 Hz and 1 s of time duration. Every cycle of hammering certainly shows individual patterns of impacts and the amplitudes of vibrations depending on the current state of the treated material (granite). In addition, reinforced concrete samples were used in the experiments.

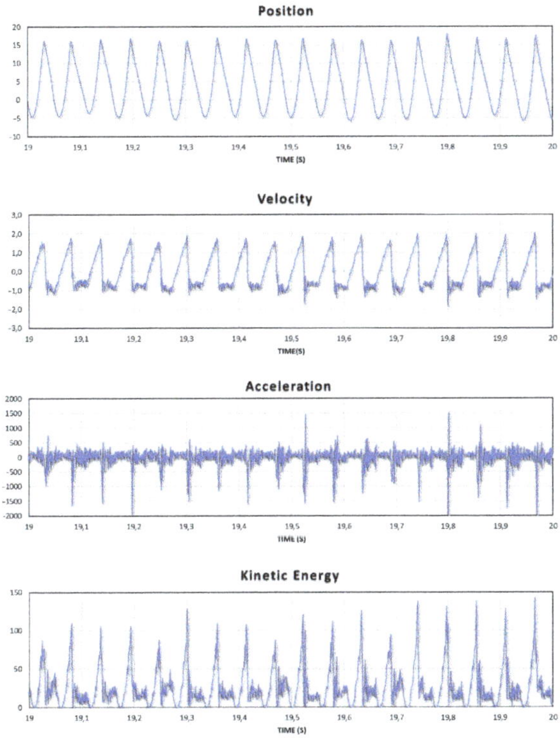


**Figure 17.** Dynamic parameters of the electric-hammer testing at 18 Hz: chisel position (mm), velocity (m/s), acceleration (m/s$^2$), and kinetic energy (J) [114].

From these represented graphs of dynamical parameters, it is noted that after the impact on the material, the chisel rebounds, which produces corresponding peaks in velocity and acceleration signals. However, these high-frequency jumps of small amplitude are not visible in the position signal due to a double integration of the original signal. Since the kinetic energy is calculated by the velocity, this signal also exhibits certain signs of chisel rebounds from the treated material surface.

The time-domain graph of power consumption by the electric hammer at the impact frequency of 18 Hz is shown in Figure 18. The power-consumption graphs show that the power consumption of the inverter is constant when the moving-part position goes up and decreases quickly to zero as the mover is accelerated rapidly downwards onto the tool.

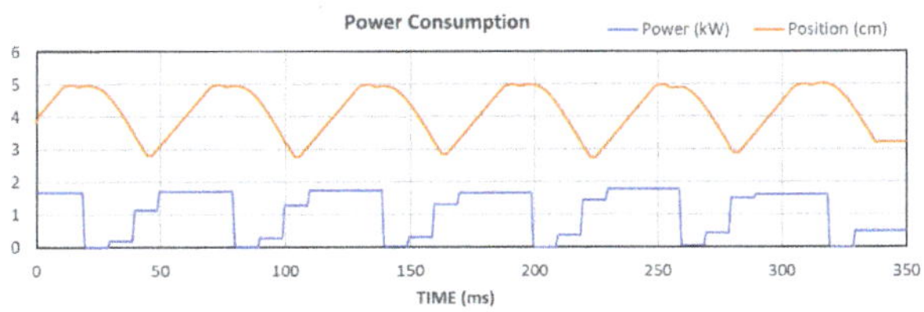


**Figure 18.** The power consumption of the electric hammer at the impact frequency of 18 Hz (Mayer [114]).

The kinetic energy of the actuator, as determined by the vibration measurements at the top of the hammer, is highest at frequency settings of 0.9 and 5 Hz (see Figure 19).

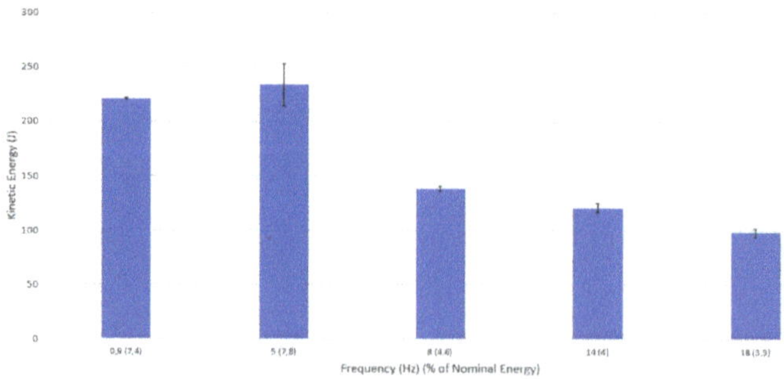


**Figure 19.** Dependence of kinetic energy on impact frequency (Mayer [114]).

Depending on the frequency, type of processed material, and shape of the crushing tool (cone, blunt, wedge), the time to fracture varies significantly (see Figure 20).

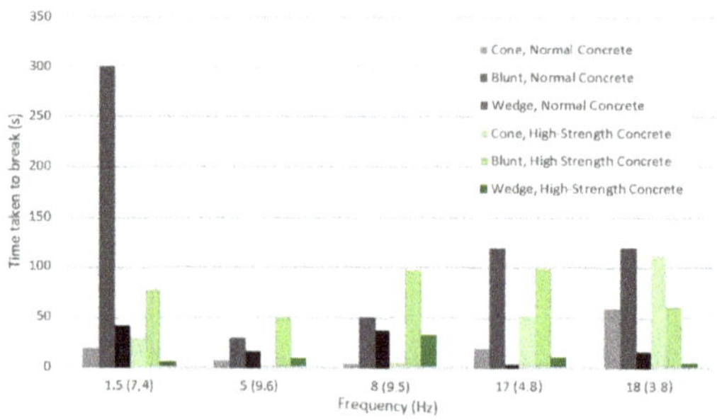


**Figure 20.** Dependence of time to break on impacts frequency (Mayer [114]).

## 6. Discussion

Many technological operations in mining, oil and gas, raw-materials processing, construction (demolition), and other heavy industries demand certain solutions to increase productivity, energy efficiency, and personnel safety along with a reduction in maintenance costs, and environmental impact (exhaust gases emissions, wastes utilization). To satisfy these frequently contradictory requirements, innovative technical solutions based on ubiquitous digitization, automation, and the electrification of machines have been proposed and implemented in recent years.

For many types of technological machines, where, traditionally, hydraulic drives were used including hammers, designers are trying to modify them with electrical counterparts (Wagner [119]). However, large-scale electric hammers are not yet enough widely used in industry due to a lack of producers and on-site investigations and comparison with hydraulic analogs of the same power and productivity.

For the correct comparison of the hydraulic breaking hammer and linear motor electric hammer in industrial conditions, the complexity of the process and the uneven conditions in the experiment should be taken into account. Many uncertainties have resulted from uncontrolled factors (e.g., human actions, inhomogeneity of the rock, and the content of oversized pieces). However, being aware of the limitations and adopting certain simplifying assumptions, the comparison can be conducted on the basis of:

- Two separate experiments for both types of hammers in similar working conditions;
- The same mass of spoil (1–5 Mg), grain size (>50 cm), rock type (the same part of the deposit);
- Hammer settings (frequency, energy, etc.) taking into account hydraulic hammer limitations (no parameters adjustment);
- Preferably the same experienced operator in both cases;

- Video recordings of both experiments.

Electric power supply for new hammers in the on-surface and underground applications for stationary manipulators (dumping points, sieving screens) is not a problem. For mobile machines, additional cabling should be provided such as in reinforced bolting installation and blast-hole drilling machines.

Although electric hammers will be subjected to excessive impacts, their durability is higher and maintenance costs are lower than the hydraulic units. Additional savings are expected from the lower amount of waste hydraulic oil for utilization. In the case of diesel-engine-powered machines, fewer exhaust-gas (NOx) emissions and, simultaneously, higher energy efficiency and fuel consumption are provided.

The main advantage of electric hammers is the possibility to control (adjust) the impact power and working frequency depending on the hardness and morphology of the treated rock material. This direction needs further development from the viewpoint of appropriate sensors and methods of material properties detection. Using visual methods for hammer control may improve tele-operation and, finally, create efficient autonomous solutions for personnel safety and comfort in harsh environmental conditions.

Probably, the only application where linear electric motor actuators are not quite suitable for hydraulic-cylinder replacement is the large-scale roof supports subjected to constant high static and dynamic loads (Szurgacz [120,121]). Instead, the combination of hydraulic cylinders and linear electric motor actuators may bring new functional benefits in the case of meeting explosion safety regulations in underground coal mines.

## 7. Conclusions

The conducted review of available literature and some preliminary experiments on the linear electric motor hammer have shown that this solution has many advantages and certain complications for implementation in the industry.

In this paper, we considered the most critical problems related to the design, performance, control, dynamics, and materials of electric hammers.

We claim that hydraulic hammers' substitution with fully electric hammers, at least in mining and construction (structures demolition), is a prospective approach allowing a significant reduction in the energy consumption and maintenance costs of appropriate machines. Certainly, the replacement of hydraulic hammers should be performed close to their lifetime. Nevertheless, the short period of return on investments (1–2 years) makes electric hammers beneficial, especially in enterprises with a big fleet of machines. The only additional investment required for electric-hammer adaptation is the cables and probably power supply points, which can also be used in other operations.

The significance of the comparison of the two types of hammers is in achieving a more clear understanding of what benefits can be achieved not only in energy saving and environmental-impact reduction but also in the overall performance improvement of raw-materials processing. Namely, preferably, the automatic or manual control of shock energy, the amplitude of stroke, and frequency of impacts depending on the rock properties, size of pieces, and surrounding conditions. To increase the effect of breaking-machine digitization, additional sensors are quite easy to integrate into an electric hammer, more than in a hydraulic counterpart.

We mainly deal here with important problems from a practical perspective; however, a lot of scientific tasks can be formulated aimed at further improvement of electric hammers' design. For example, an investigation of higher temperatures, humidity, and intensive impacts in underground mines where machine maintenance is complicated. Durability is a critical parameter of any new mechanism intended for implementation, which will be considered first by the technical staff of enterprises.

**Author Contributions:** Conceptualization, J.P., T.M. and R.Z.; methodology, P.K., A.W. and R.Z.; software, T.M.; formal analysis, R.Z.; investigation, A.W. and P.K.; resources, J.P.; data curation, A.W.; writing—original draft preparation, P.K., A.W. and R.Z.; writing—review and editing, P.K., A.W. and R.Z.; visualization, P.K. and T.M.; supervision, R.Z. All authors have read and agreed to the published version of the manuscript.

**Funding:** This activity has received funding from the European Institute of Innovation and Technology (EIT), a body of the European Union, under the Horizon 2020, the EU Framework Programme for Research and Innovation. This work is supported by EIT RawMaterials GmbH under Framework Partnership Agreement No. 21016 (ECHO: Electrical Computerised Hammering Operator).

**Data Availability Statement:** The measurement data presented in this study are not publicly available due to restrictions of privacy.

**Conflicts of Interest:** The authors declare no conflict of interest.

## References

1. Global Rock Breaker Market. Competition Forecast and Opportunities, 2013–2023. Available online: https://www.techsciresearch.com/report/rock-breaker-market/3530.html (accessed on 24 October 2022).
2. Global Hydraulic Hammer Market 2018 by Manufacturers, Regions, Type and Application, Forecast to 2023. Available online: https://www.absolutereports.com/global-hydraulic-hammer-market-13679807 (accessed on 24 October 2022).
3. Tatiya, R.R. *Surface and Underground Excavations: Methods, Techniques and Equipment*; CRC Press: Boca Raton, FL, USA, 2005. [CrossRef]
4. Zhou, H.; Xie, X.; Feng, Y. Rock breaking methods to replace blasting. *IOP Conf. Ser. Mater. Sci. Eng.* **2018**, *322*, 022014. [CrossRef]
5. Moray, S.; Throop, N.; Seryak, J.; Schmidt, C.; Fisher, C.; D'Antonio, M. Energy efficiency opportunities in the stone and asphalt industry. In Proceedings of the Twenty-Eighth Industrial Energy Technology Conference, New Orleans, LA, USA, 2006; pp. 71–83.
6. Vukovic, M.; Leifeld, R.; Murrenhoff, H. Reducing Fuel Consumption in Hydraulic Excavators—A Comprehensive Analysis. *Energies* **2017**, *10*, 687. [CrossRef]
7. Ficarella, A.; Giuffrida, A.; Laforgia, D. The Effects of Distributor and Striking Mass on the Performance of a Hydraulic Impact Machine. In Proceedings of the Commercial Vehicle Engineering Congress & Exhibition, Rosemont, IL, USA, 7–9 October 2008. [CrossRef]
8. Gieras, B.F. *Linear Electric Motors. Electric Power Generation, Transmission, and Distribution: The Electric Power Engineering Handbook*, 3rd ed.; CRC Press: London, UK, 2012; Chapter 34.
9. Budig, P.K. The application of linear motors. In Proceedings of the IPEMC 2000. Third International Power Electronics and Motion Control Conference (IEEE Cat. No.00EX435), Beijing, China, 15–18 August 2000; Volume 3, pp. 1336–1341. [CrossRef]
10. Shen, Y.; Lu, Q.; Ye, Y. Double-Stator Air-Core Tubular Permanent Magnet Linear Motor for Vehicle Active Suspension Systems. In Proceedings of the 2016 IEEE Vehicle Power and Propulsion Conference (VPPC), Hangzhou, China, 17–20 October 2016; pp. 1–6.
11. Gysen, B.L.J.; Paulides, J.J.H.; Janssen, J.L.G.; Lomonova, E.A. Active Electromagnetic Suspension System for Improved Vehicle Dynamics. *IEEE Trans. Veh. Technol.* **2010**, *59*, 1156–1163. [CrossRef]
12. van Casteren, D.; Gysen, B.; Kessels, J.; Paulides, J.; van den Bosch, P.; Lomonova, E. Non-Linear Full-Car Modeling and Sky-Hook Control for a Direct-Drive Active Suspension System. *SAE Int. J. Passeng. Cars Mech. Syst.* **2013**, *6*, 252–268. [CrossRef]
13. Lo, D.S.; Amara, Y.; Barakat, G.; Chabour, F. Cogging force reduction in linear tubular flux switching permanent-magnet machines. *Open Phys.* **2018**, *16*, 243–248. [CrossRef]
14. Govindpure, N.; Hipparagi, R.; Kumar, A.; Talange, D.B.; Bhole, V. Basic Design equations of Linear Electric Machines. In Proceedings of the 2018 IEEE International Conference on Power Electronics, Drives and Energy Systems (PEDES), Chennai, India, 18–21 December 2018; pp. 1–6. [CrossRef]
15. Grinchenko, V. Justification of the basic design of a linear electric motor. *Theor. Appl. Sci.* **2013**, *7*, 58–60. [CrossRef]
16. Brittain, J.E.; Laithwaite, E.R. A history of linear electric motors. *Technol. Cult.* **1990**, *31*, 337. [CrossRef]
17. Kurilin, S.P.; Rubin, Y.B.; Dli, M.I.; Denisov, V.N. Models and methods of designing linear electric motors for non-ferrous metals industry applications. *Non-Ferr. Met.* **2021**, *11*, 83–90. [CrossRef]
18. Gieras, J.F. Linear electric motors in machining processes. *J. Int. Conf. Electr. Mach. Syst.* **2013**, *2*, 380–389. [CrossRef]
19. Korendiy, V.; Kachur, O.; Zakharov, V.; Kuzio, I. Studying the Dynamics of a Vibratory Finishing Machine Providing the Single-Sided Lapping and Polishing of Flat Surfaces. *Eng. Proc.* **2022**, *24*, 9. [CrossRef]
20. Korendiy, V.; Kachur, O.; Zakharov, V.; Kuzio, I.; Hurey, I.; Predko, R. Experimental study of the lap motion trajectory of vibratory finishing machine. *Vibroengineering PROCEDIA* **2022**, *46*, 1–7. [CrossRef]
21. Tiunov, V.V. Combined linear electric motors for robotic systems. *Russ. Electr. Eng.* **2017**, *88*, 745–749. [CrossRef]
22. Rudzki, P.; Krot, P. Dynamics control of powered hydraulic roof supports in the underground longwall mining complex. *IOP Conf. Ser. Earth Environ. Sci.* **2021**, *942*, 012014. [CrossRef]

23. Nasar, S.; Xiong, G.; Fu, Z. Eddy-current losses in a tubular linear induction motor. *IEEE Trans. Magn.* **1994**, *30*, 1437–1445. [CrossRef]
24. Kumar, A.; Supare, C. Design, Analysis and Realization of Tubular Linear Induction Motor for Hammering Application. In Proceedings of the 2020 IEEE International Conference on Power Electronics, Drives and Energy Systems (PEDES), Jaipur, India, 16–19 December 2020; pp. 1–7. [CrossRef]
25. Kumar, A.; Hasan, M.; Akhtar, M.J.; Parida, S.; Behera, R. Design optimization of linear induction motor. In Proceedings of the 2015 6th International Conference on Power Electronics Systems and Applications (PESA), Hong Kong, China, 15–17 December 2015; pp. 1–5. [CrossRef]
26. Bonkobara, Y.; Umemura, H.; Kawano, Y.; Koyanagi, A.; Hamahata, T.; Kondou, T. Development of an electric hammer using self-synchronization phenomena. *Proc. Mech. Eng. Congr. Jpn.* **2016**, *2016*, G1000204. [CrossRef]
27. Harada, M. Vibration damping structure for electric hammer. *J. Acoust. Soc. Am.* **1997**, *101*, 3233. [CrossRef]
28. Goman, V.; Prakht, V.; Dmitrievskii, V.; Sarapulov, F. Analysis of coupled thermal and electromagnetic processes in linear induction motors based on a three-dimensional thermal model. *Mathematics* **2021**, *10*, 114. [CrossRef]
29. Peltola, T.; Peltola, J.; Pyrhonen, J. A Hammer Device. IPC E02F3/96; E01C23/12; E02F5/30; E02F5/32; B25D17/00. International Patent Application No. WO2019068958A1, 11 April 2019.
30. Peltola, T.; Peltola, J.; Juha, P. A Linear Electric Machine. IPC H02K41/03; B25D11/06; H02K1/04; H02K5/02. International Patent Application No. WO2020058565A1, 26 March 2020.
31. Peltola, J.; Peltola, T. A Brake for Linear Movement and a Hammer Device Comprising the Same. IPC F16D59/00; F16D63/00; B25D17/24. International Patent Application No. WO2022129680A1, 23 June 2020.
32. Peltola, T.; Peltola, J.; Pyrhonen, J. A Hammer Device. IPC B25D 11/00. European Patent Application No. EP3692217A1, 10 August 2022.
33. Milanesi, F. Design Optimization and Control Strategies for PM Multiphase Tubular Linear Actuators. Ph.D. Thesis, University of Bologna, Bologna, Italy, 2009.
34. van Zyl, A. Design, Construction and Evaluation of a Modified Tubular Linear Synchronous Motor. Ph.D. Thesis, University of the Witwatersrand, Johannesburg, South Africa, 2006.
35. Wang, J.; Jewell, G.; Howe, D. Design optimisation and comparison of tubular permanent magnet machine topologies. *IEE Proc. Electr. Power Appl.* **2001**, *148*, 456–464. [CrossRef]
36. Jiao, Z.; Cao, Y.; Yan, L.; Li, X.; Zhang, L. Design and Analysis of Novel Linear Oscillating Loading System. *Appl. Sci.* **2019**, *9*, 3771. [CrossRef]
37. Giuffrida, A.; Laforgia, D. Modelling and Simulation of a Hydraulic Breaker. *Int. J. Fluid Power* **2005**, *6*, 47–56. [CrossRef]
38. Gorodilov, L.V. Investigation into Characteristics of Working Cycles of Hydraulic Percussive Machines with Ideal Distributor. *J. Min. Sci.* **2002**, *38*, 74–79. [CrossRef]
39. Neyman, L.A.; Neyman, V.Y.; Markov, A.V. Mathematical model of the technological vibratory unit with electromagnetic excitation. *J. Phys. Conf. Ser.* **2020**, *1661*, 012063. [CrossRef]
40. Neyman, L.; Neyman, V.N. Generalized model of a single-coil synchronous impact electromagnetic machine. *Proc. Russ. High. Sch. Acad. Sci.* **2019**, *2*, 56–71. [CrossRef]
41. Herisanu, N.; Marinca, B.; Marinca, V. Dynamics of the Vibro-Impact Nonlinear Damped and Forced Oscillator under the Influence of the Electromagnetic Actuation. *Mathematics* **2022**, *10*, 3301. [CrossRef]
42. Chen, K.; Zhang, G.; Wu, R.; Wang, L.; Zheng, H.; Chen, S. Dynamic analysis of a planar hydraulic rock-breaker mechanism with multiple clearance joints. *Shock Vib.* **2019**, *2019*, 4718456. [CrossRef]
43. Song, C.; Kim, D.J.; Chung, J.; Lee, K.W.; Kweon, S.S.; Kang, Y.K. Estimation of impact loads in a hydraulic breaker by transfer path analysis. *Shock Vib.* **2017**, *2017*, 8564381. [CrossRef]
44. Ummaneni, R.B.; Jaillot, C.; Nilssen, R.; Brennvall, J. Experimental characterisation of linear permanent magnet actuator with gas springs. In Proceedings of the 2009 IEEE International Electric Machines and Drives Conference, Miami, FL, USA, 3–6 May 2009; pp. 369–372. [CrossRef]
45. Agnello, G.; Caruso, M.; Di Dio, V.; Miceli, R.; Nevoloso, C.; Spataro, C. Speed control of tubular linear induction motors for industrial automated applications. In Proceedings of the 2016 IEEE International Conference on Renewable Energy Research and Applications (ICRERA), Birmingham, UK, 20–23 November 2016; pp. 1196–1201. [CrossRef]
46. Wang, W.; Tian, W.; Wang, Z.; Hua, W.; Cheng, M. A Fault Diagnosis Method for Current Sensors of Primary Permanent-Magnet Linear Motor Drives. *IEEE Trans. Power Electron.* **2021**, *36*, 2334–2345. [CrossRef]
47. Yoon, B.J.; Lee, K.S.; Lee, J.H. A Novel Predictable Rock Breaker Using Intelligent Hydraulic Control with ICT Convergences. *Appl. Sci.* **2019**, *9*, 3333. [CrossRef]
48. Szewczyk, K.; Walasek, T. Dynamic diagnostics of moving ferromagnetic material with the linear induction motor. *ITM Web Conf.* **2017**, *15*, 07003. [CrossRef]
49. Ruiz-Carcel, C.; Starr, A. Data-Based Detection and Diagnosis of Faults in Linear Actuators. *IEEE Trans. Instrum. Meas.* **2018**, *67*, 2035–2047. [CrossRef]
50. Ficarella, A.; Giuffrida, A.; Laforgia, D. Investigation on the impact energy of a hydraulic breaker. In *Proceedings of the SAE Technical Paper Series*; SAE International: Warrendale, PA, USA, 2007; Number 2007-01-4229.

51. Ficarella, A.; Giuffrida, A.; Laforgia, D. Numerical Investigations on the Working Cycle of a Hydraulic Breaker: Off-Design Performance and Influence of Design Parameters. *Int. J. Fluid Power* **2006**, *7*, 41–50. [CrossRef]
52. Park, J.W.; Kim, H.E. Development of the test system for measuring the impact energy of a hydraulic breaker. *Proc. JFPS Int. Symp. Fluid Power* **2005**, *2005*, 75–79. [CrossRef]
53. Wu, T.; Tang, Y.; Tang, S.; Li, Y.; He, W.; Chen, E. Design and analysis of a new down-the-hole electromagnetic hammer driven by tube linear motor. *IET Electr. Power Appl.* **2017**, *11*, 1558–1565. [CrossRef]
54. Wu, T.; Fu, K.; Zhu, J.; Lei, G. The influence analysis of thrust and gap magnetic field of a down-to-hole tubular permanent magnet linear hammer due to high temperature in deep hole. In Proceedings of the 2017 20th International Conference on Electrical Machines and Systems (ICEMS), Sydney, NSW, Australia, 11–14 August 2017; pp. 1–5. [CrossRef]
55. Zhang, S.; Norum, L.; Nilssen, R. Analysis of tubular linear permanent magnet motor for drilling application. In Proceedings of the 2009 International Conference on Electric Power and Energy Conversion Systems, (EPECS), Sharjah, United Arab Emirates, 10–12 November 2009; pp. 1–5.
56. Wodecki, J.; Góralczyk, M.; Krot, P.; Ziętek, B.; Szrek, J.; Worsa-Kozak, M.; Zimroz, R.; Śliwiński, P.; Czajkowski, A. Process Monitoring in Heavy Duty Drilling Rigs—Data Acquisition System and Cycle Identification Algorithms. *Energies* **2020**, *13*, 6748. [CrossRef]
57. Petit, P. Electric rock drilling system for in-stope mining in platinum operations. In Proceedings of the International Platinum Conference 'Platinum Surges Ahead, Sun City, South Africa, 8–12 October 2006.
58. Zhang, G.; Thuro, K.; Konietzky, H.; Menschik, F.M.; Kasling, H.; Bayerl, M. In-situ investigation of drilling performance and bit wear on an electrical drill hammer. *Tunn. Undergr. Space Technol.* **2022**, *122*, 104348. [CrossRef]
59. Antonucci, A.; Barr, A.; Martin, B.; Rempel, D. Effect of bit wear on hammer drill handle vibration and productivity. *J. Occup. Environ. Hyg.* **2017**, *14*, 640–649. [CrossRef]
60. Rempel, D.; Antonucci, A.; Barr, A.; Cooper, M.R.; Martin, B.; Neitzel, R.L. Pneumatic rock drill vs. electric rotary hammer drill: Productivity, vibration, dust, and noise when drilling into concrete. *Appl. Ergon.* **2019**, *74*, 31–36. [CrossRef]
61. SCHULER. First Linear Hammer with ServoDirect Technology. Available online: https://www.schulergroup.com/major/download_center/broschueren_forging/download_forging/forging_flyer_linearhammer_servodirekt_e.pdf (accessed on 5 December 2022).
62. Bembenek, M.; Zięba, A.; Kopyściański, M.; Krawczyk, J. Analysis of the Impact of the Consolidated Material on the Morphology of Briquettes Produced in a Roller Press. *J. Mater. Eng. Perform.* **2020**, *29*, 3792–3799. [CrossRef]
63. Korendiy, V.; Kachur, O.; Hurey, I.; Predko, R.; Palash, R.; Havrylchenko, O. Modelling and experimental investigation of the vibratory conveyor operating conditions. *Vibroengineering Procedia* **2022**, *47*, 1–7. [CrossRef]
64. METSO Crushing and Screening Handbook. Available online: https://www.mogroup.com/insights/e-books/crushing-and-screening-handbook (accessed on 16 November 2022).
65. Makarov, L.N.; Denisov, V.N.; Kurilin, S.P. Designing and modeling a linear electric motor for vibration-technology machines. *Russ. Electr. Eng.* **2017**, *88*, 166–169. [CrossRef]
66. Gursky, V.; Kuzio, I.; Krot, P.; Zimroz, R. Energy-Saving Inertial Drive for Dual-Frequency Excitation of Vibrating Machines. *Energies* **2021**, *14*, 71. [CrossRef]
67. Bardzinski, P.; Walker, P.; Krol, R.; Kawalec, W. Simulation of Random Tagged Ore Flow through the Bunker in a Belt Conveying System. *Int. J. Simul. Model.* **2018**, *17*, 597–608. [CrossRef] [PubMed]
68. Bardzinski, P.; Jurdziak, L.; Kawalec, W.; Król, R. Copper Ore Quality Tracking in a Belt Conveyor System Using Simulation Tools. *Nat. Resour. Res.* **2019**, *29*, 1031–1040. [CrossRef]
69. Bardzinski, P.; Krol, R.; Jurdziak, L. Empirical Model of Discretized Copper Ore Flow within the Underground Mine Transport System. *Int. J. Simul. Model.* **2019**, *18*, 279–289. [CrossRef]
70. Gursky, V.; Krot, P.; Korendiy, V.; Zimroz, R. Dynamic Analysis of an Enhanced Multi-Frequency Inertial Exciter for Industrial Vibrating Machines. *Machines* **2022**, *10*, 130. [CrossRef]
71. Gursky, V.; Krot, P.; Dilay, I.; Zimroz, R. Optimization of the Vibrating Machines with Adjustable Frequency Characteristics. In *Nonstationary Systems: Theory and Applications, Proceedings of the 13th Workshop on Nonstationary Systems and Their Applications, Grodek nad Dunajcem, Poland, 3–5 February 2020*; Chaari, F., Leskow, J., Wylomanska, A., Zimroz, R., Napolitano, A., Eds.; Springer International Publishing: Cham, Switzerland, 2022; pp. 352–363. [CrossRef]
72. Krot, P.; Zimroz, R.; Michalak, A.; Wodecki, J.; Ogonowski, S.; Drozda, M.; Jach, M. Development and verification of the diagnostic model of the sieving screen. *Shock Vib.* **2020**, *2020*, 8015465. [CrossRef]
73. Aipov, R.; Linenko, A.; Badretdinov, I.; Tuktarov, M.; Akchurin, S. Research of the work of the sieve mill of a grain-cleaning machine with a linear asynchronous drive. *Math. Biosci. Eng.* **2020**, *17*, 4348–4363. [CrossRef]
74. Doroszuk, B.; Król, R. Analysis of conveyor belt wear caused by material acceleration in transfer stations. *Min. Sci.* **2019**, *26*, 189–201. [CrossRef]
75. Król, R. Studies of The Durability of Belt Conveyor Idlers with Working Loads Taken into Account. *IOP Conf. Ser. Earth Environ. Sci.* **2017**, *95*, 042054. [CrossRef]
76. Góralczyk, M.; Krot, P.; Zimroz, R.; Ogonowski, S. Increasing Energy Efficiency and Productivity of the Comminution Process in Tumbling Mills by Indirect Measurements of Internal Dynamics—An Overview. *Energies* **2020**, *13*, 6735. [CrossRef]

77. Bortnowski, P.; Gładysiewicz, L.; Król, R.; Ozdoba, M. Energy efficiency analysis of copper ore ball mill drive systems. *Energies* **2021**, *14*, 1786. [CrossRef]
78. Kujundžić, T.; Klanfar, M.; Korman, T.; Briševac, Z. Influence of Crushed Rock Properties on the Productivity of a Hydraulic Excavator. *Appl. Sci.* **2021**, *11*, 2345. [CrossRef]
79. Krot, P.; Zimroz, R.; Sliwinski, P.; Gomolla, N. Safe Operation of Underground Mining Vehicles Based on Cyclic Fatigue Monitoring of Powertrains. In *Structural Integrity and Fatigue Failure Analysis*; Lesiuk, G., Szata, M., Blazejewski, W., Jesus, A.M.d., Correia, J.A., Eds.; Springer International Publishing: Cham, Switzerland, 2022; pp. 283–292.
80. Gorodilov, L.V.; Maslov, N.A.; Korovin, A.N. Evaluation of parameters of hydraulic impact devices of active bucket with direct connection to the hydraulic system of II grade excavator. *Interexpo GEO-Sib.* **2020**, *2*, 45–51. [CrossRef]
81. Gorodilov, L.V.; Korovin, A.N. Analysis of active bucket designs of open-pit and construction excavators. *Interexpo GEO-Sib.* **2021**, *2*, 171–179. [CrossRef]
82. Mazur, M. Determination of crushing energy during vibratory crushing. *New Trends Prod. Eng.* **2019**, *2*, 287–294. [CrossRef]
83. Bond, F.C. The Third Theory of Comminution. In *Transactions of the American Institute of Mining, Metallurgical and Petroleum Engineers*; American Institute of Mining, Metallurgical, and Petroleum Engineers: New York, NY, USA, 1952; Volume 193, pp. 484–494.
84. Kabachkov, Y.F.; Vainer, B.M.; Lesnikov, V.V. Use of an electric hammer for splitting oversize magnesite blocks. *Refractories* **1982**, *23*, 468–469. [CrossRef]
85. Minaev, A.G.; Shakhtarin, R.A.; Kabachkov, Y.F. Use of electric hammers to break up concrete foundations in the repair of production equipment. *Metallurgist* **1986**, *30*, 376–377. [CrossRef]
86. Krot, P.; Śliwiński, P.; Zimroz, R.; Gomolla, N. The identification of operational cycles in the monitoring systems of underground vehicles. *Measurement* **2020**, *151*, 107111. [CrossRef]
87. Stefaniak, P.; Wodecki, J.; Jakubiak, J.; Zimroz, R. Development of Test Rig for Robotization of Mining Technological Processes—Oversized Rock Breaking Process Case. *IOP Conf. Ser. Earth Environ. Sci.* **2017**, *95*, 042028. [CrossRef]
88. KGHM ZANAM. Stationary Rock Breaker Type URB/Klim. Available online: https://www.kghmzanam.com/wp-content/uploads/2020/06/URB_Klim_PL.pdf (accessed on 5 December 2022).
89. Siwulski, T.; Warzynska, U.; Panowska, K.; Wolter, M. Improving the efficiency of a rock breaker hydraulic working system by changing the structure of the hydraulic system. *MM Sci. J.* **2022**, *2022*, 5738–5747. [CrossRef]
90. Krauze, K.; Rączka, W.; Sibielak, M.; Konieczny, J.; Kubiak, D.; Culer, H.; Bajus, D. Automated transfer point URB/ZS-3. *Mininig Inform. Autom. Electr. Eng.* **2017**, *2*, 80. [CrossRef]
91. Duff, E.; Caris, C.; Bonchis, A.; Taylor, K.; Gunn, C.; Adcock, M. The development of a telerobotic rock breaker. In *Field and Service Robotics*; Springer: Berlin/Heidelberg, Germany, 2010; pp. 411–420. [CrossRef]
92. Cárdenas, D.; Parra-Tsunekawa, I.; Leiva, F.; Ruiz-del Solar, J. Automatic Determination of Rock-Breaking Target Poses for Impact Hammers. *Energies* **2022**, *15*, 6380. [CrossRef]
93. Correa, M.; Cárdenas, D.; Carvajal, D.; Ruiz-del Solar, J. Haptic Teleoperation of Impact Hammers in Underground Mining. *Appl. Sci.* **2022**, *12*, 1428. [CrossRef]
94. Takahashi, H.; Sano, K. Automatic detection and breaking system for boulders by use of CCD camera and laser pointer. *Fragblast* **1998**, *2*, 397–414. [CrossRef]
95. Lampinen, S.; Niu, L.; Hulttinen, L.; Niemi, J.; Mattila, J. Autonomous robotic rock breaking using a real-time 3D visual perception system. *J. Field Robot.* **2021**, *38*, 980–1006. [CrossRef]
96. Bieniawski, Z.T. *Engineering Rock Mass Classifications: A Complete Manual for Engineers and Geologists in Mining, Civil, and Petroleum Engineering*; John Wiley & Sons: Nashville, TN, USA, 1989.
97. Aksoy, C.O. Review of rock mass rating classification: Historical developments, applications, and restrictions. *J. Min. Sci.* **2008**, *44*, 51–63. [CrossRef]
98. Mezentsev, I.V. Influence of Design Factors on the Efficiency of Hydraulic Hammers. *J. Min. Sci.* **2003**, *39*, 400–404. [CrossRef]
99. Aksoy, C.O.; Ozacar, V.; Safak, S. An updated formula and method to predict the performance of impact hammers. *Int. J. Rock Mech. Min. Sci.* **2013**, *61*, 289–295. [CrossRef]
100. Ismael, M.; Abdelghafar, K.; Sholqamy, M.; Elkarmoty, M. Performance prediction of hydraulic breakers in excavation of a rock mass. *Min. Geol. Pet. Eng. Bull.* **2021**, *36*, 107–119. [CrossRef]
101. Kucuk, K.; Aksoy, C.; Basarir, H.; Onargan, T.; Genis, M.; Ozacar, V. Prediction of the performance of impact hammer by adaptive neuro-fuzzy inference system modelling. *Tunn. Undergr. Space Technol.* **2011**, *26*, 38–45. [CrossRef]
102. Tumac, D.; Hojjati, S. Predicting performance of impact hammers from rock quality designation and compressive strength properties in various rock masses. *Tunn. Undergr. Space Technol.* **2016**, *59*, 38–47. [CrossRef]
103. Li, S.; Tian, S.; Li, W.; Yan, T.; Bi, F. Research on the Resonance Characteristics of Rock under Harmonic Excitation. *Shock Vib.* **2019**, *2019*, 6326510. [CrossRef]
104. Chiang, L.E.; Elías, D.A. A 3D FEM methodology for simulating the impact in rock-drilling hammers. *Int. J. Rock Mech. Min. Sci.* **2008**, *45*, 701–711. [CrossRef]
105. Hogan, J.D.; Rogers, R.J.; Spray, J.G.; Boonsue, S. Dynamic fragmentation of granite for impact energies of 6–28J. *Eng. Fract. Mech.* **2012**, *79*, 103–125. [CrossRef]

106. Kumano, A.; Goldsmith, W. Projectile impact on soft, porous rock. *Rock Mech. Felsmech. Mec. Des Roches* **1982**, *15*, 113–132. [CrossRef]
107. Kahraman, S.; Bilgin, N.; Feridunoglu, C. Dominant rock properties affecting the penetration rate of percussive drills. *Int. J. Rock Mech. Min. Sci.* **2003**, *40*, 711–723. [CrossRef]
108. Zhang, P.; Wu, Z.; Sun, J.; Liu, Y.; Chu, Z. Experimental and numerical studies of the impact breakage of granite with high ejection velocities. *PLoS ONE* **2022**, *17*, e0266241. [CrossRef]
109. Gorodilov, L.V.; Efimov, V.P.; Kudryavtsev, V.G. Modeling the striking head-impact tool-rock mass interaction. *J. Min. Sci.* **2013**, *49*, 618–624. [CrossRef]
110. Efimov, V.; Gorodilov, L. Experimental investigation of shock pulses in the piston-bit system in interaction with rock mass. *IOP Conf. Ser. Earth Environ. Sci.* **2021**, *773*, 012041. [CrossRef]
111. Gorodilov, L.V.; Kudryavtsev, V.G. Hydraulic Impactor Control Methods and Charts. *J. Min. Sci.* **2022**, *58*, 52–64. [CrossRef]
112. Corkum, A.G.; Asiri, Y.; Naggar, H.E.; Kinakin, D. The Leeb Hardness Test for Rock: An Updated Methodology and UCS Correlation. *Rock Mech. Rock Eng.* **2017**, *51*, 665–675. [CrossRef]
113. *ASTM D5873-14*; Standard Test Method for Determination of Rock Hardness by Rebound Hammer Method. ASTM: West Conshohocken, PA, USA, 2005. Available online: https://www.astm.org/d5873-14.html (accessed on 16 November 2022).
114. Mayer, T. *Investigating Parameter's of Lekatech's Electric Hammer*; Technical Report; Lekatech: Kausala, Finland, 2021. Available online: https://www.lekatech.fi/site/assets/files/3535/investigating_parameters_of_lekatechs_electric_hammer.pdf (accessed on 11 December 2022).
115. Korhonen, S. Reducing Energy Consumption of Hammering With Electric Excavators. Master's Thesis, University of Oulu, Oulu, Finland, 2006.
116. Ukkola, J. CE-marking of the electric breaker hammer and product safety requirements for the US and Canadian markets. Master's Thesis, Lappeenranta–Lahti University of Technology (LUT), Lappeenranta, Finland, 2022.
117. Lekatech. Linear Electric Technology for Hammering Applications. 2022. Available online: https://www.lekatech.fi/ (accessed on 8 December 2022).
118. LEKATECH Electric Hammer Data Sheet. Available online: https://echo.pwr.edu.pl/wp-content/uploads/2022/11/LekaTech_Datasheet.pdf (accessed on 5 December 2022).
119. Wagner Edward, P.S. Electric Excavator. International Patent Application No. IPC E02F3/32. International Patent Application No. F15B2211/20515; U.S. Patent Application No. US2021062459A1, 4 March 2021.
120. Szurgacz, D. Dynamic Analysis for the Hydraulic Leg Power of a Powered Roof Support. *Energies* **2021**, *14*, 5715. [CrossRef]
121. Szurgacz, D.; Brodny, J. Analysis of the Influence of Dynamic Load on the Work Parameters of a Powered Roof Support's Hydraulic Leg. *Sustainability* **2019**, *11*, 2570. [CrossRef]

energies

MDPI

*Article*

# The Development of Coal Mine Methane Utilization Infrastructure within the Framework of the Concept "Coal-Energy-Information"

**Arina Smirnova [1,2], Kirill Varnavskiy [1], Fedor Nepsha [1,*], Roman Kostomarov [1] and Shaojie Chen [2]**

1 Mining Industry Digital Transformation Lab, Mining Institute, T.F. Gorbachev Kuzbass State Technical University, 650000 Kemerovo, Russia
2 State Key Laboratory of Mining Disaster Prevention and Control Co-Founded by Shandong Province and The Ministry of Science and Technology, Shandong University of Science and Technology, Qingdao 266590, China
* Correspondence: nepshafs@kuzstu.ru

**Abstract:** The operation of coal mines is intricately linked with emitting a large quantity of coal mine methane, and in most cases, this methane releases into the atmosphere. In total, according to statistics, coal mining enterprises emit 8% of anthropogenic methane, determining a contribution to greenhouse gas emissions to the amount of 17%. There are various means for coal mine methane utilization. In this study, the concept "Coal-Energy-Information" is proposed. This concept implies both the construction of data processing centers on the industrial sites of coal mines and the usage of coal mine methane. Coal mine methane can be used as a primary energy source for the energy supply of data processing center consumers as well as coal mine consumers with necessary energy resources (electricity, heat, and cooling). Within the framework of the proposed concept, three options of coal mine methane utilization are considered. The first option is the use of gas genset for electrical and thermal energy generation (cogeneration) and their usage for coal mine and constructed data processing centers and consumers' power supply. The second option is absorption refrigerator usage (with coal mine methane direct burning) for cooling the IT equipment of constructed data processing centers. The last one is the use of a gas genset and absorption refrigerator (trigeneration) for constructed data processing centers' and coal mine consumers' energy supplies (electricity, heat, and cooling). In conclusion, it is noted that proposed concept is closely correlated with the program for the development of the coal industry in Russia for the period up to 2035, since it allows creating a base for the implementation of innovative technologies based on digital platforms that ensure the development of coal mining technology without the constant presence of personnel in underground mining facilities.

**Keywords:** coal mines; data processing centers; trigeneration; coal mine methane utilization

check for updates

**Citation:** Smirnova, A.; Varnavskiy, K.; Nepsha, F.; Kostomarov, R.; Chen, S. The Development of Coal Mine Methane Utilization Infrastructure within the Framework of the Concept 'Coal-Energy-Information". *Energies* **2022**, *15*, 8948. https://doi.org/10.3390/en15238948

Academic Editor: Dameng Liu

Received: 31 October 2022
Accepted: 20 November 2022
Published: 26 November 2022



## 1. Introduction

According to the results and reports of 2021, the coal production in Russia almost reached the level of 2019 and was more than 435 million tons, virtually 10% higher than that in 2020. Therefore, the dynamics of the huge surge in coal demand in the world market has been noticed in the extraction volume and national coal costs over the period of 2021.

Figure 1 provides current data about the coal costs per ton of three global futures [1]. According to the given information, the behavior of steam coal costs for the period from 2021 to present was ambiguous. It is shown that in the second half of the year, coal costs remained approximately stable and were USD 220 per ton. By the end of 2021, the Rotterdam and Newcastle Coal Futures prices began to increase rapidly. The reason for this was the abolishing of anti-COVID restrictions in the world and the growth in demand for primary energy sources. However, containment measures are still being taken occasionally in China. Due to that fact, the significant decline in production in China is reflected in the steady cost of coal.

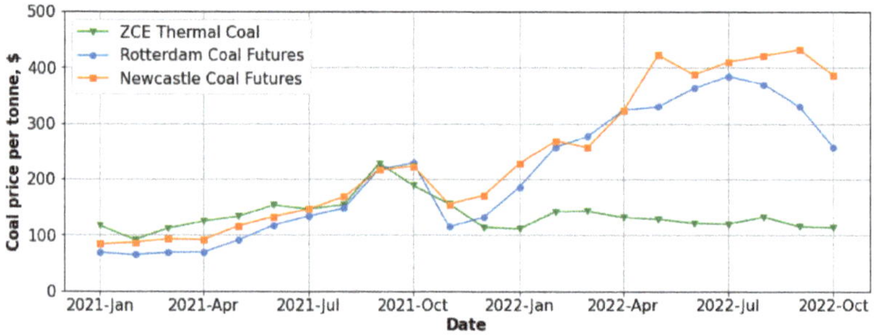


**Figure 1.** Trends and forecasts of world coal prices [1].

There was a spike in coal costs because of aggravation of the international energy crisis in the first half of 2022. Therefore, the peak of Newcastle Coal Futures prices was reached in September and was USD 420 per ton, and for Rotterdam Coal Futures, its peak was in July with a USD 390 per ton price.

The ambiguous dynamics of coal prices have a negative impact on the economic stability of coal mining enterprises. In this regard, an urgent task arises: finding alternative ways to ensure economic sustainability using the existing infrastructure of coal mining enterprises.

According to most forecasts, the consumption of coal will not grow significantly anymore. Moreover, it will slow down in the future, and it seems logical to use the current period of high prices for coal to search for new techniques for business diversification of coal mining enterprises, because the obtained windfall can be directed to the creation and implementation of new technologies.

The most obvious way to introduce technologies for coal deep processing is to use the industrial site of coal mining enterprises whenever possible. However, this method requires significant capital investments, while the markets for such products are not always clear.

Another promising option is the implementation of various technologies within the "Coal-Gas-Electricity" concept [2], with the production of electrical energy at the output.

This option certainly has great prospects for practical implementation, but according to the authors, it can be supplemented and expanded, namely by the creation of technologies within the framework of the "Coal-Energy-Information" concept. In practice, this implies the use of infrastructure and production facilities of coal mining enterprises for the construction and provision of the engineering infrastructure of data processing centers (DPCs) with the necessary energy resources.

Currently, there is constant growth in the IT industry. The volume of processed data is increasing, and consequently, the volume of electricity consumption of DPCs is also growing [3]. Therefore, the growing demand for DPCs leads to a construction cost increase.

Energy consumption by DPCs is significant, and it can be said that DPCs belong to the energy-intense industrial energy consumers. The share of electrical energy costs is about 40% of the total OPEX. Taking into consideration the trend to use clean energy sources, in [4], the need to use renewable energy sources to create "green" DPCs is mentioned. However, the use of "traditional" renewable energy sources in Kuzbass (the main coal-producing region in Russia) is far from constant cost effectiveness.

However, for Kuzbass, it is relevant to use one the most ecologically clean sources from the hydrocarbon ones: coal bed methane (CBM). The coal deposits in Kuzbass contain large amounts of CBM comprising methane from 80% to 95%, with lower contents of such heavier hydrocarbons as ethane and propane and nonhydrocarbon gases such as nitrogen and carbon dioxide [5].

The Kuznetsk Basin, situated in the southern part of Western Siberia, is the most mature basin for CBM exploration and development in Russia, with prospective original in-place hydrocarbon reserves of 13.1 trillion $m^3$ [6]. Thus, CBM can be simultaneously not only a hazardous industrial factor in coal mining but also a valuable co-product for recovery and utilization [7].

CMM is a type of CBM released as a direct result of the physical process during underground mining works and coal seam extraction. According to the authors of [8], coal mines emit 8% of their anthropogenic methane emissions during extraction and processing activities. In the first half of 2022, Kuzbass mining enterprises extracted 38 million tons with underground methods, while the average methane emissions per 1 ton of extracted coal was 17 $m^3/t$ [9]. The rapid development of the coal mining industry led to the increase in mining depth, which remarkably affects environmental pollution with methane emissions. Because of the high CMM quantities emitted into the environment, methane utilization is an advanced and relevant direction for coal region development. For greenhouse gas emissions mitigation, methane could be captured and used for power generation [10].

Therefore, the use of one more alternative sources of energy (CMM) for power generation is of growing interest. According to the authors of [11], the cost of power generation using CMM is 30–50 percent lower than the cost of power generation produced by wind power plants. At the same time, the investments to eliminate 1 ton of equivalent annual carbon dioxide emissions when using CMM are USD 34, which is four times lower than the CAPEX of wind farms (USD 100–142).

In general, three main factors determine the feasibility of CMM utilization: (1) reduction of the number of MAM explosions during underground coal deposit development and improving the safety of mining activities; (2) new job formation at gas fields and gas processing enterprises and improving the economic performance of coal mines; (3) improving the environmental situation in coal mining regions by the reduction of greenhouse gas emissions, one of which is methane. (The global warming potential of methane is 21 times greater than that of carbon dioxide.)

There are many studies devoted to the issue of CMM processing [8,12–19]. In general, they are aimed at ensuring the maximum use of CMM released from coal seams with efficiency of up to 95%. For this purpose, a gas genset operating in a trigeneration cycle can be used. The trigeneration cycle allows providing an electrical power supply to a coal mine and DPC, as well as providing a heat supply to a coal mine and cooling to the IT equipment of a DPC. In the future, the deep processing of CMM can provide production of carbon dioxide, which can be used in agriculture.

Within the framework of this article, the authors considered three options for the utilization of CMM:

1. The use of a gas genset for generation of electrical and thermal energy (cogeneration);
2. The use of a gas genset and absorption refrigerator (AR) for an energy supply to a DPC constructed on the industrial site of a coal mine (trigeneration);
3. The use of an AR (with direct burning of CMM) for the cooling of IT equipment of constructed DPCs.

This article is structured as follows. In Sections 2–4, options for the utilization of CMM are considered. Sections 5 and 6 provide a comparison of three methods of CMM utilization and considers the stakeholders interested in the development of the CMM utilization sector and the deployment of DPCs. Section 7 provides the conclusion of the article.

## 2. Energy Supply System Based on a CMM-Fired Gas Genset

### 2.1. Common Information

In contrast to traditional gasfields, CBM is an unconventional methane resource, and methane is not in a free form in the porous medium but in a bound (sorbed) form, which is stored in the natural fractures (cleats) and coal micropores [20,21]. The coal seam methane content is the volume of methane contained in a unit volume of coal (in $m^3/t$) on a dry ash-free (daf) basis. The values for the coal seam methane content in the Kuznetsk Basin area reach more than 30 m$^3$/t on a daf basis and rise with increasing coal deposit depths and decreasing heat and humidity [16]. Currently, the normative and technical documentation obliges Russian users of subsurface resources to carry out coal seam degassing work when the natural gas content exceeds 13 m$^3$/t on a daf basis. Therefore, CMM utilization is becoming required because of this rule.

The volume of methane released during mining operations in a unit of time is characterized by the methane-bearing capacity in m$^3$/min. According to the authors of [22], the yearly average relative methane-bearing capacity in Kuzbass coal mines is from 11 to 150 m$^3$/min while mining coal at a rate of 1600–13,074 t/day. The values of the relative methane-bearing capacity of the most dangerous mines in Kuzbass are shown in Figure 2 in descending order [23].

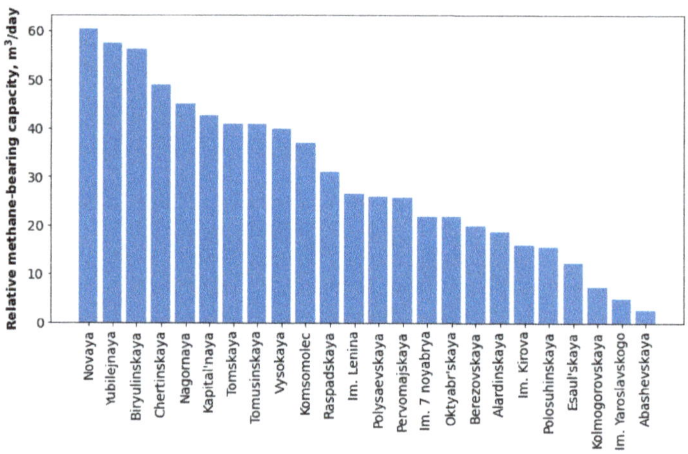


**Figure 2.** Relative methane-bearing capacity of the Kuzbass coal mines [23].

## *2.2. Analysis of the Composition of GAM*

Figure 3 presents a monthly graph of the amount of a methane-air mixture removed from underground mine operations, as well as the concentration of methane from one of the coal mines of Kuzbass. The case of this coal mine is considered in this paper.

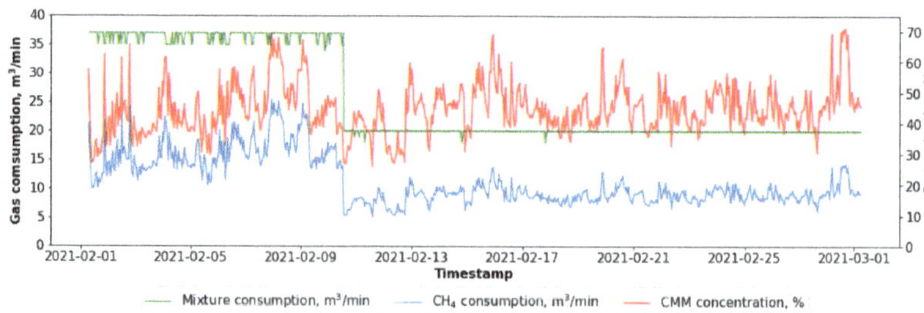


**Figure 3.** Monthly graph of changes in methane concentration and volumes of GAM removed from the mine.

These graphs show that the methane concentration did not drop below 30%, which indicates that there is no need to enrich GAM for use as a fuel for a gas genset. At the same time, it is necessary to note the sharply variable nature of the change in concentration, which was caused by a change in the concentration of methane in the process of coal mining.

A structural diagram of a power system with a gas genset is presented in Figure 4.

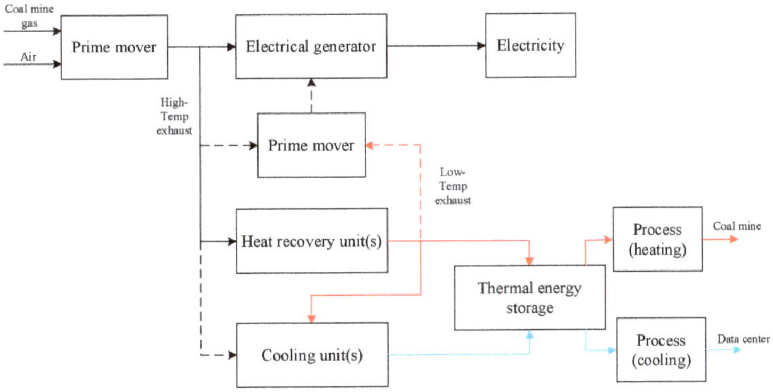


**Figure 4.** Block diagram of CMM utilization using gas genset.

The diagram in Figure 4 allows one to determine the approach to the choice of GPU: (1) analysis of the composition of the gas-air mixture (GAM); (2) assessment of the calorific capacity of the gas mixture; (3) assessment of amount of generated electrical and thermal energy; and (4) the choice of gas genset.

### 2.3. Algorithm of Choosing the CMM Utilization Scheme

This paper assumes the use of a mine infrastructure for the construction of DPCs, which will be the main consumer of energy resources produced using CMM utilization. Figure 5 provides the algorithm for choosing the optimal CMM utilization method with this approach.

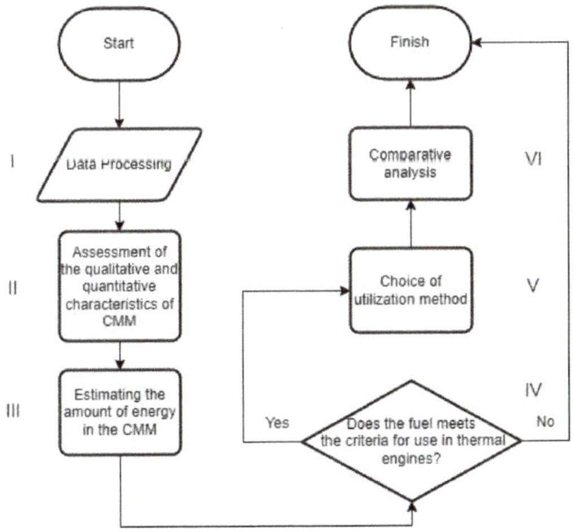


**Figure 5.** Algorithm for choosing a method of utilization of CMM.

The algorithm in Figure 5 includes the following stages:

*I.* Obtaining data on the volume and composition of the gas-air mixture and the CMM concentration.

*II.* Analysis the composition of the gas-air mixture for the needed filtration and dehumidification. In general, the CMM must meet the main criteria specified in Table 1. If the permissible values are exceeded, then filtering and dehumidification equipment must be used.

**Table 1.** The main criteria for the qualitative composition of CMM. The data in the table are given in accordance with the state standards of the Russian Federation.

| Index | Unit | Admissible Value |
|---|---|---|
| Sulfur | $g/m^3$ | 0.036 |
| $H_2O$ | $mg/m^3$ | 9 |
| Solid impurities | $mg/m^3$ | 1 |
| Particle size of solids | micron | 10 |

*III*. At this stage, the calorific value of the captured gas-air mixture is calculated.

*IV*. Determine the possibility of using CMM as a fuel for thermal engines. The main evaluation criteria are the calorific value, the concentration of methane in the mixture (not less than 25%) and the volume of the captured gas-air mixture. Based on the above data, it is possible to determine the type and capacity of equipment for the utilization of CMM.

*V*. Next, the choice of utilization method for CMM is carried out. In this paper, three ways to utilize CMM are considered:

- DPC with cogeneration energy center (CEC): The CEC consists of a gas genset operating on a cogeneration cycle and providing the DPC and coal mine with electrical and thermal energy. In this case, excess heat can be given off to meet the mine's own needs.
- DPC with refrigeration from an AR: In this case, it is proposed to use a direct-combustion AR to provide cooling to the DCP.
- DPC with a trigeneration energy center (TEC): The TEC combines the advantages of a CEC and AR and provides the DPC with electricity, heat and cooling.
- At this stage, the selection of equipment and a feasibility study for each specific case are also carried out.

*VI*. Conduct a comparative analysis to determine the most effective method.

Each of the stages is discussed in more detail below.

### 2.4. Assessment of the Calorific Capacity of GAM

For determination of the gas genset power when operating on CMM, it is necessary to estimate the potential amount of energy contained in cubic meters in the GAM which ventilates from the mine. The drive engine of a gas genset is an internal combustion engine, and therefore, the analysis of the energy potential of GAM must be carried out based on the assessment of the calorific value of GAM and its debit [24].

To assess the calorific value of GAM, it is proposed to use Equation (1) of D.I. Mendeleev:

$$LCV_f = LCV_1 \cdot a_1 + LCV_2 \cdot a_2 + \ldots LCV_n \cdot a_n = \sum_{i=1}^{n} LCV_i \cdot a_i \quad (1)$$

where $LCV_f$ is the lower heating value of fuel ($MJ/m^3$), $LCV_i$ is the lower heating value of gases in the composition of GAM ($MJ/m^3$), and $a_i$ is the share of *i*-gas in GAM.

### 2.5. Assessment of the Amount of Generated Electrical and Thermal Energy

The lower heating value of GAM can be calculated by the following Equation (2):

$$LCV_v = 60 \cdot R \cdot LCV_f \quad (2)$$

where $R$ is the debit of GAM ($m^3$), $LCV_f$ is the lower heating value of fuel ($MJ/m^3$), and $LCV_v$ is the lower heating value of GAM ($MJ/m^3$).

From the heat balance in Equation (3) and approximate heat balance values (Table 2), it can be seen that no more than 28% of the fuel energy in a gas internal combustion engine can be transformed into effective power and then into electrical energy:

$$Q_0 = Q_e + Q_c + Q_{exh} + Q_{icf} + Q_{unc} \quad (3)$$

where $Q_e$ is the heat equivalent to efficient engine operation (MJ/h), $Q_c$ is the heat given off to the cooling medium (MJ/h), $Q_{exh}$ is the heat carried away from the engine with exhaust gases (MJ/h), $Q_{icf}$ is the heat lost due to incomplete combustion of fuel (MJ/h), and $Q_{unc}$ is the unaccounted heat loss (MJ/h).

**Table 2.** Approximate heat balance of internal combustion engines (%) [25].

| Engine | $Q_e$ | $Q_{exh}$ | $Q_c$ | $Q_{icf}$ | $Q_{unc}$ |
|---|---|---|---|---|---|
| Petrol | 21–28 | 30–55 | 12–20 | 0–45 | 3–8 |
| Gas | 23–28 | 35–45 | 20–25 | 0–5 | 5–10 |
| Diesel | 29–45 | 25–45 | 15–35 | 0–5 | 2–5 |

*2.6. The Choice of Genset Type and Feasibility Study*

According to the obtained values of $LCV_e$, it is possible to select the type and composition of a gas genset.

For the feasibility study, Equations (4)–(9) were used.

The fuel costs in USD/kW·h are expressed as

$$C_{fuel} = \frac{R_{gas} \cdot C_{gas} \cdot K_{up}}{P_{unit}}, \quad (4)$$

where $R_{gas}$ is the gas consumption ($m^3$/h), $C_{gas}$ is the gas cost, (when using CMM, it is conventionally taken as equal to USD 0), $P_{unit}$ is the power of the gas genset (kW), and $K_{up}$ is the capacity factor.

The maintenance costs in USD/kW·h are expressed as

$$C_{oil} = \frac{V_{oil} \cdot C_{oil1}}{T_m \cdot P_{unit}}, \quad (5)$$

where $V_{oil}$ is the oil volume (l), $C_{oil1}$ is the cost of 1 L of oil (USD), and $T_m$ is the frequency of maintenance in running hours.

The service costs, including overhaul, in USD/kW·h are expressed as

$$C_m = \frac{C_{m1}}{T_{oh} \cdot P_{unit}}, \quad (6)$$

where $C_{m1}$ is the service costs, including overhaul (USD), and $T_{oh}$ is the frequency of overhaul in running hours.

The oil burning costs in USD/kW·h are expressed as

$$C_{ob} = \frac{R_{ob} \cdot C_{oil}}{P_{unit}}, \quad (7)$$

where $R_{ob}$ is the oil consumption for burning (gr/kWh).

The costs for spare parts, including overhaul, in USD/kW·h are expressed as

$$C_{sp} = \frac{C_{sp1}}{T_{oh} \cdot P_{unit}}, \quad (8)$$

where $C_{sp1}$ is the cost of consumables and spare parts, including overhaul (USD). ($C_{sp1}$ is accepted to be equal to 50% of the total cost of installation.)

Amortization in USD/kW·h is expressed as

$$C_{dep} = \frac{C_{unit}}{T_{oh} \cdot P_{unit}}, \quad (9)$$

where $C_{unit}$ is the full cost of the gas genset and auxiliaries' systems (USD).

Significantly increasing the efficiency of a gas genset is possible using an exhaust gas heat and coolant recovery system.

When calculating the thermal correction, it is accepted that there will be a replacement of an existing heat source in a gas heat and coolant recovery system. The capacity factor at this stage is neglected.

The thermal correction ($/kWh) is the cost of 1 kW of thermal energy produced from an existing source (boiler room). In the case of a cogeneration cycle, it is replaced by the waste heat of the exhaust gases of the gas genset. This can be determined according to Equation (10):

$$C_{heat} = \frac{R_{fuel} \cdot C_{fuel}}{Q_{unit}}, \tag{10}$$

where $C_{fuel}$ is the cost per $m^3$ (ton) of fuel for the current heat source (USD), $Q_{unit}$ is the thermal capacity of a heat source (kW), and $R_{fuel}$ is the fuel consumption.

### 2.7. Calculation of Greenhouse Gas Emissions

For calculation of the reduction of greenhouse gas emissions, it is necessary to normalize to the amount of emissions of greenhouse gases into units (global warming potential). The *GWP* of methane is 25 units.

Thus, reducing emissions from CMM utilization based on gas gensets can be calculated by Equation (11):

$$CO_{2reduced} = \frac{0.6682 \cdot R_{cmm} \cdot \%CH_4}{1000} \cdot GWP, \tag{11}$$

where 0.6682 is the methane density ($kg/m^3$), $R_{cmm}$ is the gas consumption of a gas genset ($m^3/h$), and $\%CH_4$ is the percentage of a greenhouse gas (methane) in GAM.

In accordance with [26], the average cost of $CO_2$ emissions in 2021 was USD 58.97/ton, and the calculation of savings on quotas is produced by Equation (12):

$$ETS_{year} = CO_{2reduced} \cdot P_{ETS} \tag{12}$$

where $P_{ETS}$ is the cost of $CO_2$ emissions.

The cost calculation in USD/kWh is shown in Equation (13):

$$C_{kW} = C_{gas} + C_{oil} + C_m + C_{ob} + C_{sp} + C_{dep} - C_{heat}. \tag{13}$$

$C_\Delta$ in USD/kWh is calculated as shown in Equation (14):

$$C_\Delta = C_{kW} - Rate_e, \tag{14}$$

where $C_\Delta$ is the difference in the cost of 1 kW of energy produced by a cogeneration gas genset and the current electricity rate and $Rate_e$ is the electricity rate (USD/kWh).

The economic effect in USD per year is calculated as shown in Equation (15):

$$E = C_\Delta T_{eng} \cdot P_{unit} \cdot K_{up}, \tag{15}$$

where $T_{eng}$ is the running hours.

The payback period in years is calculated as shown in Equation (16):

$$S = \frac{C_{unit}}{E} \tag{16}$$

The calculation results are shown in Table 3. The calculations were made according to the average values of the data for the volume and concentration of methane presented in Figure 3.

**Table 3.** Calculation results.

| Parameter Name | Symbol | Unit | Value |
|---|---|---|---|
| Calorific capacity | $LCV_f$ | $MJ/m^3$ | 13.24 |
| Amount of energy | $LCV_v$ | $MJ/m^3$ | 30,200.88 |
| Effective power | $Q_e$ | MJ/h | 1812.05 |
| Maintenance costs | $C_{oil}$ | USD/kW·h | 0.0016 |
| Service costs, including overhaul | $C_m$ | USD/kW·h | 0.0055 |
| Oil burning costs | $C_{ob}$ | USD/kW·h | 0.00052 |
| Costs for spare part, including overhaul | $C_{sp}$ | USD | 329,676.34 |
| Amortization | $C_{dep}$ | USD/kW·h | 0.01 |
| The thermal correction | $C_{heat}$ | USD/kW·h | 0.01 |
| Savings on quotas | $ETS_{year}$ | USD | 70,302.49 |
| Cost price | $C_{kW}$ | USD/kW·h | 0.0076 |
| Economic effect | $E$ | USD/year | 360,798.50 |
| Payback period | $S$ | year | 5 |

### 2.8. *Effects of the Considered Option*

The use of a gas genset for the generation of electrical and thermal energy (cogeneration) allows achieving the following:

1. Reduction of payments for electricity consumed from the grid (by reducing the maximum power and reducing the consumption of electrical energy);
2. Operation of a gas genset in cogeneration mode allows providing heating and a hot water supply to industrial site consumers (replacement or alternative to a boiler room);
3. Reduction of fees for greenhouse gas emissions (methane);
4. Increasing a power supply's reliability.

## 3. DPC Construction in Addition to the Gas Gensets

To improve the reliability of the DPC and coal mine energy supply, CMM utilization can be used. In this case, the coal mine energy supply structure is as shown in Figure 6.

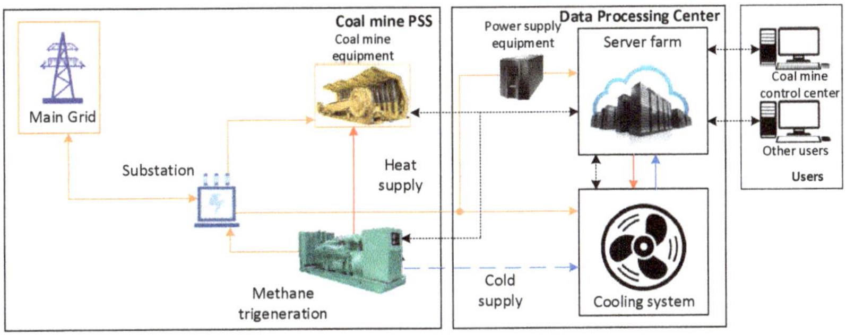


**Figure 6.** Structure of power supply system of coal mine and DPC.

In our case, it was assumed that the DPC equipment was located on the surface. However, it is possible to place the IT equipment in the underground part of the coal mine. One such example is the DPC placed in Norway's Lefdal mine, where olivine, a mineral used to make heat-resistant glass, was previously mined [27]. It must be noted that the DPC equipment can be used to build applied control systems for the technological systems of a coal mine [28].

The cooling system is the second consumer in the DPC after computer equipment. Therefore, it is quite useful to use gas piston units in trigeneration mode. In this case, they can provide refrigeration within a coal mine and significantly reduce the Power Usage

Efficiency (PUE), which is a ratio that describes how efficiently a computer DPC uses energy in comparison with cooling and other overhead.

The share of other consumers such as lighting, alarm, control, and monitoring systems depends on the DPC size and is about 6% of the total consumption of the DPC.

### 3.1. Options for the Use of DPC by Enterprises of the Mineral Resources Sector

According to the authors of [29], DPC leases are grouped into four categories:

1. *Wholesale or powered shell leases* are the provision of premises and infrastructure for the power supply of a DPC building. These leases are typically long-term contracts due to the significant investment made by the tenant.
2. *Turnkey or enterprise leases*, where the landlord builds the shell building and installs all infrastructure needed, including the raised floor. The tenant has a server room ready for operation at his disposition.
3. *Colocation agreements* consist of providing a fully equipped server room, including the server racks.
4. *Cloud agreements (IaaS)* offer IT services to a wide range of tenants. The tenant rents computing power and is not involved in the operation of DPC equipment.

The second and third options are preferable for coal mining enterprises. The DPC construction can be entrusted to an external, more experienced company when implementing the second option. During the transition to the third option, it is assumed that the enterprise will have personnel serving the life support system of the DPC. In the future, it is possible to switch to the fourth option, where an enterprise will be able to provide cloud services. These services can be related to the automation of business processes of other regional enterprises. This includes platform services (IaaS) and application components (SaaS).

### 3.2. DPC Construction Cost Calculation

The DPC construction cost calculation (CAPEX) is a rather complicated process. Therefore, for this purpose, a special service Data Center Capital Cost Calculator was used [30].

The initial parameters for the calculation are presented below:

1. *DPC environment*

Location of data center: Europe, Bulgaria. (The choice of this country was justified by the lack of Russia in the Data Center Capital Cost Calculator. Therefore, it was decided to choose the country with the closest value for the average gross monthly wage (Russia = USD 1046, Bulgaria = $885 in 2022).

Data center design capacity: 1000 kW.

Cooling system: computer room air handler (CRAH).

UPS architecture: traditional, non-scalable UPS.

Power distribution type: basic wall-mount panelboards.

Power density: 5 kW.

Cost of work for one person hour: USD 22.5/h.

CAPEX: USD 3,000,000.

2. *Redundancy level*

DPC redundancy is typically described in four groups with increasing resilience to component failure: N, N + 1, 2N, and 2N + 1. The N set-up is configured to have just the number of components it needs to function, meaning that whenever one component fails, the entire system fails. The N + 1 configuration indicates that there is one extra component on site regardless of the size of N.

The price includes the costs of the racks, raised floor, fire suppression, switchgear, and dropped ceiling.

As a result, the projected data center included 200 racks and was located on an area of 664 $m^2$, while the building area was 1135 $m^2$. At the same time, the construction cost was EUR 3 million.

According to the authors of [29], the operating costs were USD 225/$m^2$ per month.

### 3.3. Determination of DPC Profitability

Colocation agreements are considered when determining the yield. Here, 5 kW racks were leased, and the Tier 3 reliability level was provided. According to the authors of [29], the rental price was USD 800 per month. Therefore, it was assumed that 70% of the 200 racks were always leased. In this case, the annual OPEX will be USD 255,375, and the rental income from the racks was USD 1,920,000. The payback period at a discount rate of 12.5% (for the Russian Federation) is about 4 years.

### 3.4. Effects of the Considered Option

In addition to the previously considered option, the following effects will be obtained:

1. Business diversification toward the IT sector with the possibility of generating additional income;
2. The ability to provide cooling to DPC IT equipment when using the gas-generating set in trigeneration mode;
3. The ability to achieve the maximum efficiency of the generating unit up to 85%;
4. Development of the regional IT industry by building a new DPC to use local deposits and employment creation.

## 4. DPC Construction Together with Absorption Refrigerator

IT equipment generates a lot of heat during operation. Therefore, the cooling of the DPC premises, in which the IT equipment is located, is the most important task to ensure the reliability of the DPC operation. In turn, energy consumption for the cold supply is, on average, 40–45% of the total energy consumption.

The use of AR in the cooling system of a DPC located on the territory of a coal mine will significantly reduce energy costs for cooling IT equipment. In this case, it is proposed to use AR with direct combustion while using CMM as a primary fuel.

The power supply structure of a coal mine and DPC when using an AR is shown in Figure 7.

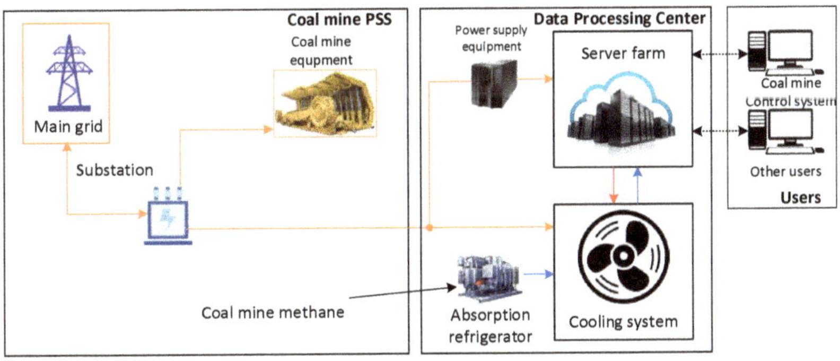


**Figure 7.** Power supply structure of coal mine and DPC when using ARs.

With direct combustion of the primary fuel, it is most efficient to use two-stage ARs. The cooling capacity of an AR can be determined by Equation (17):

$$Q_{cool} = Q_{heat} \cdot COP, \tag{17}$$

where $COP$ is the refrigerating factor (for a two-stage AR, COP can be taken to be equal to 1.2).

The heat from the burner to the AR can be calculated by Equation (18) [13]:

$$Q_{bur} = LCV_{CMM} \cdot Rate_{CMM}, \tag{18}$$

On average, according to the catalog, the consumption of AR (with direct combustion) with a refrigerating capacity of 1.1 MW was approximately 85 $m^3/h$. However, the use of natural gas was assumed. Preservation of the thermal power of the AR burner during the combustion of CMM implies the introduction of appropriate changes in the burner design, as well as an increase in gas consumption (the amount of gas consumed depends on its composition) [24].

Using Equation (18), the amount of heat supplied from the burner was determined when using natural heat for the considered AR $Q_{bur} = 778$ kW. When using CMM with a net calorific value of 18.38 $MJ/m^3$, a supply of 152.64 $m^3/h$ of coal mine methane (or 2.53 $m^3/min$) would be required to operate a similar AR.

The amount of methane captured from the mine in question was four times higher than that required for the operation of the selected AR and, accordingly, the cooling supply of the DPC.

During the cold season, when the cooling of IT equipment can be carried out using free cooling, an AR can be used for the heating and hot water supply of the administrative premises of the DPC and mine, but the efficiency of the AR in heating mode will decrease.

*Effects of the Considered Option*

1. Business diversification toward the IT sector, with the possibility of generating additional income;
2. The possibility to provide effective CMM utilization year-round;
3. Using AR for the heating and hot water supply of industrial site consumers (replacement for or alternative to a boiler room);
4. Reduction of fees for greenhouse gas emissions (methane).

## 5. Common Results

The CAPEX of DPC construction was USD 3,000,000. The OPEX was USD 255,000 per year. With a gas genset (cogeneration), AR, or gas genset together with an AR (trigeneration) installation, it became possible to decrease operational expenditures. Table 4 presents the comparison of CMM utilization options.

**Table 4.** Comparison of CMM utilization options.

| Parameter Name | DPC with Cogeneration Energy Center (CEC) | AR + DPC | DPC with Trigeneration Energy Center (TEC) |
|---|---|---|---|
| Additional CAPEX (USD) | 672,000 | 345,000 | 1,140,000 |
| CAPEX decrease (USD) | - | 200,000 | 200,000 |
| Additional OPEX (USD) | 67,000 | 10,000 | 77,000 |
| Profitability or Savings (USD) | 200,000 | 60,000 | 260,000 |
| Total OPEX (USD) | 122,000 | 205,000 | 72,000 |
| Payback period of CMM utilization equipment | ≈5 years | ≈7 years | ≈6 years |

Table 4 shows that all proposed CMM utilization options allowed decreasing the total OPEX by lowering electricity and heating bills or providing cooling for the computer room. The AR and trigeneration energy center replaced some parts of the cooling system, such as the chiller and cooling tower. Therefore, the CAPEX was lowered by USD 200,000. The choice of CMM utilization option depends on the geological conditions of the coal mine and on electricity and heat tariffs. For the considered use case, the third option yielded the best result.

## 6. Discussion

The following benefits can be achieved through implementation of proposed CMM utilization options:

1. Improving labor safety at underground coal mines.
2. The possibility of using industrial sites of closed (or mothballed) underground coal mines and partial preservation of jobs.
3. An increase in tax revenues for the regional budget from organizations that own DPCs, as well as service companies that maintain and operate DPCs.
4. The creation of additional job opportunities in one of the most advanced industries: the IT sector.
5. Improving the environmental situation by reducing greenhouse gas emissions.
6. Development of the regional IT sector through the creation of infrastructure facilities, namely DPCs with increased energy efficiency in the local resource base.
7. The possibility of creating environmentally friendly heat supply facilities for the population (with the usage of heat removed from server farms of DPCs).
8. Improving the efficiency of the development of coal deposits through an integrated approach (the use of coal and CMM) and increasing the profitability of coal mining.
9. Diversification of the business of coal companies with the possibility of generating additional income at reduced operating costs for DPCs (compared with the standard solutions for DPCs).
10. The opportunity to save on electrical energy and heating costs at existing coal mines.

However, in doing so, there are several risks and issues in implementation of the proposed CMM utilization options.

First, it should be noted that in each case, when considering CMM as a primary energy source, it is necessary to consider the variability of the CMM well rate and, if necessary, organize additional fuel reserves. This can lead to additional CAPEXs and a rise in the need for the creation of a special department, thus increasing the OPEX.

In addition, an important component of the implementation of the "Coal-Energy-Information" concept is the need to amend standards, technical regulations, and other regulatory documents.

Finally, there is one more difficulty, which is the global deficit of semiconductors and the related limitation of server hardware supplies. This problem can increase the deadlines for the delivery of DPCs and influence the architecture of server hardware.

Nevertheless, despite the mentioned difficulties, it is obvious that at present, there is a trend toward decarbonization in the world. However, not all countries imply the complete rejection of the use of coal. This means that for the coal mining industry, it is advisable to implement modern technologies and concepts that allow making coal mining as environmentally friendly as possible and allow diversifying the business of coal mining enterprises to ensure their stabile operation and financial sustainability.

Yet, over the coming 10–20 years, a significant decrease in coal production is not forecasted, which in turn determines some predictability of income for coal mining enterprises. It is logical to use part of this income for the implementation of different approaches for comprehensive development of coal deposits, assuming the simultaneous production of coal and extraction of CMM, which will allow achieving the goals of coal mining enterprises for environmental friendliness improvement and economic efficiency increases. In particular, the concept of "Coal-Energy-Information" can be considered one of such approaches.

## 7. Conclusions

Three options in the framework of the proposed concept (1 = DPC with CEC, 2 = DPC with AR, and 3 = DPC with TEC) which allow improving the reliability and efficiency of power supply services, decreasing the negative influence on the environment, and diversifying the business of coal mining enterprises were shown in the paper. All three options assumed construction of the DPC on the industrial site of a coal mining enterprise, with different solutions for the energy supply of this DPC and the technological equipment of the coal enterprise by all necessary types of energy (electricity, heat, and cooling), using CMM as the primary source of energy.

Each proposed option implies the creation of its own configuration of energy supply systems with the necessary power equipment, but implementation of any of the three options, on one hand, allows for reducing $CH_4$ emissions, contributing to respecting the environment, and on the other hand, makes it possible for coal mining enterprises to offer the market IT services as a new "product" and thus diversify production activities.

For each option, a primary feasibility study was conducted. CAPEX, OPEX, and profitability and savings calculations were made, and the payback periods of investments (five years for option 1, seven years for option 2, and six years for option 3) were determined. This paper also mentioned the major risks and issues inherent in the implementation of the proposed options, namely the necessity to consider the variability of the CMM wells rate, which determines the need to organize additional fuel reserves in some cases. Another potential difficulty is the global deficit of semiconductors and the related limitation of server hardware supplies.

Finally, it was pointed out that the construction of DPCs based on coal mines makes it possible to stimulate mining enterprises toward efficient CMM utilization and therefore improve the safety of mining activities, to increase economic efficiency, as well as to give impetus to the development of the IT industry in mining cities. In addition, it is noted that the proposed concept is in close correlation with the program for the development of the coal industry in Russia for the period up to 2035, since it allows one to create a base for the implementation of innovative technologies based on digital platforms that ensure the development of coal mining technology without the constant presence of personnel in underground mining workings, or "unmanned" coal mines.

**Author Contributions:** Methodology, F.N.; Formal analysis, A.S., K.V., F.N. and R.K.; Resources, A.S.; Writing—original draft, A.S., K.V., F.N. and R.K.; Writing—review & editing, S.C.; Visualization, F.N.; Project administration, K.V. All authors have read and agreed to the published version of the manuscript.

**Funding:** This research was supported by a state assignment of the Ministry of Science and Higher Education of the Russian Federation (No. 075-03-2021-138/3).

**Institutional Review Board Statement:** Not applicable.

**Informed Consent Statement:** Not applicable.

**Data Availability Statement:** Not applicable.

**Conflicts of Interest:** The authors declare no conflict of interest.

## References

1. Investing. Com-Stock Market Quotes & Financial News. Available online: https://www.investing.com/ (accessed on 17 November 2022).
2. Puchkov, L.A.; Vorobjev, B.M.; Vasyuchkov, Y.F. *Advanced Coal-Energy Complexes (Future Vision)*; Publishing House of Moscow State Mining University: Moscow, Russia, 2007.
3. The Importance of Green Data Centres | News | BroadGroup. Available online: https://www.broad-group.com/data/news/documents/b1m0b65jb01p34/the-importance-of-green-data-centres?page=1 (accessed on 5 December 2021).
4. Rahmani, R.; Moser, I.; Cricenti, A.L. Modelling and Optimisation of Microgrid Configuration for Green Data Centres: A Meta-heuristic Approach. *Future Gener. Comput. Syst.* **2020**, *108*, 742–750. [CrossRef]
5. Mazanik, E.V.; Mogileva, E.M.; Kolikov, K.S. Coal bed methane utilization: The state of the art, objectives and future considerations. *Russ. Min. Undustry* **2014**, *113*, 59–64.
6. Trofimova, G.I.; Cheremisina, V.G. Organization of field production of coal-bed methane in Kuzbass as a modern efficient task. *Int. Sci. J. Symb. Sci.* **2015**, *9*, 23–37.
7. Zhang, Y.; Kolesnik, Y.I. Modern technologies for coal bed methane production in China: Trends and development prospects. *Baikal Res. J.* **2022**, *13*. [CrossRef]
8. Karacan, C.Ö.; Ruiz, F.A.; Cotè, M.; Phipps, S. Coal Mine Methane: A Review of Capture and Utilization Practices with Benefits to Mining Safety and to Greenhouse Gas Reduction. *Int. J. Coal Geol.* **2011**, *86*, 121–156. [CrossRef]
9. Shubina, E.A.; Lukyanov, V.G. Trends in coal mining laws and taxation aimed at stimulating commercial coal seam methane production. *Bull. Tomsk. Polytech. Univ. Geo Assets Eng.* **2015**, *326*, 131–139.
10. Bakkhaus, K.; Kasyanov, V.V. Analysis of technologies for the use of coal mine methane. *J. Min. Sci.* **2012**, *15*, 169–182.

11. *A Strategy for Coal Bed Methane (CBM) and Coal Mine Methane (CMM) Development and Utilization in China. Energy Sector Management Assistance Program*; Formal Report 326/07; The International Bank for Reconstruction and Development, World Bank: WA, DC, USA, 2007; p. 146.
12. Cheng, Y.-P.; Wang, L.; Zhang, X.-L. Environmental Impact of Coal Mine Methane Emissions and Responding Strategies in China. *Int. J. Greenh. Gas Control* **2011**, *5*, 157–166. [CrossRef]
13. Zhou, F.; Xia, T.; Wang, X.; Zhang, Y.; Sun, Y.; Liu, J. Recent Developments in Coal Mine Methane Extraction and Utilization in China: A Review. *J. Nat. Gas Sci. Eng.* **2016**, *31*, 437–458. [CrossRef]
14. Meybodi, M.A.; Behnia, M. Australian Coal Mine Methane Emissions Mitigation Potential Using a Stirling Engine-Based CHP System. *Energy Policy* **2013**, *62*, 10–18. [CrossRef]
15. Hummel, J.A.; Ruiz, F.A.; Kelafant, J.R. Quantifying the Benefits of Coal Mine Methane Recovery and Use Projects: Case Study on the Application of in-Mine Horizontal Pre-Drainage Boreholes at Gassy Coal Mines in India and the Optimization of Drainage System Design Using Reservoir Simulation. *Environ. Technol. Innov.* **2018**, *10*, 223–234. [CrossRef]
16. Uddin, N.; Blommerde, M.; Taplin, R.; Laurence, D. Sustainable Development Outcomes of Coal Mine Methane Clean Development Mechanism Projects in China. *Renew. Sustain. Energy Rev.* **2015**, *45*, 1–9. [CrossRef]
17. Yin, F.; Nie, B.; Wei, Y.; Lin, S. Co-Production System Based on Lean Methane and Biogas for Power Generation in Coal Mines. *Atmosphere* **2022**, *13*, 803. [CrossRef]
18. Wang, X.; Zhou, F.; Ling, Y.; Xiao, Y.; Ma, B.; Ma, X.; Yu, S.; Liu, H.; Wei, K.; Kang, J. Overview and Outlook on Utilization Technologies of Low-Concentration Coal Mine Methane. *Energy Fuels* **2021**, *35*, 15398–15423. [CrossRef]
19. Yutyaev, A.E.; Belyaev, V.V.; Agafonov, V.V. Cogeneration of Resource-Saving Technologies in the Development of Coal Deposits. *Min. Inf. Anal. Bull. (Sci. Tech. J.)* **2013**, *6*, 69–74.
20. Puchkov, L.A. Modern problems of coal bed methane. *Min. Inf. Anal. Bull.* **1997**, *6*, 3–16.
21. Flores, R.M. Chapter 4—Coalification, Gasification, and Gas Storage. In *Coal and Coalbed Gas*; Flores, R.M., Ed.; Elsevier: Boston, MA, USA, 2014; pp. 167–233, ISBN 978-0-12-396972-9.
22. Zaburdyaev, V.S. Methane hazard of coal mines. *Occup. Saf. Ind.* **2013**, *8*, 60–64.
23. Nenasheva, R.I. *Mining and Geological Conditions in the Development of Coal Seams in Kuzbass*; Kuzbassvuzizdat: Kemerovo, Russia, 2012; ISBN 978-5-202-01058-3.
24. Redko, A.; Redko, I. Numerical Investigation of the Low-Caloric Gas Burning Process in a Bottom Burner. *Probl. Reg. Energetics* **2017**, *2*, 72–81.
25. Sharoglazov, B.A.; Farafontov, M.F.; Klementiev, V.V. *Internal Combustion Engines: Theory, Modeling and Calculation of Processes*; South Ural State University: Chelyabinsk, Russia, 2005.
26. EU Carbon Permits-2022 Data-2005–2021 Historical-2023 Forecast-Price-Quote. Available online: https://tradingeconomics.com/commodity/carbon (accessed on 26 June 2022).
27. Case Study: In the Hall of the Mountain Data King. Available online: https://images.forbes.com/forbesinsights/StudyPDFs/IBM-InTheHallOfTheMountainDataKing-REPORT.pdf (accessed on 29 October 2022).
28. Nepsha, F.; Voronin, V.; Kostomarov, R.; Varnavskiy, K.; Ermakov, A. *Employing the Digital Platform for Control System Development in a Coal Mining Enterprise*; IOP Publishing Ltd.: Bristol, UK, 2021, p. 012012.
29. Louwerens, T. *Data Centre Investment: An Investment Model & Associated Risk-Return Profile*; Delft University of Technology: Delft, The Netherlands, 2014.
30. Data Center Capital Cost Calculator. Available online: https://www.se.com/ww/en/work/solutions/system/s1/data-center-and-network-systems/trade-off-tools/data-center-capital-cost-calculator/ (accessed on 7 November 2021).

energies

Article

# Design of an Infrared Image Processing Pipeline for Robotic Inspection of Conveyor Systems in Opencast Mining Sites

**Mohammad Siami [1], Tomasz Barszcz [2,*], Jacek Wodecki [3] and Radoslaw Zimroz [3]**

1 AMC Vibro Sp. z o.o., Pilotow 2e, 31-462 Kraków, Poland
2 Department of Robotics and Mechatronics, AGH University of Science and Technology, Al. Mickiewicza 30, 30-059 Kraków, Poland
3 Department of Mining, Faculty of Geoengineering, Mining and Geology, Wrocław University of Science and Technology, 50-370 Wroclaw, Poland
* Correspondence: tbarszcz@agh.edu.pl

**Abstract:** Conveying systems play an essential role in the continuous horizontal transportation of raw materials in mining sites. Regular inspections of conveyor system structures and their components, especially idlers, are essential for proper maintenance. Traditional inspection methods are labor-intensive and hazardous; therefore, robot-based thermography can be considered a quality assessment tool for the precise detection and localization of overheated idlers in opencast mining sites. This paper proposes an infrared image processing pipeline for the automatic detection and analysis of overheated idlers. The proposed image processing pipeline can be used for the identification of significant temperature anomalies such as hotspots and hot areas in infrared images. For the identification of such defects in idlers, firstly, the histogram of captured infrared images was analyzed and improved through the pre-processing stages. Afterward, the location of thermal anomalies in infrared images was extracted. Finally, for the validation of segmentation results, the shapes and locations of segmented hot spots were compared with RGB images that were synchronized by captured infrared images. A quantitative evaluation of the proposed method for the condition monitoring of belt conveyor idlers in an open-cast mining site shows the applicability of our approach.

**Keywords:** overheated idlers detection; maintenance; inspection robots; IR images; hot spot detection



**Citation:** Siami, M.; Barszcz, T.; Wodecki, J.; Zimroz, R. Design of an Infrared Image Processing Pipeline for Robotic Inspection of Conveyor Systems in Opencast Mining Sites. *Energies* **2022**, *15*, 6771. https://doi.org/10.3390/en15186771

Academic Editor: Chunhua Liu

Received: 31 July 2022
Accepted: 13 September 2022
Published: 16 September 2022



## 1. Introduction

Conveyors have been developed and used as the most common system for conveying all forms of material in the mining industry. For decades, conveyors have been used for transporting raw materials due to their efficiency and relatively straightforward design. Despite the conveyor advantages, there are still significant challenges for conducting regular inspections to guarantee their operation under harsh environmental conditions in mines [1–7].

Idlers are important parts of the conveyors that support the belt to carry the material along its full length [8,9]. Idlers can be damaged by friction, tear, wear, jamming, or seizure. Faulty idlers can become overheated and cause belt damage; thus, the temperature, noise emissions, and vibrations of idlers should be constantly monitored through regular inspections. Idlers are located along the conveyor, and the typical length of conveyors in mining tunnels could reach a kilometer [10]. Human inspections of idlers by walking along the belt is time-consuming, costly, and hazardous, as even a small conveyor of 150 m consists of nearly 450 carrying rollers and 50 return rollers that should be inspected individually [11].

Monitoring the surface temperature of idlers is a key to finding faulty idlers because the abnormal temperature rise is an important characterization of idler failures on conveyor systems. The detection of overheated idlers focuses on identifying areas in IR images with a significantly higher temperature than other elements in a image. However, the automatic identification of overheated idlers in IR images is difficult due to the presence of sunlight reflection or non-informative objects (from the hot idler detection perspective) [12,13].

To summarize, according to the current status of condition monitoring (CM) methods for conveyor systems, there is a challenging need to minimize the presence of humans by the automatization of the inspection processes. Direct monitoring methods, such as IR thermography, are capable of detecting and diagnosing defects in idler modules. In this paper, IR image processing techniques alongside shape detection algorithms are experimentally assessed for their applicability for CM of idlers. The proposed techniques are applied to IR images of idler modules captured by a mobile robot during several field IR thermographic measurements. The direct relation between idlers surface temperature and their health status in terms of efficiency was the main topic for investigation.

The paper is organized as follows. First, the problem becomes increasingly explored (as predictive maintenance and inspection robotics are discussed by many authors); thus, a comprehensive literature review is provided. It has been divided into several paragraphs, as a few perspectives need to be mentioned. Then, an original procedure for overheated idler detection is proposed. Next, we describe the experimental trials and data acquired by the inspection robot in the real environment, and finally, the results of the proposed methodology applied to real data are presented and discussed.

## 2. Literature Review

### 2.1. Application of IR Thermography for Diagnosing Industrial Infrastructures

Critical infrastructures are almost always equipped with many sensors and supervisory systems. However, in some situations, as is considered in this paper, CM systems can be applied on a limited scale only. Drive units (engine, gearbox, etc.) are usually monitored by supervisory control and data acquisition (SCADA) [14] but the rest of the conveyors (for example, a typical conveyor is 1 km length), namely the moving belt, hundreds of rotating idlers, etc. are difficult to monitor by stationary installations and need to be inspected by maintenance staff [15]. Unfortunately, the mining environment is very harsh for that reason; therefore, there is a general tendency to minimize the presence of humans and atomization of inspection processes by intelligent robots.

IR thermography is categorized as a non-destructive CM technique that can be used for analyzing the temperature patterns in objects based on utilizing IR radiations that are emitted from an object surface [16,17]. The simple analysis of IR images can give us information about the surface temperature of machines, while by further analysis, we can find possible thermal emission abnormalities. Together with extracted features from IR images, the degree of deterioration can be evaluated by analyzing the thermodynamics and physical characteristics of inspected machines [18].

Several types of faults and conditions in rotating machinery such as coupling looseness, rotor imbalance, misalignment, rolling element bearing damage, and lubricant inadequacy can be detected in IR images [2,19–21].

It is worth mentioning that while IR imaging applications in the identification of overheated modules in industrial infrastructures have already been discussed in controlled laboratory environments, robot-based IR imaging methods for the identification of abnormal temperature in real case experiments so far have been rarely discussed and their results rarely presented.

In [22], researchers developed a method for the CM of rotating machinery using IR image processing techniques. The authors discussed the advantages of IR imaging-based machine health monitoring over vibration-based methods. However, in their proposed method, they only used IR images as a reference for the identification of faulty modules; furthermore, they used a stationary IR camera system for conducting their research in a controlled environment. Similarly, in [23], the authors investigated a fault detection method using thermography techniques for identifying air leakages in the pipeline in a laboratory environment.

In [4,12,15], different methods for CM of conveyor systems in mines are discussed. In [4], researchers propose a method for analyzing the thermal state of a belt conveyor in an underground mine. However, the thermal images were captured by a human inspector on a limited scale. Inspired by the same problem, ref. [15] proposed a fault analysis method for the identification of faulty idlers in conveyor systems. However, they conducted their experiments in a controlled environment, and thermal images present a few idlers in high resolution; furthermore, the authors did not propose a solutions for images with complex backgrounds that should be considered in real case scenarios.

Particularly, Dabek et al. [12] suggest an automatic robot-based IR imaging method for the identification of overheated idlers in an open-cast mining site. The authors propose an efficient diagnostic procedure for the detection of overheated idlers, such as defining regions of interest (ROIs) on captured images to reduce the redundant data as well as the color-based segmentation method. For improving the proposed method by Dabek et al., firstly, we discuss a new ROI estimation technique for removing the non-ROIs areas. Furthermore, we propose a histogram-based technique for improving the overall quality of captured extracted IR frames. Therefore, we could accurately track the location of idlers in capturing IR and RGB frames and improve the overall accuracy of segmentation results.

Automated Diagnostic Methods

During the measurements, the inspection robot was able to capture the sequence of IR images without information about the true temperature of the conveyor elements. Due to the automatic scaling of colors to the temperature range, "hard" thresholding based on the predefined value of temperature was not possible.

The hotspot areas in grayscaled IR images can be extracted using an automatic thresholding method where the maximum gray pixel value determines the maximum temperature in the defined region of interest. The problem with general IR image processing pipelines is they do not provide accurate results in identifying the objects in a complex background, as IR images tend to be over-segmented. Some examples of IR image segmentation results using automatic thresholding methods are shown in Figure 1. One can notice that most of the segmented images tend to be over-segmented in comparison to the ground truth image, which leads to some parts of the equipment or components to merged with the background image.

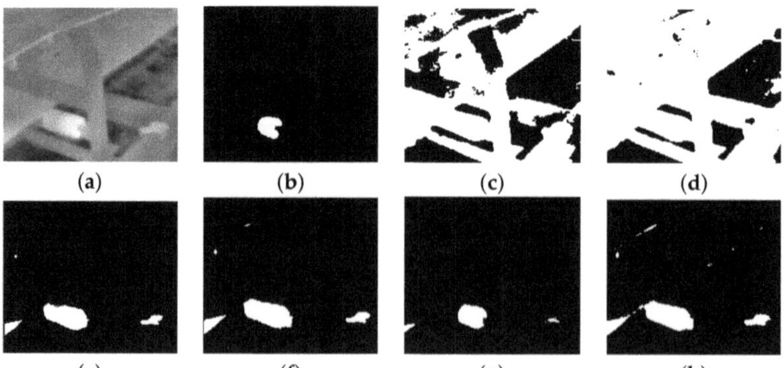

(a) (b) (c) (d)

(e) (f) (g) (h)

**Figure 1.** Example results of infrared image segmentation using thresholding method: (**a**) original infrared image, (**b**) ground truth segmentation, (**c**) Shanbhag [24], (**d**) Otsu [25], (**e**) Intermode [26], (**f**) Triangle [27], (**g**) Yen [28], (**h**) Maximum Entropy [29].

In this paper, our detection strategy was focused on firstly proposing techniques for improving the general characteristics of extracted IR frames histograms and in the next step proposing an outlier-based automatic segmentation method together with shape detection algorithms for the identification of overheated idlers.

The overall quality of IR images can be improved based on histogram enhancement techniques. In [30], the authors propose techniques for the enhancement of the subtle thermal signatures. They suggest noise smoothing by means of, e.g., median or Gaussian filtering, as the most common preprocessing procedures. The application of both median and Gaussian filters to IR images with poor or insufficient information about the health of PV modules is discussed by [31]. The combination of Contrast Limited Adaptive Histogram Equalization (CLAHE) with Gamma correction can be considered an effective way for improving the overall quality of digital images [32,33]. In our work, an automatic gamma correction method together with CLAHE and median filter method has been used for improving the contrast and reducing the noises in captured IR images.

Due to the nature of IR images which are quite different in comparison to visual light images, extracting the hot regions within an IR image is a very challenging task [34]. The distribution of pixel intensities in IR images is based on the heat distribution of an object. Low-intensity contrast and over-centralized intensity distributions in IR images are important factors that bring some difficulties to automatic segmentation methods.

For addressing the mentioned issues in this paper, we exploit the concept of outliers. If in a given IR image, any hot element will appear in the distribution of pixels, the right tail of image histograms related to "hot" colors will be heavier than for "normal temperature" elements. Our target is to detect really hot elements in the conveyor, i.e., significantly higher temperatures than other elements in the picture. In this context, refs. [35–39] proposed IR image histogram analysis techniques for the identification of thermal anomalies in industrial infrastructures.

## 3. Automatic Procedure for Detection of Overheated Idlers in IR Images

### 3.1. General Concept

The aim of this section is to describe the key elements of the methodology. A summarized flowchart of the proposed procedure is presented in Figure 2.

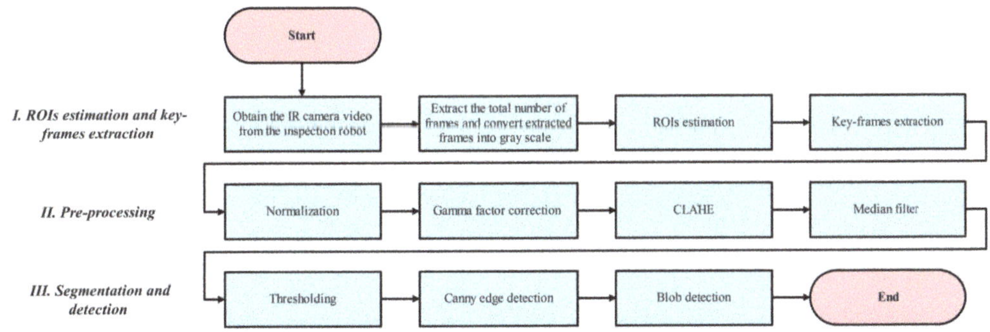


**Figure 2.** Simplified flowcharts of proposed procedure.

The proposed method started by loading the captured data by the inspection robot. The captured data during the experiment were download in a local computer and were further processed offline. Afterward, the total number of captured frames was extracted from loaded IR and RGB videos. The extracted IR frames were converted into 8-bit grayscaled images for further analysis. The camera system captured wide-angle videos from the mining site; therefore, the extracted videos contained many non-informative objects that were not related to the conveyor system. For reducing the number of non-informative objects in the captured frames, ROIs were defined on IR and RGB image data sets.

During the measurements, several data acquisition sessions were performed. It is worthwhile to notice that the environmental conditions are time-varying (even if it is in a kind of indoor condition). A critical issue is that during the experiment, the true temperature of the conveyor element automatically adjusts the colors to a given temperature range producing many complicated images. By additionally assuming the linear relation between the intensity of the brightest pixel (hottest area) to the darkest pixel (coldest area), we could apply statistical methods to segment overheated idlers from the background in grayscaled IR images.

To analyze and assess the quality of idlers in conveyor systems with respect to thermal defects, we first detect the overheated idler modules that have significantly higher temperatures in comparison to other modules. The overheated idlers appeared in the ROI as areas with an average temperature higher than their surroundings. We follow a statistical, data-driven approach that consists of the following steps: (1) normalization (2) correction and refinement, (3) thresholding, (4) canny edge detection implementation, and finally (5) blob detection algorithm for detecting the overheated idlers. The proposed pipeline was developed in the Python language using the OpenCV library for computer vision algorithms.

### 3.2. IR Image Histogram Analysis

A histogram of IR images acts as a graphical representation of the color or intensity distribution of pixels. For a gray-level image, the intensity value of pixels refers to discrete temperature values. The statistical analysis of an IR image histogram is a practical way of indicating anomalies in temperature patterns in captured scenes. The mean value, variance, and standard deviation are the statistical-based features that describe the intensity distribution of pixels in IR images. For a grayscaled IR image, the first-order histogram probability P(g) is defined as follows [40]:

$$P(g) = \frac{L(g)}{M} \tag{1}$$

where $M$ is the total number of pixels and $L(g)$ describes the number of gray levels $g$. In gray-level images, the total number of the intensity level of pixel L spans into [0, 256]. As the tonal distribution represents the thermal distribution in captured scenes, gray-level infrared images can be processed based on tonal intensities. The general brightness of images is defined by the mean value:

$$\bar{g} = \sum_{g=0}^{L-1} g \cdot P(g) \tag{2}$$

Furthermore, the dispersion of a set of data points around their mean value is defined by variance and is given by the following equation:

$$\sigma_g^2 = \sum_{g=0}^{L-1} (g - \bar{g})^2 \cdot P(g) \tag{3}$$

The standard deviation or the square root of the variance has described the spread in IR image data and can give us information about the contrast of IR images. As temperature distribution is a key index of possible defects, the standard deviation can be considered as an important factor for identifying overheated idlers.

### 3.3. Adaptive Region of Interest Estimation

The original captured videos contain both information on the target and redundant areas; therefore, it would be an advantage if non-ROI regions can be removed before precise (final) analysis. This reduction has to be performed in a way such that no idlers data are lost while the computational burden is reduced. The ROI analysis is defined by a set of techniques that can be used for selecting areas of an image from which the individual or average pixel values are extracted for further analysis. The ROI can be defined manually or by automated methods. The first one is faster but less precise, whereas the second method is more time-consuming but in general more accurate.

The purpose of ROI generation is to automatically extract regions of interest (ROI) from the extracted IR frames that only contain pixels that are related to idlers. During the examination, for capturing RGB and IR videos, the inspection mobile robot was navigated through the free spaces between belt conveyors. The camera system's point of view (POV) was fixed and pointed toward the belt conveyor. As long as the mobile robot followed a straight line alongside the conveyor belt, the changes in the camera system POV were neglectable.

There are many different methods for tracking objects in the sequence of the frame. Since the camera system POV changes during the inspection were neglectable, we could accurately estimate the approximate location of the idler in the sequence of extracted frames, as shown in Figure 3. By the consideration of small changes in the camera system POV, a rectangular region of the predetermined size, 400 × 600 pixels surrounding the estimated location of idlers with fixed positions, was defined on extracted IR and RGB frames. The size of the defined ROI was large enough to only capture the idler, while it was considerably smaller than the original captured frames; therefore, we could effectively reduce the computational burden.

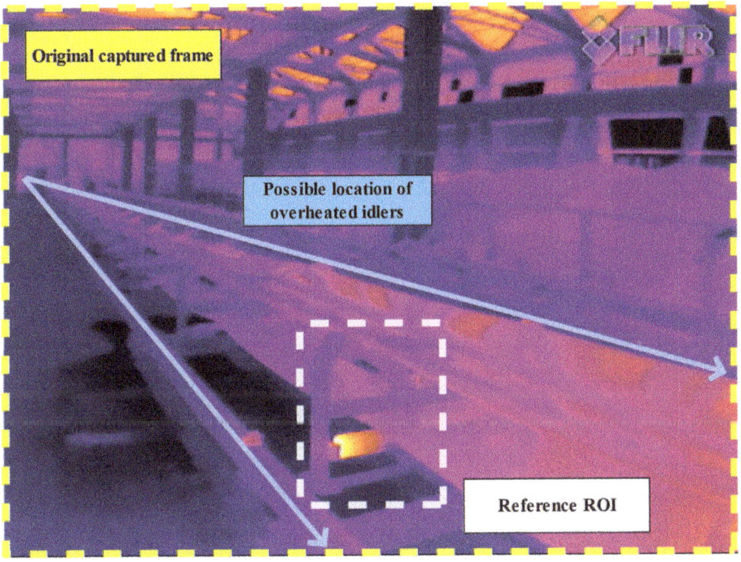


**Figure 3.** Selection of ROI on the captured IR images.

### 3.4. The Key Frame Extraction Method

Through the conducted experiment, the inspection mobile robot captured continuous thermal videos from different conveyor systems. Using every individual frame for identification of the overheated idlers is unnecessary, as many frames are almost repeated in a certain time interval. Therefore, in our research, we chose a key frame extraction method to summarize and reduce the size of the extracted IR frames.

Mean Square Error (MSE) is one of the two error metrics which can be used for calculating the cumulative squared error between the reference image and the target image. In this paper, the global similarity between each reference and target frame in relation to their pixel intensities has been measured by the MSE method, where frame $f_i$ and $f_{i-1}$ are the target and reference frames, and $m$ and $n$ are coordination of each pixel in compared frames [41].

$$\text{MSE} = \frac{1}{MN}\sum_{n=0}^{M}\sum_{m=1}^{N}[f_{i-1}(n,m) - f_i(n,m)]^2 \quad (4)$$

For calculating the degree of similarity between each of the two IR frames, firstly, the reference and target $f_i$ and $f_{i-1}$ are loaded. Afterward, histograms of both frames are taken, and the mean square difference between two histograms is calculated. Through the examination, the parameter MSE $\geq 5$ turned out to be sufficient and considered as a threshold for the selection of the keyframes; therefore, when the difference between compared frames is greater than the threshold, the target frame is declared as the next keyframe. This process is repeated until there are no frames left for processing.

### 3.5. Normalization

The constant changes in color scheme adjustments by the camera can only affect the calibration factor regarding absolute temperature values but not affect the distribution pattern of the temperature. As long as pixel intensity was predefined by the camera with respect to the hottest and coldest point in original frames, we need to normalize the intensity value of a pixel with respect to the brightest and darkest pixel in defined ROIs.

The intensity value of infrared images was normalized to a constant range with respect to pixel intensity distribution in defined ROI. Normalizing the temperature pattern allows us to define a set of parameters that works well for analyzing ROIs with varying temperature ranges. The normalization of ROIs with different seasons can rescale the radiant temperature to the same level between the hottest and lowest in defined ROIs and thus reduce the seasonal difference. Accordingly, the extracted ROIs were normalized using the following equation:

$$N_i = \frac{TS_i - TS_{\text{min}}}{TS_{\text{max}} - TS_{\text{min}}} \tag{5}$$

In Equation (5), $N_i$ is the normalized value of pixel i, where $TS_i$ is the intensity value of pixel i. Furthermore, $TS_{max}$ and $TS_{min}$ can be defined as the maximum and minimum values of pixel intensity in a ROI.

### 3.6. Correction and Refinement

The automatic thresholding method cannot correctly exclude background regions from the foreground due to their high variance. Therefore, some data normalization steps are usually required before thresholding. To avoid the loss of subtleties in the extracted ROIs, sets of pre-processing techniques were used for improving the general characteristics of frames. The correction and refinement incorporate CLAHE, and Gamma correction was followed by median filtering.

#### 3.6.1. CLAHE Method

Histograms in normalized ROIs would be skewed toward the lower end of the grayscale, and all the image detail can be compressed into the dark end of the IR image histogram. For addressing this issue, histogram-based methods can be used to improve image quality and adjust the contrast.

Histogram equalization (HE) is a simple method for enhancing the contrast of the image by spreading out the intensity range of the image or stretching out the most frequent intensity value of the image. Stretching the intensity values changes the natural brightness of the input image and introduces some undesirable noises [42]. To improve the HE method, Adaptive Histogram Equalization (AHE) [43] was proposed. In the AHE method, the input image is split into smaller images, which are called tiles. The noise, however, often increases when the histogram slope is steep.

CLAHE is an effective contrast enhancement method that effectively enhances the contrast of the image. CLAHE is an improved version of the AHE method that works precisely in the same way, but it clips the histogram at specific values for limiting the amplification before computing the cumulative distributive function (CDF). This change reduces the noise because clipping prevents a CDF from being steep. The computation of CLAHE is performed as:

$$p = (p_{\max} - p_{\min}) * P(f) + p_{\min} \tag{6}$$

where $p$ represents the pixel value after applying CLAHE, $p_{\max}$ and $p_{\min}$ represent the maximum and minimum pixel value of an image, respectively, and $P(f)$ represents the cumulative probability distribution function [44].

#### 3.6.2. Gamma Correction

Gamma correction can be used to control the overall brightness of images. It is responsible for performing a nonlinear calculation of the pixels intensity of the input image and thereby adjusting the saturation of the image. It is necessary to determine the optimal gamma value; therefore, it should neither be too minimum nor maximum. For enhancing the contrast of the ROIs, an adaptive gamma correction (AGC) technique was applied to the image data set. For having uniform distribution in an image histogram, the optimal value for gamma factor by consideration of $\bar{I} = 0.5(L-1)$ as mean intensity can be defined as follows [45]:

$$\gamma = \frac{\log\left(\frac{I}{(L-1)}\right)}{\log(0.5)} \tag{7}$$

#### 3.6.3. Median Filtering

In extracted frames where sun reflections were captured on the belt, some unwanted variations were observed within the ROIs. These variations appear in an area with the presence of sun reflection on the belt surface in the form of bright spots that can be wrongly segmented as hot areas. Obviously, it is not possible to exclude them through the ROI estimation process, as they appear at the center of the camera POV. A possible solution to this obstacle is to apply median filtering.

The normalized ROIs after double enhancement undergo median filtering to prepare the images before the thresholding process. The median filter is a nonlinear digital filtering method and is employed to eliminate salt-and-pepper noises in pre-processed ROIs. The median filter can reduce the noise without diminishing the sharpness of the image.

Figure 4 shows the original ROI (on the left) with the unwanted variations and the same ROI after being modified through the pre-processing stages (on the right).

### 3.7. *Thresholding*

Overheating in idlers can be recognized as a hotspot in certain areas of ROIs. In IR images with uniform backgrounds, the number of pixels belonging to background or cold areas is much larger than the number of pixels belonging to foreground or overheated objects. We know that an overheated idler's surface always looks relatively brighter than the background in captured IR images. For distinguishing pixels that are related to the background (cold pixels) and pixels that are related to the foreground (overheated idlers), we classify them based on their distance to the mean value. Our automatic segmentation method worked a base on the outlier detection technique. The main advantage of the proposed method is that we can accurately detect abnormal pixels that lie far away from other observation values. Therefore, in ROIs containing a very complex background and low signal-to-noise ratio (SNR), we can precisely find abnormal pixels and prevent the results to be over or under-segmented.

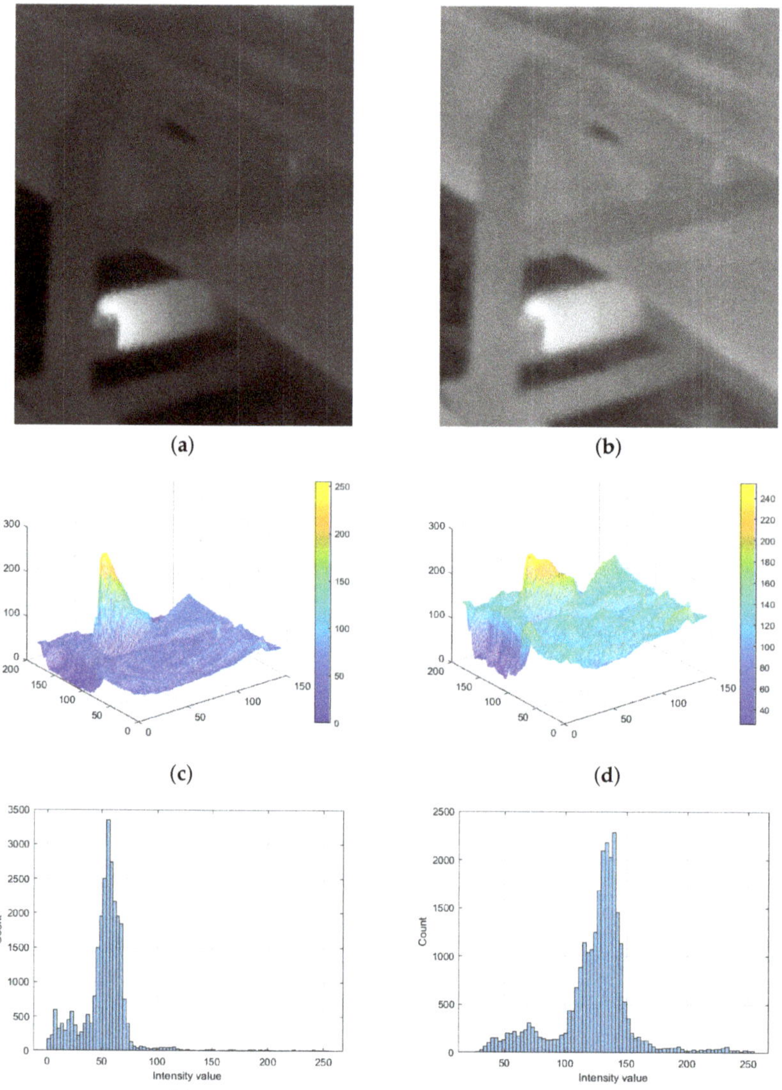


**Figure 4.** Comparison of pre-processed and orginal IR image after modification through pre-processing stages. (**a**) Original ROI, (**b**) Pre-processed ROI, (**c**) Three-dimensional (3D)-map of original ROI, (**d**) Three-dimensional (3D)-map of pre-processed ROI, (**e**) Histogram of orginal ROI, (**f**) Histogram of Pre-processed ROI.

Outlier detection is a problem of finding patterns in data that are not in the range of normal behavior. In this paper, hot spots in IR images are considered anomalous patterns and treated as outliers. An outlier will thus indicate a temperature abnormality. To identify defective idlers, we apply the IQR method to the extracted histogram features $\mathcal{F}_i := \{f_1, f_2, \ldots, f_{255}\}$.

IQR is a technique that helps to find outliers in the data which are continuously distributed. IQR is the difference between the first quartile and the third quartile: $IQR = Q3 - Q1$ where, $Q1$ and $Q3$ can be defined by Equation (8) [46,47].

$$\begin{aligned} Q1 &= \bar{g} - 0.675\sigma; \\ Q3 &= \bar{g} + 0.675\sigma \end{aligned} \tag{8}$$

The following thresholds (referred to as fences) are required to be defined for classifying the outliers in two different classes. The outliers can be defined as values that are either among inner and outer fences (mild outliers) or beyond outer fences (extreme outliers). The lower and upper inner fences can be computed as Q1 − 1.5 IQR and Q3 + 1.5 IQR, while the lower and upper outer fences can be defined as Q1 − 3 IQR and Q3 + 3 IQR, respectively [48,49].

Let $\alpha$ represent the pre-processed ROI and $\beta$ represent the extracted binary image from $\alpha$ and T as the threshold value by the proposed method. Furthermore, $W$ is ROI width and $H$ is image height. Since this captured IR is a digital image, $x$ and $y$ are indenting the coordination of pixels.

$$\begin{aligned} &\beta(x,y) = \begin{cases} 1 & \text{if } \alpha(x,y) > T \\ 0 & \text{if } \alpha(x,y) \leq T \end{cases} \\ &\forall 0 \leq x < W, 0 \leq y < H \end{aligned} \tag{9}$$

Likewise, all the pixel values of pre-processed $\alpha$ are set to 1 when the pixel values are greater than the computed T. On the other hand, the other pixel values are set to 0 when they are less than the defined T. In order to obtain object image $\gamma$ from image $\alpha$ and $\beta$, the following formula is used:

$$\begin{aligned} &\gamma(x,y) = \begin{cases} \alpha(x,y) & \text{if } \beta(x,y) = 1 \\ 0 & \text{if } \beta(x,y) = 0 \end{cases} \\ &\forall 0 \leq x < W, 0 \leq y < H \end{aligned} \tag{10}$$

After the segmentation process, $\gamma$ is the thermal image of the inspected equipment after removing the background.

Through examination, we found out that overheated idlers cannot always be defined as mild outliers. Accordingly, for increasing the chance of true detection, the value of extreme outliers is extracted for each frame and considered as an optimal threshold value for segmenting the overheated idlers Figure 5.

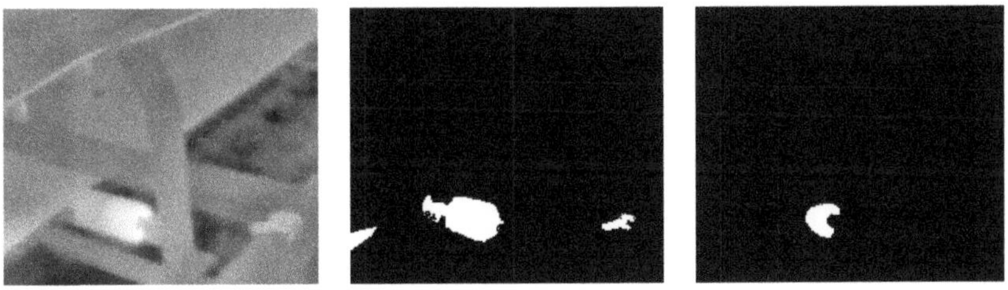

(a) (b) (c)

**Figure 5.** Comparison of extreme and mild outliers in segmentation of an overheated idlers. (**a**) Pre-processed ROI before segmentation, (**b**) Segmentation results of mild outliers detection, (**c**) Segmentation results of extreme otliers detection.

### 3.8. Canny Edge Detection Implementation

In the previous section, we could accurately partition the input image into homogeneous hotspots and backgrounds using an adaptive thresholding method. The post-segmentation processing algorithms including canny edge detection and blob detection are designed to differentiate the hotspots edges from each. Our objective is to extract the boundary of the segmented hotspots and classify them as separate sources of heat.

The aim of performing edge detection, in general, is to significantly reduce the amount of data in an image, which will increase the computation speed of the approach while preserving the structural properties of the image [50–52]. The correctness of detected edges is an important factor that affects the blob extraction step; therefore, both the edge detection and blob extraction stages are tightly connected. To find the shape and size of the extracted hot spots, firstly, we used the canny edge detection method for detecting the boundaries of the segmented hotspots.

The canny edge detector is the most common method for detecting a wide range of edges in images. It uses a multi-stage algorithm consisting of five separate steps: smoothing, gradient finding, non-maximum suppression, double thresholding, and edge tracking by hysteresis for detecting boundaries in images. It determines the spots in images more accurately than other operators. It convolves the segmented frames with a Gaussian filter for reducing the noises and then computes the gradient and gradient direction for each possible edge. Furthermore, the detected image gradients undergo double thresholding to remove edge-like noise [53].

### 3.9. Blob Detection

Now, because all boundaries of the segmented hotspots are detached, they can be detected and counted using techniques such as blob detection. A blob can be defined as a region inside the calculated boundaries in which the pixels are considered to be similar to each other, while they should be different from the surrounding neighborhoods. Blobs are defined as interest points or interest regions. The interest points are referred to as local extremes in scale-location spaces, which will indicate circular or square regions.

The blob detection method is commonly used in many applications that are related to measuring the object's shape, location, diameter, etc. For providing complementary information about the number of hot spots which are not obtained from edge detectors, the blob detection algorithm is applied to the canny edge detection results.

We used a blob detector method based on the Laplacian of the Gaussian (LoG). Therefore, an image is convolved by a Gaussian kernel. Furthermore, a multi-scale blob detector with automatic scale selection is then computed using a scale normalized Laplacian operator (see Figure 6). After labeling circular shapes, we could detect and count the different overheated elements in ROIs [54,55].

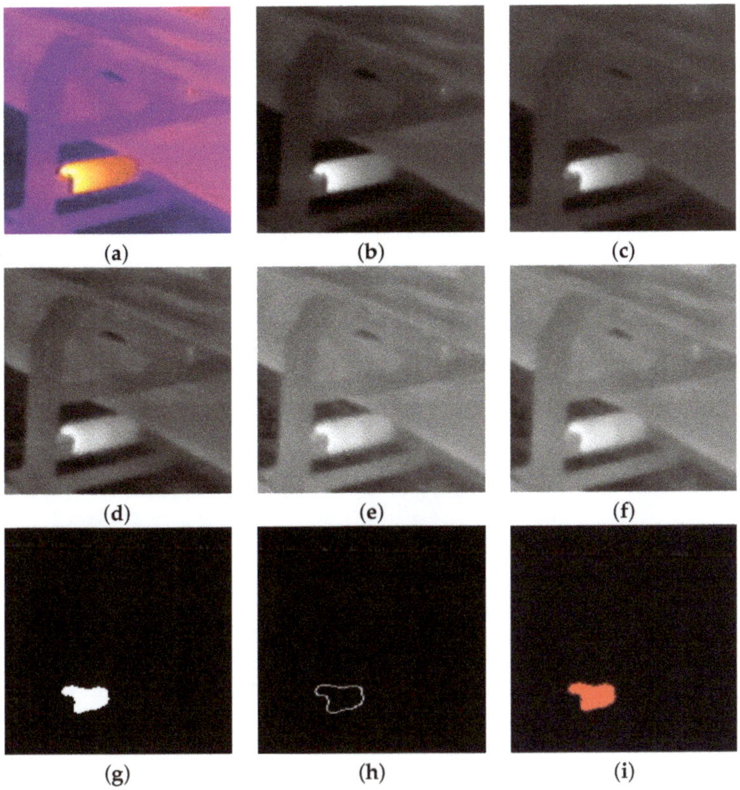

**Figure 6.** Intermediate and final results of presented overheated idlers module detection approach. (**a**) Original IR image, (**b**) Grayscaled, (**c**) Normalization, (**d**) CLAHE, (**e**) Gamma correction, (**f**) Median filter, (**g**) Thresholding, (**h**) Canny edge detection, (**i**) Blob detection algorithm.

*3.10. Performance Metrics*

Six performance metrics—Sensitivity, Specificity, Precision, Accuracy and Matthew's correlation coefficient (MCC)—were used as evaluation metrics where TP, FP, TN, and FN are true positive, false positive, true negative, and false negative, respectively.

$$\text{Sensitivity} = \frac{(TP)}{(TP + FN)} \tag{11}$$

$$\text{Specificity} = \frac{(TN)}{(TN + FP)} \tag{12}$$

$$\text{Precision} = \frac{(TP)}{(TP + FP)} \tag{13}$$

$$\text{Accuracy} = \frac{(TP + TN)}{(TP + FN) + (FP + TN)} \tag{14}$$

$$\text{MCC} = \frac{\text{TP} \times \text{TN} - \text{FP} \times \text{FN}}{\sqrt{(\text{TP} + \text{FP})(\text{TP} + \text{FN})(\text{TN} + \text{FP})(\text{TN} + \text{FN})}} \tag{15}$$

In this paper, true positives refer to the frames where overheated idlers were correctly segmented. On other hand, false positive represents the number of frames where other thermal sources were wrongly segmented as overheated idlers. Furthermore, true negative cases are referred to as the frames where no overheated idler was neither present nor detected. In contrast, a false negative can be described as the number of frames where faulty idlers were not segmented due to the underestimation.

## 4. Experiments and Data Description

The experiments were carried out by capturing data from a conveyor system located in an opencast mine close to the bunker where transported material was stored.The mining site is located in Jaroszów, 50 km to the west from Wrocław. The basis of our measurement system is a remote-controlled mobile robot with an extensive navigation system figure. The robot is custom built for the Wrocław University of Science and Technology as a universal mobile platform for inspections, as shown in Figure 7 with main features explained in Table 1. During the inspection mission, the robotic platform captured various types of data, including RGB images, IR images, sound, LiDAR data, etc. The camera system with a fixed point of view was directed toward the conveyor to better cover the ROI. The main specifications are mentioned as follows:

**Figure 7.** View of the robot during inspection.

**Table 1.** Inspection robot main characteristics.

| | |
|---|---|
| Locomotion type | Wheeled, skid steering |
| Navigation system | Autonomous (Internal computer) |
| | Manual (Pilot using remote computer conection) |
| Internal software | Robot Operating System (ROS) |
| Power system | Internal battery, 24 V |
| Robot gross weight | 65 kg |
| Maximum payload capacity | 75 kg |

The analyzed conveyor system in this paper is a mechanical system used for the continuous horizontal transport of raw materials including raw clays, milled clays, and chamotte from the mine pit to the bunker. The considered conveyor was the last section that ends the entire series of conveyors. The analyzed section of the conveyor system always runs without the material, because the material is dumped into appropriate silos just before this section. The belt itself carries the material, which is important for thermal reasons, but the idlers do not experience the additional weight of the material.

The investigated conveyor system was several hundred meters long, and it operates in harsh environmental conditions. The design of the conveyor is classical, as shown in

Figure 8. The key problem was to identify overheated idlers as a potential source of the fire. It is worth noting that inspection had not started at the very beginning of the belt conveyor.

**Figure 8.** A general picture of the raw materials storage with belt conveyor to transport raw materials.

*Data Description*

Based on the manual analysis of the acquired data (additional information in Table 2), we have selected several interesting situations that could be problematic during automatic image analysis. Below, we present some of such "difficult to analyze" pictures. In general, one may group them into several classes, namely: images without a hot idler, images with a hot idler, images with sunlight reflection without a hot idler, images with sunlight reflection with a hot idler, etc., see the examples presented below Figure 9.

**Table 2.** (**a**) The camera system parameter, (**b**) Belt conveyor parameters.

| (a) | |
|---|---|
| **Parameter** | **Value** |
| Resolution | 640 × 480 pixles |
| Frames per second | 25 fps |
| Observation angle | 45° |
| Mounting height | 100 cm above shelf |
| **(b)** | |
| **Parameter** | **Value** |
| Conveyor length | 150 m |
| Idler diameter | 133 mm |
| Idler spacing | 1.45 m |
| Belt width | 800 mm |

Despite the advantage of IR imaging methods, there are still different factors that need to be considered even when conducting an indoor inspection. Generally, the precision of a thermographic measurement is directly related to specific background parameters, i.e., the environmental conditions, the optical properties of the target material, and the possible presence of any nearby object. Objects with a high emissivity value such as greasy, black or reflective objects have quite a high emissivity value, typically as high as 0.97; therefore, they can strongly reflect the IR radiations [56]. In Figure 10, one can notice that in the raw IR images, there are many non-informative heating sources that are not related to the conveyor system: for example, windows or sunlight reflection on the belt (marked by arrows).

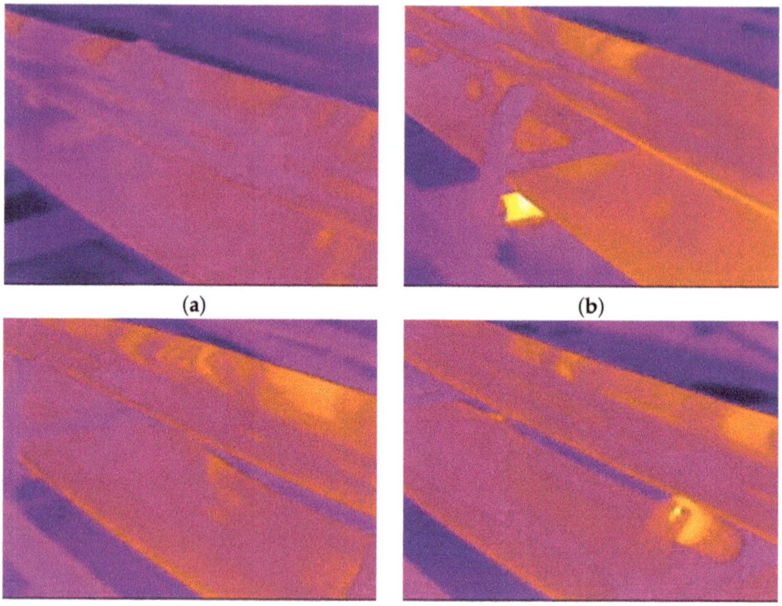

(a) (b) (c) (d)

**Figure 9.** Four classes of images that were difficult to analyze. (**a**) Image without hot idler, (**b**) Image with hot idler, (**c**) Image with sunlight reflections, (**d**) Image with sunlight reflections and a hot idler.

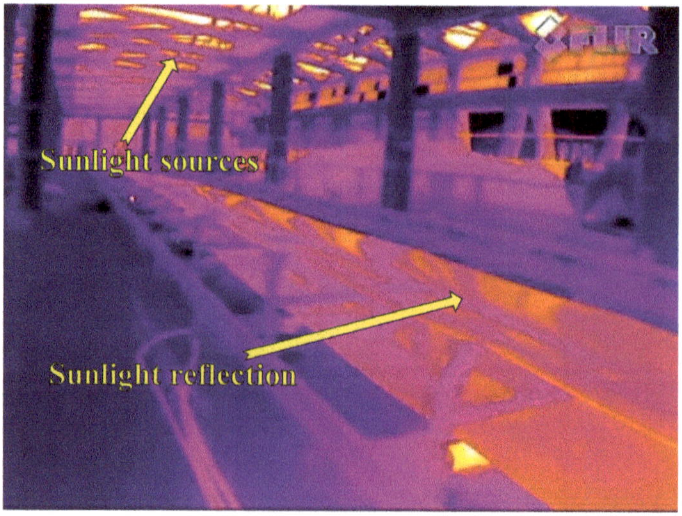


**Figure 10.** Location of sunlight sources and sun reflections on belt that captured in a raw IR image.

The experiment was conducted on a sunny day. In a real scenario, it is impossible to predict what the weather will be like, and since the measurement session was organized in a real mine ahead of time, there was no way of predicting the weather several weeks ahead. However, we managed to provide a solution that can be applied in every condition.

Solar radiation can heat the equipment, especially those with high absorption of the sun's energy, which can make small thermal differences. In our case (inspecting the idlers), the solar radiation was mostly blocked by the celling; therefore, their effects on the surface temperature of idlers were neglectable. However, one can notice that the belt surface is black, smooth, and shiny, which makes a very probable sunlight reflection problem. Therefore, the IR camera measured the reflected temperature instead of measuring the temperature of the belt itself. This will make image processing difficult, and this is the main reason to limit the analyzed area to conveyor-related only (defining the ROIs).

In Figure 11, one can see another example of hot spots (marked by frames) not related to idlers. It shows that the application of real industrial data is always difficult due to unpredictable sources of noise/unwanted components.

As result, the methodology will provide a false detection of sunlight recognized as a hot area. Examples presented here with a detailed discussion on the detection efficiency and understanding of the source of the problem are necessary before the automatic processing of hundreds of images. In the next sections, we will discuss global efficiency with some indicators of detection quality. Even if there are some problematic examples that are hard to recognize, other known techniques appear much less efficient. Moreover, the proposed technique is automatic and provides results in an objective way that is better than the subjective opinion performed by experts.

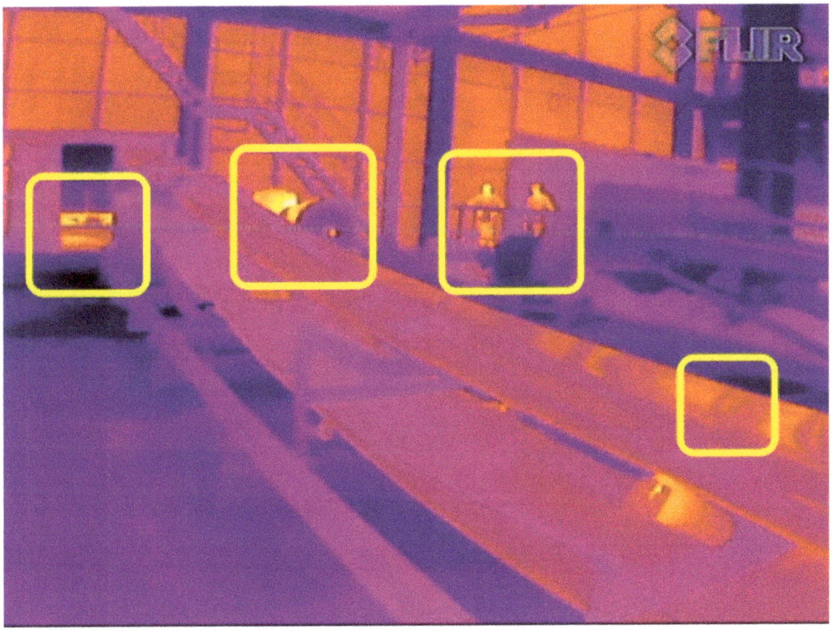


**Figure 11.** Examples of hotspots that are not related to idlers.

## 5. Results and Data Validation

The performance of the method in the identification of overheated idlers was tested on image data sets that were captured from a conveyor systems. The inspection robot moved alongside the conveyor system two times (back and forth) and captured data. Initially, 6275 frames from data set 1 (moving forward) and 6135 frames from data set 2 (moving backward) were extracted from captured RGB and IR videos. After applying the keyframe extraction, we could reduce the data set size up to 40%; therefore, 2470 frames from data set 1 and 2205 frames from data set 2 were chosen for further analysis. Furthermore, we compared the performance of the proposed method in identification of overheated idlers with Maximum entropy, Yen and Minimum method [26].

### 5.1. Validation of Detection Results Based on Manual Analysis

The test images were hand labeled. The evaluation was performed by comparing the final segmentation mask through the visual interpretation process. By consideration of the number of thermal sources that can be detected as an overheated idler alongside the conveyor, validation of the segmentation results is important, as other thermal sources can be wrongly segmented as overheated idlers. Therefore, through the validation process, the shape and location of the segmented hotspots in segmented frames were compared to RGB images, as shown in Figure 12.

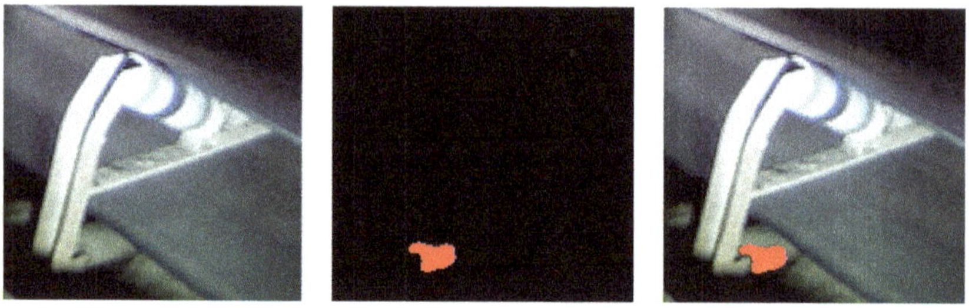

(a) (b) (c)

**Figure 12.** Fusion of thermal and RGB images for validaiton and localization of detected hotspots. (**a**) RGB image, (**b**) The blob detection, (**c**) RGB and blob detecion fusion results.

### 5.2. Results

The specificity and accuracy value for both data sets were about 0.98. The specificity value indicates the performance of the proposed method regarding identifying true positives and true negatives, while the accuracy value discussed the proportion of correctly predicted samples among the total number of the processed samples. The precision value determines the ratio of correctly predicted positive observations to the total predicted positive observations, which were above 0.66 for both data sets.

We additionally computed the F1-score for both data sets (see Tables 3 and 4). The F1-score combines the precision and recall of a classifier into a single metric by taking their harmonic mean and measure of accuracy incorporating both the precision and recall. It can be used as a single performance test for positive classifications, which were 0.76 for data set 1 and 0.78 for data set 2. The F1-score of the proposed method was 80% higher than the compared methods, which indicates the low performance of other thresholding methods in the true detection of overheated idlers in the studied data sets.

**Table 3.** Comparison of the performance factors of the proposed method with other selected methods—data set 1.

| Measures | Proposed Method | Maximum Entropy | Yen | Minimum Method |
|---|---|---|---|---|
| Sensitivity | 1 | 1 | 1 | 1 |
| Specificity | 0.98 | 0 | 0 | 0 |
| Precision | 0.62 | 0.05 | 0.03 | 0.03 |
| Accuracy | 0.98 | 0.05 | 0.03 | 0.03 |
| F1-score | 0.76 | 0.1 | 0.05 | 0.06 |

Table 4. Comparison of the performance factors of proposed method with other selected method—data set 2.

| Measures | Proposed Method | Maximum Entropy | Yen | Minimum Method |
|---|---|---|---|---|
| Sensitivity | 0.94 | 1 | 1 | 1 |
| Specificity | 0.98 | 0 | 0 | 0 |
| Precision | 0.66 | 0.05 | 0.05 | 0.01 |
| Accuracy | 0.98 | 0.05 | 0.05 | 0.01 |
| F1-score | 0.78 | 0.1 | 0.1 | 0.03 |

## 6. Concluding Remarks

Analyzing the intensity distribution of pixels from the acquired IR images by means of ROI analysis, canny edge detection, blob detection methods and the fusion of segmented IR and RGB frames showed an evident correlation between abnormalities in pixel intensity (temperature patterns) and the existence of hotspots on the surface of the faulty idlers. Furthermore, the proposed diagnostic approach gave promising results, succeeding to diagnose four out of four defective idlers in the studied conveyor system.

In conclusion, robotic-based IR thermography with the combination of thermal image processing techniques was proved to be a potential and reliable method for CM and fault diagnosis of idler modules. The presented approach gave easily interpreted results and fast detection of faulty idlers, utilizing both qualitative and quantitative data from the processed thermal images from the two conveyor systems.

The most important limitation of the proposed method is the fact that any specular object present in the background could cause unwanted gray-level variations that may be conflicting with the actual variations related to hotspots that may cause false alarms. In order to reduce unwanted gray-level variations, sets of pre-proposing algorithms, including CLAHE, adaptive gamma correction, and median filter, were applied to ROIs.

In grayscaled IR images, the changes in temperature are indicated with changes in pixel intensity; therefore, they depict some degree of smoothness during the traversal from one pixel to another, and there is a lack of sudden and sharp change. Hence, most of the conventional automatic methods have different performances in the automatic thresholding of IR images as compared to their performances with other types of images.

There are further limitations referring to emissivity uncertainties, the presence of sunlight reflection on the belt surface of the conveyor systems, and the presence of other heat sources in mining sites that have to be always taken into account for field measurements.

Additional investigation in improving the detection results can be completed based on the definition of adaptive ROI on extracted IR and RGB frames. Through the conducted examination, we found out that in some cases, as long as the mobile robot was moved through the harsh surfaces, obstacle avoidance maneuvers and sudden changes in the robot paths were necessary. The camera system position was fixed during the inspection; therefore, the camera's point of view (POV) was changed due to sudden modifications in the mobile robot path. For addressing sudden changes in camera POV and improving the detection results, a mapping technique that fuses the camera system with the LiDAR data for estimating the optimal region of interest on captured videos is of further interest to the current research team.

Characterizing idler defects, e.g., bearing defects, can be completed based on a micro-scale analysis combining the current experience with active thermography approaches or with the fusion of acoustic and thermal imaging approaches for a more accurate CM plan. In addition, a more thorough understanding of degradation mechanisms and failure modes in idler modules is critical to improvements in idler design and reliability. Performing robotic-based IR imaging in large-scale conveyor systems with new fusion techniques can be considered in future research.

**Author Contributions:** Conceptualization, R.Z.; methodology, M.S.; software, M.S.; validation, R.Z., J.W. and T.B.; formal analysis, M.S.; investigation, M.S. and J.W.; resources, R.Z. and J.W.; data curation, J.W. and M.S.; writing—original draft preparation, M.S.; writing—review and editing, M.S., J.W., R.Z. and T.B.; visualization, M.S.; supervision, R.Z. and T.B.; project administration, R.Z.; funding acquisition, R.Z. All authors have read and agreed to the published version of the manuscript.

**Funding:** Part of this work was supported by the European Commission via the Marie Sklodowska Curie program through the ETN MOIRA project (GA 955681)—Mohammad Siami. This activity has received funding from the European Institute of Innovation and Technology (EIT), a body of the European Union, under the Horizon 2020, the EU Framework Programme for Research and Innovation. This work is supported by EIT RawMaterials GmbH under Framework Partnership Agreement No. 19018 (AMICOS. Autonomous Monitoring and Control System for Mining Plants). Scientific work was published within the framework of an international project co-financed from the funds of the program of the Minister of Science and Higher Education titled "PMW" 2020-2021; contract no. 5163/KAVA/2020/2021/2.

**Data Availability Statement:** Archived data sets cannot be accessed publicly according to the NDA agreement signed by the authors.

**Acknowledgments:** The authors (M. Siami) gratefully acknowledge the European Commission for its support of the Marie Sklodowska Curie program through the ETN MOIRA project (GA 955681). Support was also provided by the Foundation for Polish Science (FNP)—Jacek Wodecki.

**Conflicts of Interest:** The authors declare no conflict of interest.

## References

1. Uth, F.; Polnik, B.; Kurpiel, W.; Kriegsch, P.; Baltes, R.; Clausen, E. An innovative person detection system based on thermal imaging cameras dedicate for underground belt conveyors. *Min. Sci.* **2019**, *26*, 263–276. [CrossRef]
2. Błazej, R.; Sawicki, M.; Kirjanów, A.; Kozłowski, T.; Konieczna, M. Automatic analysis of thermograms as a means for estimating technical of a gear system. *Diagnostyka* **2016**, *17*, 43–48.
3. Kozłowski, T.; Wodecki, J.; Zimroz, R.; Błazej, R.; Hardygóra, M. A diagnostics of conveyor belt splices. *Appl. Sci.* **2020**, *10*, 6259. [CrossRef]
4. Szurgacz, D.; Zhironkin, S.; Vöth, S.; Pokorný, J.; Sam Spearing, A.; Cehlár, M.; Stempniak, M.; Sobik, L. Thermal imaging study to determine the operational condition of a conveyor belt drive system structure. *Energies* **2021**, *14*, 3258. [CrossRef]
5. Obuchowski, J.; Wylomańska, A.; Zimroz, R. Recent developments in vibration based diagnostics of gear and bearings used in belt conveyors. *Appl. Mech. Mater.* **2014**, *683*, 171–176. [CrossRef]
6. Doroszuk, B.; Krol, R. Analysis of conveyor belt wear caused by material acceleration in transfer stations. *Min. Sci.* **2019**, *26*, 189–201. [CrossRef]
7. Król, R. Studies of the durability of belt conveyor idlers with working loads taken into account. *IOP Conf. Ser. Earth Environ. Sci* **2017**, *95*, 042054. [CrossRef]
8. Gładysiewicz, L.; Król, R.; Kisielewski, W. Measurements of loads on belt conveyor idlers operated in real conditions. *Meas. J. Int. Meas. Confed.* **2019**, *134*, 336–344. [CrossRef]
9. Król, R.; Kisielewski, W. Research of loading carrying idlers used in belt conveyor-practical applications. *Diagnostyka* **2014**, *15*, 67–74.
10. Zimroz, R.; Hardygóra, M.; Blazej, R. Maintenance of Belt Conveyor Systems in Poland—An Overview. In *Proceedings of the 12th International Symposium Continuous Surface Mining—Aachen 2014*; Niemann-Delius, C., Ed.; Springer International Publishing: Cham, Switzerland, 2015; pp. 21–30.
11. Nascimento, R.e.a. An integrated inspection system for belt conveyor rollers advancing in an enterprise architecture. In Proceedings of the ICEIS 2017—19th International Conference on Enterprise Information Systems, Porto, Portugal, 26–29 April 2017; Volume 2, pp. 190–200. [CrossRef]
12. Dabek, P.; Szrek, J.; Zimroz, R.; Wodecki, J. An Automatic Procedure for Overheated Idler Detection in Belt Conveyors Using Fusion of Infrared and RGB Images Acquired during UGV Robot Inspection. *Energies* **2022**, *15*, 601. [CrossRef]
13. Szrek, J.; Zimroz, R.; Wodecki, J.; Michalak, A.; Góralczyk, M.; Worsa-Kozak, M. Application of the infrared thermography and unmanned ground vehicle for rescue action support in underground mine—The amicos project. *Remote Sens.* **2021**, *13*, 69. [CrossRef]
14. Sawicki, M.; Zimroz, R.; Wyłomańsk, A.; Obuchowski, J.; Stefaniak, P.; Żak, G. An Automatic Procedure for Multidimensional Temperature Signal Analysis of a SCADA System with Application to Belt Conveyor Components. *Procedia Earth Planet. Sci.* **2015**, *15*, 781–790. [CrossRef]
15. Liu, Y.; Miao, C.; Li, X.; Ji, J.; Meng, D. Research on the fault analysis method of belt conveyor idlers based on sound and thermal infrared image features. *Measurement* **2021**, *186*, 110177. [CrossRef]

16. Qu, Z.; Jiang, P.; Zhang, W. Development and Application of Infrared Thermography Non-Destructive Testing Techniques. *Sensors* **2020**, *20*, 3851. [CrossRef] [PubMed]
17. Abdel-Qader, I.; Yohali, S.; Abudayyeh, O.; Yehia, S. Segmentation of thermal images for non-destructive evaluation of bridge decks. *NDT E Int.* **2008**, *41*, 395–405. [CrossRef]
18. Bagavathiappan, S.; Lahiri, B.; Saravanan, T.; Philip, J.; Jayakumar, T. Infrared thermography for condition monitoring—A review. *Infrared Phys. Technol.* **2013**, *60*, 35–55. [CrossRef]
19. Carvalho, R.; Nascimento, R.; D'Angelo, T.; Delabrida, S.; Bianchi, A.G.C.; Oliveira, R.A.R.; Azpúrua, H.; Uzeda Garcia, L.G. A UAV-Based Framework for Semi-Automated Thermographic Inspection of Belt Conveyors in the Mining Industry. *Sensors* **2020**, *20*, 2243. [CrossRef]
20. Yang, W.; Zhang, X.; Ma, H. An inspection robot using infrared thermography for belt conveyor. In Proceedings of the 2016 13th International Conference on Ubiquitous Robots and Ambient Intelligence (URAI), Xi'an, China, 19–22 August 2016; pp. 400–404.
21. Szrek, J.; Wodecki, J.; Błażej, R.; Zimroz, R. An Inspection Robot for Belt Conveyor Maintenance in Underground Mine—Infrared Thermography for Overheated Idlers Detection. *Appl. Sci.* **2020**, *10*, 4984. [CrossRef]
22. Jia, Z.; Liu, Z.; Vong, C.M.; Pecht, M. A rotating machinery fault diagnosis method based on feature learning of thermal images. *IEEE Access* **2019**, *7*, 12348–12359. [CrossRef]
23. Tong, K.; Wang, Z.; Si, L.; Tan, C.; Li, P. A novel pipeline leak recognition method of mine air compressor based on infrared thermal image using IFA and SVM. *Appl. Sci.* **2020**, *10*, 5991. [CrossRef]
24. Shanbhag, A. Utilization of Information Measure as a Means of Image Thresholding. *CVGIP Graph. Model. Image Process.* **1994**, *56*, 414–419. [CrossRef]
25. Otsu, N. A threshold selection method from gray-level histograms. *IEEE Trans. Syst. Man Cybern.* **1979**, *9*, 62–66. [CrossRef]
26. Prewitt, J.M.S.; Mendelsohn, M.L. The Analysis of Cell Images. *Ann. N. Y. Acad. Sci.* **1966**, *128*, 1035–1053. [CrossRef]
27. Zack, G.W.; Rogers, W.E.; Latt, S.A. Automatic measurement of sister chromatid exchange frequency. *J. Histochem. Cytochem.* **1977**, *25*, 741–753. [CrossRef]
28. Yen, J.C.; Chang, F.J.; Chang, S. A new criterion for automatic multilevel thresholding. *IEEE Trans. Image Process.* **1995**, *4*, 370–378. [CrossRef]
29. Kapur, J.N.; Sahoo, P.K.; Wong, A.K. A new method for gray-level picture thresholding using the entropy of the histogram. *Comput. Vision Graph. Image Process.* **1985**, *29*, 273–285. [CrossRef]
30. Ibarra-Castanedo, C.; González, D.; Klein, M.; Pilla, M.; Vallerand, S.; Maldague, X. Infrared image processing and data analysis. *Infrared Phys. Technol.* **2004**, *46*, 75–83. [CrossRef]
31. Vergura, S.; Falcone, O. Filtering and processing IR images of PV modules. In Proceedings of the International Conference on Renewable Energies and Power Quality (ICREPQ'11), Las Palmas de Gran Canaria, Spain, 13–15 April 2011.
32. Halim, S.A.; Manurung, Y.H.; Mohamad, S.; Morni, M.F. The Effect of CLAHE and Gamma Correction in Enhancement of Digital Radiographic Image for Weld Imperfection Detection. *Int. J. Eng. Technol.* **2018**, *7*, 36.
33. Ikhsan, I.A.M.; Hussain, A.; Zulkifley, M.A.; Tahir, N.M.; Mustapha, A. An analysis of X-ray image enhancement methods for vertebral bone segmentation. In Proceedings of the 2014 IEEE 10th International Colloquium on Signal Processing and Its Applications, Kuala Lumpur, Malaysia, 7–9 March 2014. [CrossRef]
34. Li, Y.; Mao, X. An Efficient Method for Target Extraction of Infrared Images. In *Artificial Intelligence and Computational Intelligence;* Wang, F.L., Deng, H., Gao, Y., Lei, J., Eds.; Springer: Berlin/Heidelberg, Germany, 2010; pp. 185–192. [CrossRef]
35. Tsanakas, J.; Botsaris, P. An infrared thermographic approach as a hot-spot detection tool for photovoltaic modules using image histogram and line profile analysis. *Int. J. Cond. Monit.* **2012**, *2*, 22–30. [CrossRef]
36. Jiang, L.; Su, J.; Li, X. Hot spots detection of operating PV arrays through IR thermal image using method based on curve fitting of gray histogram. *Matec Web Conf.* **2016**, *61*, 06017. [CrossRef]
37. Heriansyah, R.; Abu-Bakar, S. Defect detection in thermal image for nondestructive evaluation of petrochemical equipments. *NDT E Int.* **2009**, *42*, 729–740. [CrossRef]
38. Tsanakas, J.; Botsaris, P. On the detection of hot spots in operating photovoltaic arrays through thermal image analysis and a simulation model. *Mater. Eval.* **2013**, *71*, 457–465.
39. Khamisan, N.; Ghazali, K.H.; Almisreb, A.; Zin, A.H.M. Histogram-based of Healthy and Unhealthy Bearing Monitoring in Induction Motor by Using Thermal Camera. *J. Telecommun. Electron. Comput. Eng. (JTEC)* **2018**, *10*, 31–35.
40. Younus, A.M.; Widodo, A.; Yang, B.S. Image Histogram Features Based Thermal Image Retrieval to Pattern Recognition of Machine Condition. In *Engineering Asset Lifecycle Management;* Springer: Berlin/Heidelberg, Germany, 2010; pp. 943–949. [CrossRef]
41. Mitchell, H.B. *Image Fusion: Theories, Techniques and Applications;* Springer Science & Business Media: Berlin/Heidelberg, Germany, 2010.
42. Pizer, S.M.; Amburn, E.P.; Austin, J.D.; Cromartie, R.; Geselowitz, A.; Greer, T.; ter Haar Romeny, B.; Zimmerman, J.B.; Zuiderveld, K. Adaptive histogram equalization and its variations. *Comput. Vision Graph. Image Process.* **1987**, *39*, 355–368. [CrossRef]
43. Hummel, R. Image enhancement by histogram transformation. *Comput. Graph. Image Process.* **1977**, *6*, 184–195. [CrossRef]
44. Wong, S.L.; Yu, Y.P.; Ho, N.A.J.; Paramesran, R. Comparative analysis of underwater image enhancement methods in different color spaces. In Proceedings of the 2014 International Symposium on Intelligent Signal Processing and Communication Systems (ISPACS), Kuching, Malaysia, 1–4 December 2014; IEEE: Piscataway, NJ, USA, 2014; pp. 34–38. [CrossRef]

45. Kwok, N.M.; Ha, Q.P.; Liu, D.; Fang, G. Contrast enhancement and intensity preservation for gray-level images using multiobjective particle swarm optimization. *IEEE Trans. Autom. Sci. Eng.* **2008**, *6*, 145–155. [CrossRef]
46. Le, T.M.; Vo, T.M.; Pham, T.N.; Dao, S.V.T. A novel wrapper–based feature selection for early diabetes prediction enhanced with a metaheuristic. *IEEE Access* **2020**, *9*, 7869–7884. [CrossRef]
47. Bergamasco, M.; Della Rossa, F.; Piroddi, L. Active noise control of impulsive noise with selective outlier elimination. In Proceedings of the 2013 American Control Conference, Washington, DC, USA, 17–19 June 2013; IEEE: Piscataway, NJ, USA, 2013; pp. 4165–4170. [CrossRef]
48. Sasmal, S.; Chowdhury, S.; Kundu, S.; Politis, D.Z.; Potirakis, S.M.; Balasis, G.; Hayakawa, M.; Chakrabarti, S.K. Pre-Seismic Irregularities during the 2020 Samos (Greece) Earthquake (M = 6.9) as Investigated from Multi-Parameter Approach by Ground and Space-Based Techniques. *Atmosphere* **2021**, *12*, 1059. [CrossRef]
49. Pincus, R.; Barnett, V.; Lewis T. *Outliers in Statistical Data*, 3rd ed.; John Wiley & Sons: Hoboken, NJ, USA, 1994; Volume 37, p. 582.
50. Basu, M. Gaussian-based edge-detection methods-a survey. *IEEE Trans. Syst. Man Cybern. Part C Appl. Rev.* **2002**, *32*, 252–260. [CrossRef]
51. Awalludin, E.A.; Yaziz, M.M.; Rahman, N.A.; Yussof, W.N.J.H.W.; Hitam, M.S.; Arsad, T.T. Combination of canny edge detection and blob processing techniques for shrimp larvae counting. In Proceedings of the 2019 IEEE International Conference on Signal and Image Processing Applications (ICSIPA), Kuala Lumpur, Malaysia, 17–19 September 2019; IEEE: Piscataway, NJ, USA, 2019; pp. 308–313. [CrossRef]
52. Fabic, J.; Turla, I.; Capacillo, J.; David, L.; Naval, P.C. Fish population estimation and species classification from underwater video sequences using blob counting and shape analysis. In Proceedings of the 2013 IEEE International Underwater Technology Symposium (UT), Tokyo, Japan, 5–8 March 2013; pp. 1–6. [CrossRef]
53. Liu, H.; Jezek, K. Automated extraction of coastline from satellite imagery by integrating Canny edge detection and locally adaptive thresholding methods. *Int. J. Remote Sens.* **2004**, *25*, 937–958. [CrossRef]
54. Han, K.T.M.; Uyyanonvara, B. A survey of blob detection algorithms for biomedical images. In Proceedings of the 2016 7th International Conference of Information and Communication Technology for Embedded Systems (IC-ICTES), Bangkok, Thailand, 20–22 March 2016; IEEE: Piscataway, NJ, USA, 2016; pp. 57–60.
55. Khan, M.A.U.; Abdullah, F.; Akram, A.; Naqvi, R.A.; Mehmood, M.; Hussain, D.; Soomro, T.A. A Scale Normalized Generalized LoG Detector Approach for Retinal Vessel Segmentation. *IEEE Access* **2021**, *9*, 44442–44452. [CrossRef]
56. Harrap, M.J.; Hempel de Ibarra, N.; Whitney, H.M.; Rands, S.A. Reporting of thermography parameters in biology: A systematic review of thermal imaging literature. *R. Soc. Open Sci.* **2018**, *5*, 181281. [CrossRef] [PubMed]

energies

MDPI

Article

# A Method for Large Underground Structures Geometry Evaluation Based on Multivariate Parameterization and Multidimensional Analysis of Point Cloud Data

Adam Wróblewski, Jacek Wodecki, Paweł Trybała and Radosław Zimroz *

Department of Mining, Faculty of Geoengineering, Mining and Geology, Wrocław University of Science and Technology, 50-370 Wroclaw, Poland
* Correspondence: radoslaw.zimroz@pwr.edu.pl

**Abstract:** In underground mining, new workings (tunnels) are constructed by blasting or mechanical excavation. The blasting technique used in underground mines is supported by economic aspects, especially for deposits characterized by hard rocks. Unfortunately, the quality of the result may be different than expected in terms of the general geometry of work or the roughness of excavation surfaces. The blasting technique is also a source of vibrations that may affect other existing structures, affecting their stability. Therefore, it is of great importance to monitor both the quality of the new tunnels and changes in existing tunnels that may cause rockfall from the sidewalls and ceilings of both new and existing tunnels. The length of mining tunnels and support structures in underground mines is massive. Even if one would like to limit monitoring of tunnel geometry to those used every day for major technological processes such as transport, it is a vast amount of work. What is more, any stationary monitoring system is hard to utilize both due to everyday blasting procedures and mobile machine operation. The method proposed here is based on quick LiDAR/Terrestrial Laser Scanner measurements to obtain a cloud of points, which allows generating the spatial model of a mine's geometry. Data processing procedures are proposed to extract several parameters describing the geometry of the tunnels. Firstly, the model is re-sampled to obtain its uniform structure. Next, a segmentation technique is applied to separate the cross sections with a specific resolution. Statistical parameters are selected to describe each cross section for final 1D feature analysis along the tunnel length. Such a set of parameters may serve as a basis for blasting evaluation, as well as long-term deformation monitoring. The methodology was tested and validated for the data obtained in a former gold and arsenic mine Zloty Stok, Poland.

**Keywords:** underground mining; mining excavations; tunneling; LiDAR; terrestrial laser scanning; point cloud; 3D model; statistical features; geometry measurement and analysis; dimensionality reduction; principal component analysis



**Citation:** Wróblewski, A.; Wodecki, J.; Trybała, P.; Zimroz, R. A Method for Large Underground Structures Geometry Evaluation Based on Multivariate Parameterization and Multidimensional Analysis of Point Cloud Data. *Energies* **2022**, *15*, 6302. https://doi.org/10.3390/en15176302

Academic Editor: Krzysztof Skrzypkowski

Received: 29 July 2022
Accepted: 25 August 2022
Published: 29 August 2022



## 1. Introduction

In many underground mines worldwide, especially those operating in hard rocks, the drilling and blasting technique is the most commonly used method to excavate valuable material [1]. This technique is relatively cheap and provides high flexibility of operations. Although drilling and blasting technology has been strongly optimized over the years [2], there are still some major issues to overcome. One of them is the appropriate design of this process to prevent over- and underbreaks [3] together with the prediction of high risk of potential over- and under-excavated zones [4]. From the geometry point of view, it is not so easy to control and maintain the actual shape of the tunnel cross section defined in the blast plan in the presence of hard geological conditions, manifested in the inhomogeneity of the rock mass.

Geological and mining conditions are the main determinants of the design of the excavations [5]. The over- and underbreaks that occur after the drilling and blasting procedure are a serious problem, as they are not in line with the defined parameters selected under these conditions. The deviations from the assumed cross section in the form of underbreaks make the excavations insufficient for the proper operation of mining machines and devices required for the materials extraction processes. Moreover, limited dimensions of mine tunnels endanger the underground crew, determined by the permissible concentration of oxygen and harmful gases in the mine air, its temperature, and the intensity of its flow [6]. Underbreaks resulting from inaccurate drilling and blasting need to be removed by additional drilling and blasting, increasing excavation maintenance costs. On the other hand, overbreaks also contribute to increasing operating costs. As the part of the rock mass not intended for excavation, overbreak material is often non-usable and when blasted, degrades the ore, generating additional costs at the processing level. Enlarged dimensions of the excavations may also lead to issues with stability [7]. For these reasons, there is a strong need for sufficient control and prevention of over- and underbreaks.

Although there have been numerous approaches tested and validated that help avoid over- and underbreaks, there are still some hard to control factors, e.g., the complexity of the rock mass or human factors that result in an increase or decrease in the size of tunnels compared to the design. In some cases, the already adopted drill-and-blast technique may not be the most efficient, especially in rapidly changing geological and mining conditions. There is a strong need not only for over- and underbreak prevention but also for a quick and reliable method to measure and assess the quality of underground excavations for further decision making.

Moreover, since drilling and blasting is often outsourced, there is a need to develop a method for quick evaluation of mining tunnel geometry, in other words, work quality assessment. This is a particularly important matter when additional expenditures must be incurred for extra drilling and blasting due to underbreaks, as well as concreting or other techniques to overcome overbreaks. Furthermore, each sequence of blasting may negatively influence the stability of the tunnel directly or induce seismic shocks. Especially in deep mines, it is clear that there is additionally a convergence of mine tunnels that should be monitored on a regular basis.

As shown above, measurements of geometry are an important topic in underground mining. Shapes of tunnels, pillars, and excavations change over time, influencing the stress distribution in a rock mass [8]. This leads to the deformation of the tunnels. They must be monitored to ensure the acceptable speed of deformation development, thereby enabling safe and continuous operation of the mine. Another issue requiring geometric measurements is the excavation process. Metric methods of tracking the mining progress allow us to not only estimate the volume of extracted material [9] but also to check the actual excavation geometry compliance with the overall plan. This is especially important in the case of employing mining techniques such as drilling and blasting, which often cause unpredicted changes of the blasting site geometry and put additional stress on the rock mass. An ideal solution for those problems is frequent monitoring with modern methods for a 3D metric reconstruction of the underground sites, such as photogrammetry or laser scanning [10–12]. However, since the quality of the former is vastly influenced by the lighting conditions, the latter method is usually the preferred one in harsh deep mining environments. Laser scanning produces data in the form of an unstructured point cloud: an abundant set of points with coordinates in 3D space, usually characterized by additional variables, such as the intensity of the laser beam reflection. The problem of analyzing such data is often the enormous size of the point cloud and the lack of topological information of spatial relationships between points [13,14]. To get from the raw measurement data to a tunnel convergence estimation or a blasting quality assessment accessible for interested parties such as the geomechanical engineers or stakeholders, an automated and scalable procedure is needed. Such a procedure should allow approximation of the structure from a point cloud, analyzing its irregularities and deviations from the plan or prediction, and presenting them in an action-enabling way, conceivably with a model characterized by easily understandable parameterized metrics.

In this paper, we propose a methodology for geometry analysis based on a 3D scanning procedure and spatial data processing. In general, the idea of 3D scanning is well known in tunneling, but the key issue is related to 3D data modeling and analysis with the focus on features extraction, their analysis, and interpretation. It makes the mining workings case different from a tunnel inspection case.

The paper is organized as follows: firstly, the most important approaches used in tunnel inspection are recalled. Secondly, mining workings being analyzed and experimental works performed (scanning devices, procedure, and experiment) are described. The novelty of the paper is related to the methodology developed for 3D data analysis. Thus, the results of 3D data analysis are presented for experiments performed in the underground mine in the final part of this article.

## 2. State of the Art

Blasting is commonly used to mine hard rocks. This is a critical process from several perspectives. It is obvious that one should mention safety issues, financial aspects, quality of rock fragmentation, tunnel geometry that results from blasting, etc. In this paper, we focus on "quality of blasting" related to the planned mapping of the geometry of the tunnel into the real shape of the tunnel after blasting. Appropriate geometry of the tunnel is important for many other processes (tunnel maintenance, ventilation, transport, etc.). As the usual underground mine includes several hundred kilometers of tunnels, there is a need to provide a procedure to monitor mining workings in a long-term sense as well as to assess the compatibility of planned and received geometry just after blasting.

The over- and underbreaks can be brought about by two main reasons: by the drill-and-blast design and execution and by the geomechanical features of the rock mass [15]. Thus, many scientists in their research have raised the problem of drilling and blasting parameter selection to ensure the predefined shape of excavation [16–18].

The second main cause of over- and underbreaks together with the drill-and-blast factors have been studied more carefully. The role of geological discontinuities in causing blasting over- and underbreaks of the minor (<3 m) and major (>3 m) scale has been evaluated by [19] through joint analysis. In [20], multiple regression analysis, both linear and non-linear (LMRA and NMRA), and artificial neural network (ANN) were applied to forecast overbreaks and assess the influence of geological parameters based on 49 sets of overbreak and rock mass rating (RMR) data. In [21], the authors used data from 18 blasting experiments conducted on-site during construction of a highway tunnel in China as a feed for machine learning. To map the dependencies between the geological conditions, control indices, and the outputs of the smooth blasting parameters, the improved support vector regression (ISVR) model was implemented. In addition, an ISVR algorithm was supported by a genetic algorithm (GA) to automate the choice of optimal parameters of the ISVR model. Similarly, linear multiple regression analysis was performed in [22] to predict the overbreaks induced by blasting. Controllable, non-controllable (geological conditions), and semi-controllable blast factors were studied. Operative methodology to differentiate drill-and-blast-related overbreaks from geological ones, together with its volume estimation procedure, was presented in [23].

To estimate the size of under- and overbreaks, an appropriate measurement method that provides data high quality and easy collection is required. Maerz et al. [24] divided the excavation profile's measurement techniques into surveying ones (manual or laser) and photographic light sectioning methods. Simply stated, manual methods consist of determining the distance between the certain central point to the tunnel boundaries at fixed angular intervals. Utilized in the past, manual methods are time-consuming and strongly inefficient when the high accuracy of measurements is needed on long tunnels.

A promising method has been proposed in [25]. Cross-section measurements through a light sectioning method are performed by outlining it with a plane of light from a conical mirror. Although the procedure is much faster than the manual ones, there is still a need to manually process the photographs and estimate the size of over- and underbreaks. Nowadays, laser techniques are being exploited, ensuring high accuracy and enabling automatic post-processing.

LiDAR, defined as *Laser Induced Detection and Ranging* or *Light Detection and Ranging*, is a laser-based device allowing users to measure the distance between the sensor and other objects [26]. It is obtained thanks to its active sensor high frequency spins emitting a laser beam, which is reflected from the object and received back, allowing precise determination of the position of the object in relation to the sensor. LiDAR can operate in two different modes: sending short pulses or continuous signals. Based on this, the distance is obtained through Time-of-Flight (ToF) of the laser beam or phase-shift estimation of the electromagnetic wave that returns to the sensor relative to the output electromagnetic wave. Generally, the better sensing range is exhibited by ToF Terrestrial Laser Scanners, while phase-based devices give higher accuracy. As a result of laser scanner measurements, a precise and dense point cloud is obtained, which is the set of coordinates of points in a three-dimensional coordinate system. Numerous reflections of the same, scattered beam of different strength may be detected by the LiDAR and used for sophisticated data analysis, together with other features provided, e.g., intensity or reflectance of the beam [27].

Laser scanning is used in the urban tunnel's geometry reconstruction in the form of a 3D point cloud. Deformation estimation based on terrestrial laser scanning as a part of tunnel structural monitoring is presented in [28]. The 3D laser scanning for structural inspection of the tunnels performed under an autonomous UGV mission for concrete lining tunnel inspection is shown in [29]. Amedjoe and Agyeman [30] presented a mine excavations stability problems management approach based on cavity monitoring system data captured in the AngloGold Ashanti–Obuasi Mine. In order to reduce the number of accidents related to rock falls, Warneke et al. [31] applied 3D laser scanning technology to assess tunnels' geometry drifts and overbreaks in Stillwater Mine, Montana. Zou et al. [32] presented a smooth blasting evaluation method thanks to a mobile app utilizing geometry-related data captured by a laser profilometer.

Moreover, in recent years, 3D laser scanning techniques have been applied widely in other tunneling and mining applications [33,34]. As convergence monitoring is a well-known tunneling problem [35–40], the authors in [40] proved that data provided by a mobile LiDAR system can be used as an input for the method that automatically detects road tunnel luminaries.

The scanning technology is also very useful for specific mining applications. For example, Vanneschi et al. [41] successfully applied 3D scanning data for rock pillar degradation monitoring, while Xu et al. [42] proposed to use terrestrial laser scanning for water leakage. In [43], it is advised to use laser scanning technology on incident investigations in the mining industry. The authors in [44] considered the usage of scanning for drill and blast excavation forecasting. A review of commercial mobile mapping and surveying solutions suitable for GNSS-denied environments, such as underground mines, has been provided in [45].

Carrying out the LiDAR field measurements does not yet result in an easy to interpret model, as the data should be preprocessed using automated procedures to avoid mundane manual model preparation. The appropriate method for subsequent data analysis is also very important. In [46], the authors proposed a method for continuous extraction of subway tunnel cross sections based on terrestrial point clouds. In [47], the authors developed an automated and efficient method for extraction of tunnel cross sections using terrestrial laser scanned data. Raw 3D data from scanning are not suitable for tunnel condition evaluation, thus in [48] a procedure for feature extraction of a concrete tunnel liner from 3D laser scanning data has been proposed.

The next step is to provide a data-driven decision support system. The major issues in designing such a system, also applicable in this study, are the identification or creation of a crucial variable for decision making and providing thresholds or more sophisticated methods to interpret its values. Often, machine learning techniques are used for those purposes. A Back Propagation Neural Network (BPNN) and Multivariate Adaptive Regression Splin (MARS) machine learning algorithm has been proposed in [49,50]. It was already mentioned above that multiple regression analysis and artificial neural network (ANN) were applied to forecast the overbreaks in [20]. In [33], a state-of-the-art review focusing on segmentation and classification of mobile laser scanning point clouds is presented. For more details, one may refer to review papers by Grilli [51] or Remondino et al. [52,53].

## 3. Experiments and Data Description

### *3.1. Złoty Stok Gold Mine*

Złoty Stok is a town located in Lower Silesia Province, southwestern Poland, in the Eastern Sudetes. The city owes its name to the former gold exploitation from the 13th century. At a later stage of the mine's operation, arsenic ore was mined there until 1961.

The geological structure of the rock mass in which the tunnels are located is quite complex. The presence of the Złoty Stok—Skrzynka tectonic zone, which is a part o fthe Ladek—Śnieżnik metamorphic structure, is the reason for phenomena of cataclasis and mylonitization of varying intensity in the Złoty Stok area. Metamorphic rocks are represented by schists (mica, mica–quartz, and quartzite schists), but also gneisses, leptynites, amphibolites, as well as serpentinites and crystalline limestones gneisses are common. Rocks in the Złoty Stok-Skrzynka tectonic zone and Kłodzko–Złoty Stok Massif borders are contact-altered and cut by faults and dislocation zones [54].

From a huge complex of excavations (see Figure 1) consisting of over 300 km of underground corridors, located on 21 levels, only 2 adits are accessible since in 1996 an underground tourist route named "Kopalnia Złota" (Gold Mine) was opened. The first one, Gertruda adit, is two kilometers long, but only initial 500 m can be seen. The remaining part of this slant is deprived of lighting and flooded with water, and merely a small part can be visited with boats. The second one, Czaran adit, is also partially opened. One of its side corridors (approx. 200 m long) leads to a 25 m shaft from the 17th century, giving access to a huge chamber in which a unique, 8-meter long underground waterfall can be seen.

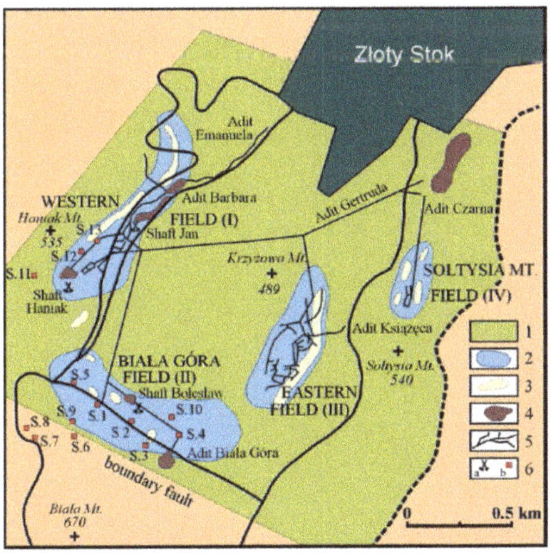


**Figure 1.** Złoty Stok post mining area [55]. Legend: 1—mining area; 2—limestones; 3—ore nests; 4—mining waste heaps; 5—slants, galleries, and adits; 6—shafts.

The deep and extensive mine excavations in the Złoty Stok Mine were made entirely by hand, then by blasting in very hard, though fractured massif of metamorphic rocks. Thus, this mine has been selected for testing purposes, firstly, because it is an old mine in which the problem with corridor geometry (roughness/unevenness of surface) is very clearly visible. Secondly, because it is more available than deep underground mines in operation. The unique characteristic of the workings is the reason for some other interesting papers related to experimental works in this mine [56,57].

*3.2. Data Collection*

For the purpose of testing the methodology developed in this study, data from the work concerning the accuracy evaluation of a LiDAR SLAM solution, namely High-Density LiDAR SLAM (HDL-SLAM [58]) have been used [59]. The procedures and details of the 3D point cloud acquisition and coregistration are included in the referenced paper. However, in our study only the part of the final point cloud acquired with the mobile LiDAR is analyzed using the proposed corridor geometry evaluation procedure.

The mobile LiDAR mounted on the UGV platform during the measurements was a Velodyne VLP-16 ('Puck') LiDAR sensor (Figure 2). It is a small and compact LiDAR that is performance and power optimized for use across a variety of applications ranging from automotive, mapping, robotics, security, and smart cities, i.e., for lower speed autonomous vehicle (AV) applications. The Puck enables real-time, surround view, 3D distance, and calibrated reflectivity measurements. The main features cover a range of 100 m, up to 600,000 points/second generation, a 360° horizontal field of view and a 30° vertical field of view, and class 1 eye-safe 905 nm technology with autonomous fleet validation [60]. The distance measurement accuracy for a single point is 3 cm ($1\sigma$).

**Figure 2.** Velodyne VLP-16 LiDAR sensor and UGV platform.

Moreover, the scanning has also been performed with the utilization of a higher accuracy device, namely Riegl VZ-400i Terrestrial Laser Scanner (TLS) Figure 3. This survey-grade device is characterized by a 3D point position determination accuracy of 5 mm ($1\sigma$ at 100 m), acquisition of up to 500,000 points/second, range from 0.5 m to 800 m, and scanning resolution of up to 0.0007° (vertical) and 0.0015° (horizontal) [61]. The scanner has a built-in Inertial Measurement Unit (IMU) used for motion estimation during changing scan positions and can be integrated with a GNSS receiver or a digital camera. Although costly and requiring the operator to carry it, the Riegl TLS can quickly obtain very dense and accurate point cloud data of a vast area.

In this paper, 3D data of the Gertruda slant (Figure 4) geometry has been processed. Renderings of the point clouds acquired with laser scanning of the Gertruda adit with the RIEGL VZ-400i TLS and a Velodyne Puck and HDL-SLAM are presented in Figure 5. Moreover, from the whole 3D point cloud representing slant geometry, an approximately 24 m long segment has been chosen (Figure 6) for analysis. Large variability of the cross sections in this part of the slant create a good test field for evaluation of the proposed diagnostic procedure. Only the point cloud acquired with SLAM has been processed using the proposed corridor geometry quality evaluation procedure to demonstrate its capabilities on a dataset obtained with an accessible, lower cost data acquisition solution. Nonetheless, a further possibility and advantages of employing it for processing highly accurate data from a survey-grade instrument are discussed in Section 6.

**Figure 3.** RIEGL VZ-400i Terrestrial Laser Scanner in the adit.

**Figure 4.** Wheeled robot during its low-speed passage through Gertruda slant.

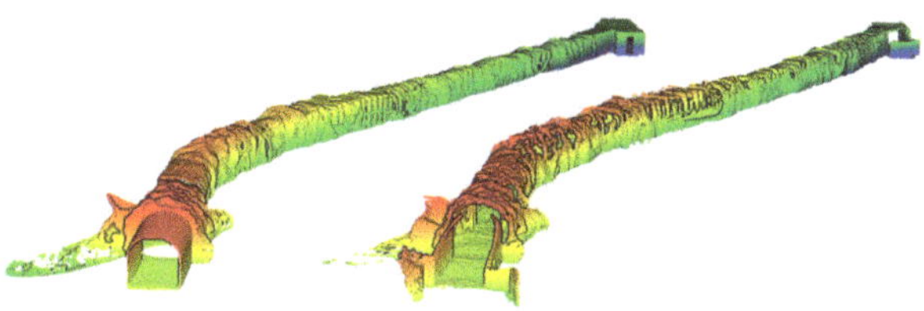

**Figure 5.** Renderings of a point cloud obtained with a RIEGL VZ-400i TLS (**left**) and a Velodyne Puck and HDL-SLAM (**right**). Point coloring by the Z coordinate.

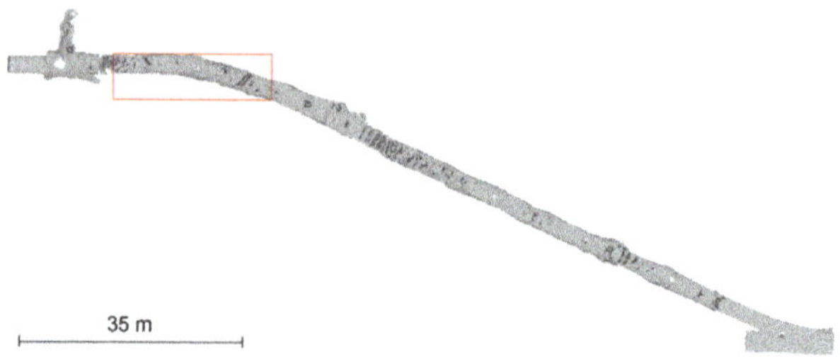


**Figure 6.** Top view of a 3D point cloud orthographic projection of the Gertruda slant with analyzed region marked in red.

## 4. Methodology

From measurement, trial one receives a spatial dataset with a massive volume of points and a specific format. As previously mentioned, raw 3D data is difficult to use for any reasoning or decision making; therefore, a method for processing is needed, which is not an easy task. If geometry evaluation is to be used in everyday practice, a simple 1D parameter is required for monitoring, but also for modeling and predicting changes in geometry.

Thus, the proposed general approach is to build a reliable 3D model from measurements, divide it into collections of cross sections, describe the 2D shape of cross sections (still multiple points) by some statistically explained features, and then again use statistical or machine learning techniques to find outliers, classes, or general patterns in tunnel geometry.

In this section, the key elements of the methodology are described. The general flow of the procedure is presented in Figure 7.

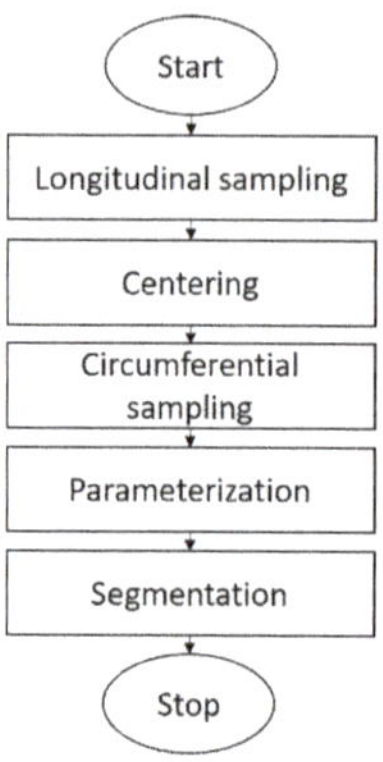


**Figure 7.** Flowchart of the procedure.

1. **Longitudinal sampling:** In the first step, the grid is defined in the dimension of tunnel length with a given resolution. Then, the original geometry produced by a scanner is manually sampled longitudinally according to the grid. In practice, when raw data are imported from the device to the computer, they are assembled from multiple files to a single point cloud. From this point cloud, running the bounding box filter of a given longitudinal interval (i.e., chosen cross-section separation), the slices of a point cloud are extracted. Projecting them on a plane perpendicular to the tunnel axis, the points establishing flat corridor cross sections (further called "profiles") of the entire geometry are generated. This way, it is possible to obtain a set of cross-sectional profiles that can be further analyzed. When a set of such profiles is obtained, they are cleaned by selecting only the edge points (see Section 4.1). This way, it is possible to get rid of unnecessary points inside the corridor.

2. **Centering:** In the next step, the profiles are centered so that their shapes can be analyzed. It is a necessary preprocessing step to counteract any changes in the directions of a path of a corridor. For example, if the corridor was excavated in a perfectly straight line, centering would not be necessary because (at least in theory) the center of each profile would lie in the same position on its plane. In the horizontal direction, the median value (see Section 4.2) of each profile is subtracted (median of horizontal coordinates of points).
In the vertical direction, profiles are first normalized to the average floor level (section of a profile describing the floor is identified, and a mean value of the vertical coordinates of this section is subtracted), and then the median value is subtracted from the entire geometry in terms of vertical coordinates, so that the projection orthogonal to the length of the corridor is centered around the origin of the coordinate system.
3. **Circumferential resampling:** Centered profiles are converted to polar coordinates. This way, their shapes are "unfolded" so that the horizontal coordinate value of each point represents the angle of a point with respect to the center of a corridor, and the vertical coordinate value of each point represents the distance from the center of a corridor (such as a unit circle converted to polar coordinates becomes a constant linear function of value 1). For each profile $P_i$ represented by a pair of vectors holding the $XY$ coordinates of individual points, the unfolded profile $Pp_i$ in a polar domain is represented by a pair of vectors holding the $R\Theta$ coordinates calculated as:

$$Pp_i = \begin{cases} R_i = \sqrt{(X^2 + Y^2)} \\ \Theta_i = tan^{-1}(Y/X) \end{cases} \quad for \quad i = 1 : N \tag{1}$$

where $N$ is the number of profiles, the coordinate $\Theta$ denotes the angle coordinate, and the coordinate $R$ denotes the radius coordinate (distance from the center).
Now the profile coordinates can be used as single-dimensional vectors $R_i$ in the domain $\Theta$, and they can be reinterpolated in the angle domain to the resolution that is common to all the profiles. First, a new domain vector $\Theta r$ is defined with $K$ evenly spaced points in range $(0, 2\pi)$. Then, all the vectors $R_i$ are resampled so that $Pp_i\{R_i, \Theta_i\}$ allows to produce $Ppr_i\{Rr_i, \Theta r\}$, where it is important to notice that the angle domain vector $\Theta r$ is common for all $Rr_i$. The resampling itself is performed using a Modified Akima cubic Hermite interpolation [62]. The interpolated values are based on a piecewise function of polynomials. This way, the profiles are described by the equal amount of points evenly spaced in the domain of the angle. At this poin,t the evenly sampled geometry of the desired grid resolution is obtained.
4. **Parameterization:** In the beginning of parameterization, the median (see Section 4.2) profile is calculated from the entire geometry which serves as a reference model. For every profile (after circumferential resampling), their vectors $Rr_i$ are arranged in a matrix $Rr$, and its median is calculated along the dimension of corridor length, which produces a new profile, also in the angle domain, such as:

$$Rr_{model} = median(Rr) \tag{2}$$

In practice, user can import additional geometry to serve as a reference model. Then, several statistics are calculated for every profile, such as:

- Total, positive, and negative deviation from the reference profile shape (see Section 4.3)—those features will be useful to describe the aspect of consistency of the excavation. Testing shows that having those 3 features together works better than using only 1 feature of total deviation, although in practice they carry the same information.
- Roughness factor (see Section 4.4)—this feature allows the user to describe the qualitative aspect of a profile in terms of how wasteful the excavation was at any given point. It is not optimal if a shape of a single profile contains a lot of variety.

- Width, height, and area of a given profile.

In total, it allows us to obtain 7 features describing the corridor along its length. Those 7 features are then used as a dataset of parameters that is used for further analysis. Those statistics can be analyzed by themselves to evaluate the geometry and draw conclusions; however, the authors propose the following method that fuses data from the statistics.

The matrix containing the statistics (Table 1) is processed using a principal component analysis (PCA) algorithm (see Section 4.5). The PCA method is known for its ability to reduce dimensionality. In practice, it means that if it is able to produce one feature that explains the vast majority of information coming from the 7-dimensional dataset, it is very practical to analyze this singular feature instead of performing 7-dimensional analysis of the data. The first component forms a diagnostic feature that describes the differences between the profiles and can be used as a working statistic for segmentation.

5. **Segmentation:** The diagnostic feature is segmented based on value thresholds. To obtain them, authors calculate a kernel density estimate of a diagnostic feature (see Section 4.6) [63] and define the thresholds as the local minima between main modes. It is performed by differential analyses of the estimated probability density function. Local minima are located at places where the first derivative is equal to 0 and the second derivative is positive. Then, profiles that belong to particular classes between those thresholds are identified. In practice, Matlab provides a function called *findpeaks* that performs this operation automatically.

**Table 1.** Statistics matrix.

| | Statistics | | | | | | |
|---|---|---|---|---|---|---|---|
| **No.** | **Positive Deviation** | **Negative Deviation** | **Total Deviation** | **Roughness** | **Total Height** | **Total Width** | **Area** |
| 1 | 13.172 | 12.881 | 26.053 | 11.239 | 2.454 | 2.949 | 6.102 |
| 2 | 4.250 | 6.120 | 10.370 | 8.160 | 2.382 | 2.831 | 5.913 |
| 3 | 2.984 | 12.852 | 15.835 | 7.871 | 2.289 | 2.878 | 5.562 |
| ... | ... | ... | ... | ... | ... | ... | ... |
| 90 | 4.016 | 14.168 | 18.184 | 11.125 | 2.384 | 2.902 | 5.570 |
| 91 | 9.192 | 25.012 | 34.204 | 10.122 | 2.312 | 2.926 | 5.405 |
| 92 | 1.304 | 13.055 | 14.359 | 8.896 | 2.336 | 2.855 | 5.469 |

### 4.1. Boundary Detection

The detection of boundary points of a flat point cloud is performed in two steps. First, the non-convex alpha shape is generated from points, and then boundary facets are determined.

The alpha shape of a set of points is a generalization of the convex hull and a subgraph of the Delaunay triangulation [64,65]. Moreover, alpha shapes allow users to control the level of detail. Varying the parameter value from 0 to infinity can produce a set of different alpha shapes for that point set.

In the second step, boundary facets (in particular—boundary edges for 2-dimensional geometries) are identified. First, edges of an alpha shape are counted individually for each triangle. Then, edges counted only once are defined as boundary edges, which allows us to define boundary points.

### 4.2. Median Calculation

The median is defined as:

$$median(X) = \begin{cases} X_{(n+1)/2} & \text{for } n\%2 = 1, \\ \frac{X_{(n/2)} + X_{(n/2)+1}}{2} & \text{for } n\%2 = 0, \end{cases} \tag{3}$$

where $X$ is a vector of values, $n$ is the number of samples, and % denotes the operation of modulo division.

### 4.3. Deviation Calculation

One of the simplest statistics that can be calculated for the profiles is the deviation from the model. In order to achieve that, unfolded profiles (see step 3 in Section 4) in the angle domain are processed by subtracting from them the model profile, such as:

$$dev_i = \left| \sum_{k=1}^{K} Rr_i - Rr_{model} \right| \tag{4}$$

In other words, for each point on the profile (each ray from the center), the distance from the center is compared with the distance of the same point for the reference profile. Now one can take the absolute value of the result to obtain the total deviation, or just take the positive or the negative part to obtain positive (outwards) or negative (inwards) deviation.

### 4.4. Roughness Factor

The idea of a roughness parameter is based on the assumption that the model profile should be relatively smooth in shape and not very jagged. In order to estimate that, the derivative of the shape of each profile is analyzed, such as:

$$RF(i) = \sum_{\theta=0}^{2\pi} |diff(P(i))| \tag{5}$$

where $|\cdot|$ stands for the absolute value, $diff()$ function denotes numerical derivative, and $P(i)$ is the ith profile in a sequence. Effectively, roughness value is a sum of amplitudes of changes between consecutive points on the outline of a profile. In practice, this calculation is performed on the unfolded form of a profile, so the value of points on the profile denotes the distance from the center.

### 4.5. Principal Component Analysis

Principal Component Analysis is a very capable analytical tool [66]. It interprets a dataset including $N$ samples over $K$ variables, as a point cloud in $K$-dimensional feature space. The aim is to rotate and translate a local coordinate system so that the variance is maximized over new dimensions, such that the first dimension displays the greatest variance, the second dimension—second greatest variance, etc.

Such a transformed system contains new values of data, which are original data but defined over a new set of dimensions (new coordinate system). Vectors representing data over the new coordinate system are known as principal components. The new feature space describes the original dataset with the most information content located within several first principal components that carry the most information. In many cases, the information contained in several first components is sufficient due to their high information content, so PCA is regarded to be a dimensionality reduction method.

Given $n$ observations of $m$-dimensional data stacked into a matrix $X \in \mathbb{R}^{n \times m}$, the principal components can be calculated using Singular Value Decomposition (SVD):

$$\frac{1}{\sqrt{n-1}} X = U \Sigma V^T, \tag{6}$$

where $U \in \mathbb{R}^{n \times n}$ and $V \in \mathbb{R}^{m \times m}$ are unitary matrices, and $\Sigma \in \mathbb{R}^{n \times m}$ contains the nonnegative real *singular values* of non-increasing magnitude ($\sigma_1 \geq \sigma_2 \geq \cdots \geq \sigma_m \geq 0$). Principal components are the orthonormal column vectors of the matrix $V$, and the variance of the $i$-th component is equal to $\sigma_i^2$.

### 4.6. Kernel Density Estimation

The distribution density is obtained using the kernel density estimator, which is the estimated empirical probability density function of a random variable [67,68]. For real values of the data $x$, the estimated distribution is given by:

$$\hat{f}_h(x) = \frac{1}{nh} \sum_{i=1}^{n} K\left(\frac{x - x_i}{h}\right), \tag{7}$$

where $x_1, x_2, \ldots, x_n$ are samples of unknown data, $K(\cdot)$ is the kernel smoothing function, $n$ is the sample size, and $h$ is the bandwidth. For this example, a Gaussian kernel is used.

The value of the bandwidth is obtained using the so-called *Silverman's rule of thumb* [68]. For the Gaussian kernel and the assumption of a Gaussian mixture, the optimal choice for $h$ (that is, the bandwidth that minimizes the mean integrated squared error) is

$$h = \left(\frac{4\hat{\sigma}^5}{3n}\right)^{\frac{1}{5}} \approx 1.06\hat{\sigma}n^{-1/5}, \tag{8}$$

where $\hat{\sigma}$ is the estimator of a standard deviation of the samples, and $n$ is the number of samples.

## 5. Results

In this section, the authors present the geometry evaluation approach on the example of a corridor section from a historical mine in Poland, shown in Figure 8. In the first place, the point cloud is segmented into slices of the same depth to acquire regular profiles. In this example, the profiles are sampled with the resolution of 25 cm; however, any different value can be chosen by the user. An example slice is shown in Figure 9a. Individual points segmented in slices are then projected onto a plane. Subsequently, Figure 9b presents an example of how those points are preprocessed in a Matlab environment. Red points are detected as boundary points, and they represent the raw form of the outlines of profiles.

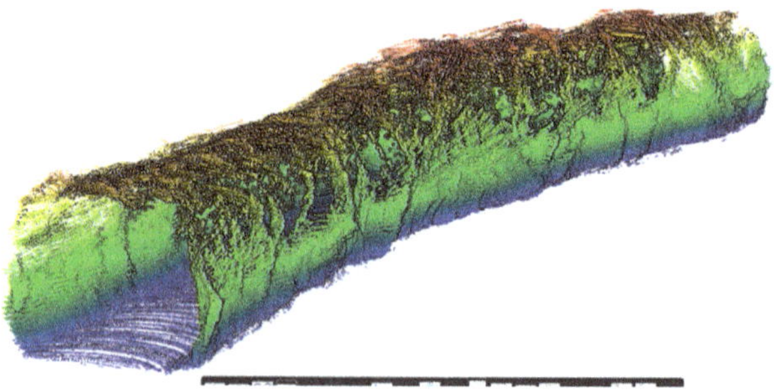

**Figure 8.** Point cloud of the analyzed corridor segment colored by elevation.

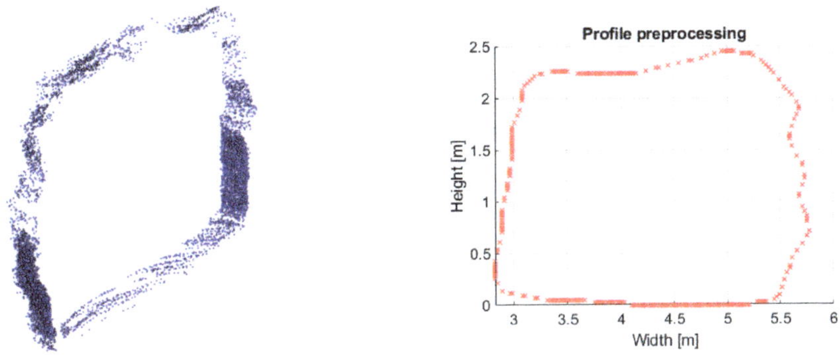


(a) (b)

**Figure 9.** Identification of boundary points for further processing: (**a**) isometric view of an example of a point cloud slice used to generate profile; (**b**) preprocessing of the slices imported from the raw geometry.

In the next step, the outline is converted to polar coordinates and resampled (see Figure 10). Figure 11 presents the shapes of outlines pulled from the original point cloud provided by the 3D scan.

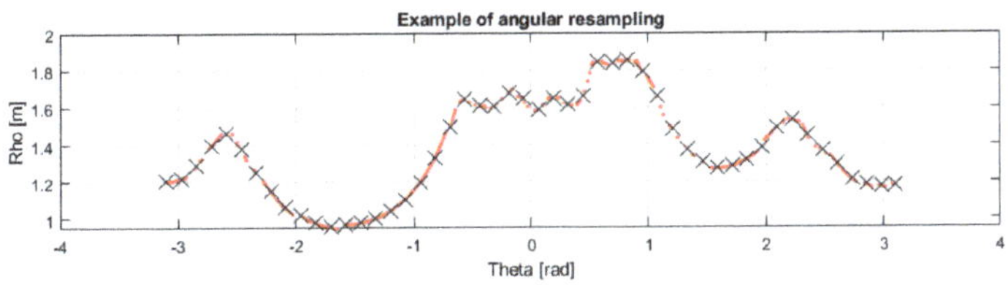


**Figure 10.** Example of angular resampling. Original boundary points from Figure 9b (red points) and evenly resampled data (black crosses). In this example, the amount of resampled points has been set to lower value for better visibility.

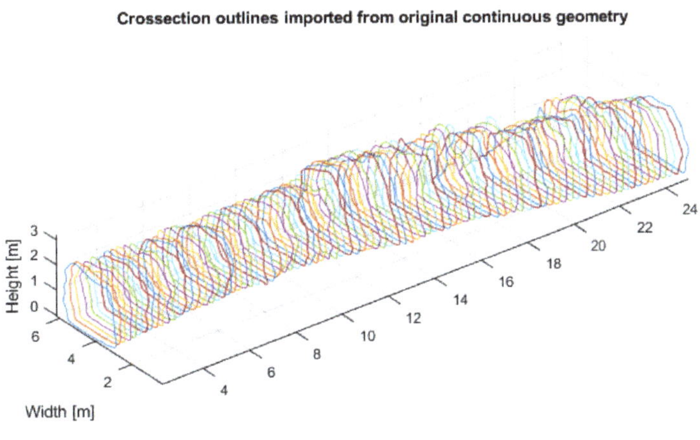


**Figure 11.** Overview of shapes of outlines. Dimension units are expressed in meters.

Afterwards, the profiles are centered (Figure 12) and resampled in the domain of angle (Figure 13). This way, every profile has the same amount of samples at the same angles, so they can be compared.

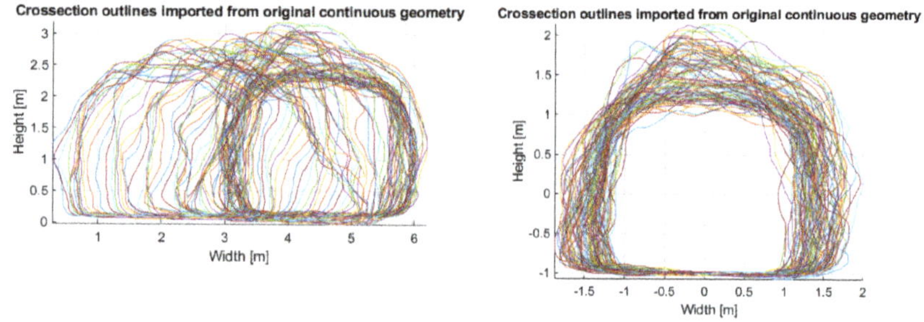


(a) (b)

**Figure 12.** Comparison of profiles between and after centering: (**a**) orthogonal projection of profiles before centering; (**b**) orthogonal projection of profiles after centering.

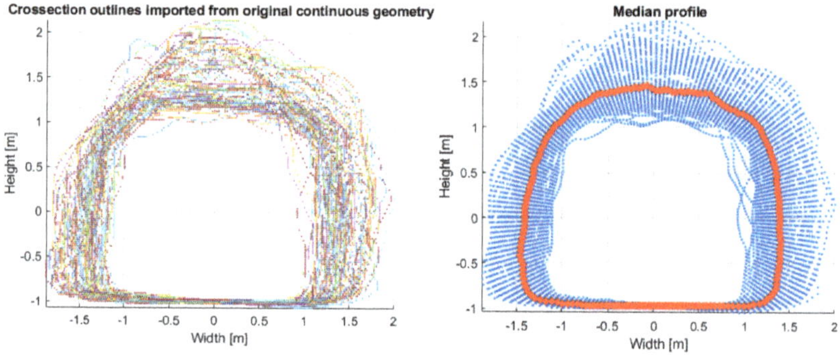


(a) (b)

**Figure 13.** Comparison of point distribution of profiles before and after resampling: (**a**) overview of raw shapes of centered outlines before angular resampling; (**b**) centered outlines after angular resampling (blue) and their median (red).

When profiles are resampled, they can be parameterized with seven statistics described in Section 4 (see Figure 14).

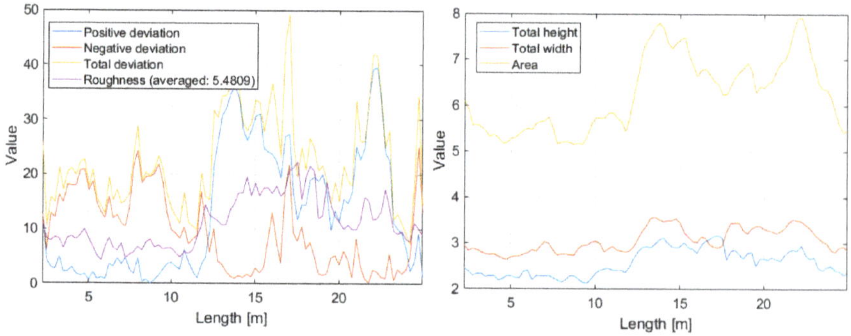


(a) (b)

**Figure 14.** Statistics calculated for the resampled profiles: (**a**) positive deviation, negative deviation, total deviation, roughness index; (**b**) cross-section height, cross-section width, cross-section area.

Statistics are then passed to the PCA, and seven principal components are calculated, while the first of them is regarded as a feature useful for further segmentation based on relative variance, which for this first component was equal to 79%.

In order to achieve that, the feature is divided into segments based on its values. First, the kernel density estimate of the feature is calculated with an automatically obtained bandwidth parameter equal to 0.12 (see Section 4.6) and the normalization mode set to PDF—probability density function. It means that the integral of a function is equal to one (see Figure 15). Then, local minima of the feature are identified. Those minima define the thresholds for feature values (see Figure 16). Then, the thresholds define regimes in the domain of corridor length (see Figure 17). Median profiles of each class and original total median profile comparison toegether with profiles selected based on the results of segmentation has ben presented in Figure 18.

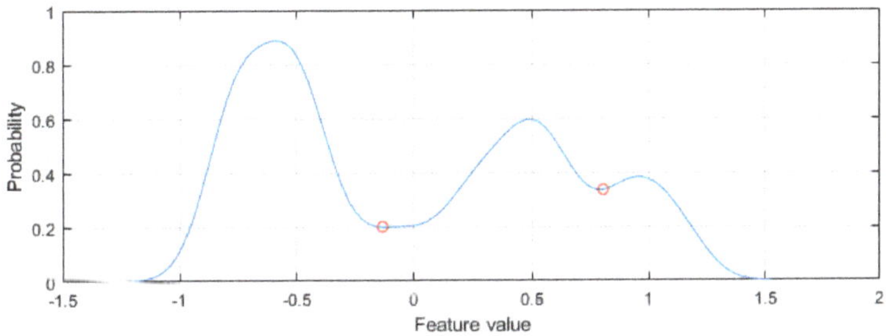


**Figure 15.** Kernel density estimate of feature values. Red circles indicate localized thresholds.

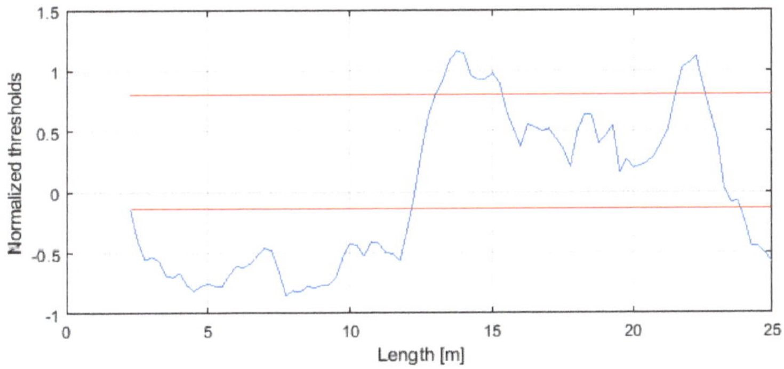


**Figure 16.** Obtained thresholds at values −0.13 and 0.81 for qualitative segmentation.

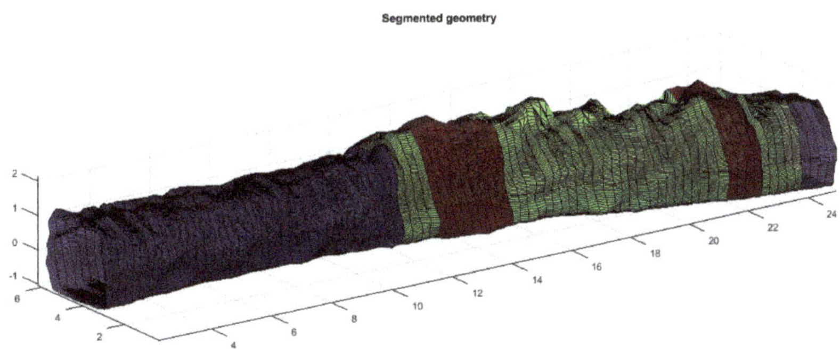


**Figure 17.** Segmented geometry. Colors denote assignment of individual profiles to classes. In this example, the blue class denotes the most "consistent and desirable" character of shape; green is the moderate one and red—the most irregular.

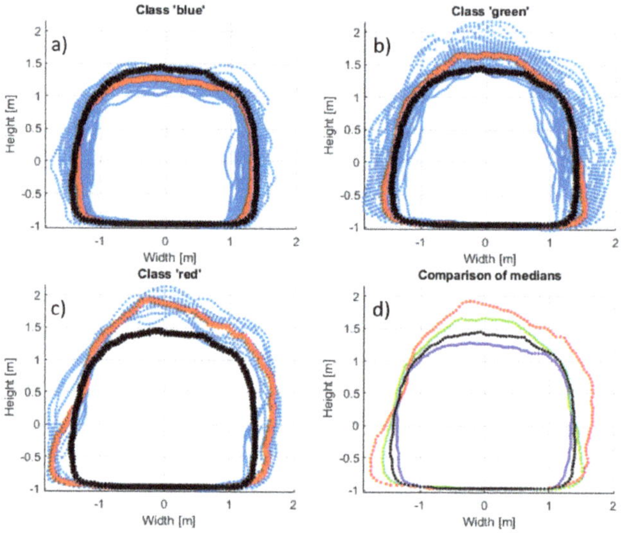


**Figure 18.** (Panels **a–c**): profiles selected based on the results of segmentation (blue points), median profile of each class (red points), and original total median profile (black points); (panel **d**): comparison of class medians (color-coded) with total median (black).

## 6. Discussion

In our study, the proposed methodology has been tested on the testbed of a closed mine corridor characterized by varied geometry due to the basic techniques employed in its creation. Such a case study could simulate a real scenario of evaluating the tunnel geometry in terms of compliance with the mining plan. In this instance, significant deviations from the plan would be in the order of tens of centimeters. As such, a point cloud obtained with a cheap SLAM solution could be used as a source of input data, despite its accuracy in the range of single centimeters. However, in cases where the desired precision of identifying deviations of the corridor geometry would be lower, e.g., during convergence monitoring, another input point cloud source of correspondingly higher measurement accuracy and resolution will be chosen. Example techniques include terrestrial laser scanning, photogrammetry (in mines with low pollution and well-lit corridors), and mobile laser scanning (for tunnels accessible by a car equipped with an MLS system).

For the proposed methodology, employing instruments of higher accuracy and resolution for measurements assures a more accurate representation of the corridor geometry mostly in the direction perpendicular to its axis. The longitudinal resolution is, however, limited by one of the initial steps of the proposed methodology—extracting slices of selected depth from the input point cloud for cross-section generation. The value of cross-section depth should be chosen individually for each use case scenario, depending on the expected or desired to be detected corridor geometry deviations regarding their dimension along the corridor. The selection of too wide cross sections may result in not detecting significant geometry disturbances or their overestimation. On the other hand, choosing too narrow point cloud slices might result in a low number of points establishing the profiles, making them an unsatisfactory representation of the real corridor geometry.

As mentioned above, geometry analysis of large underground structures consisting of dozens of kilometers of mining excavations may be of particular interest of mine maintenance services, especially geomechanical engineers or ventilation crews. The proposed methodology for spatial data utilization, combined with an appropriate statistics selection, may serve as a tool for mine ventilation optimization. As the geometry of the airways is strongly related to airflow behavior, obtained results may be strongly informative in terms of ventilation performance and indicate potential actions needed to improve the current state. On the other hand, the results may potentially contribute to reducing the aerological hazards. Moreover, a 3D point cloud from geometry measurements adequately processed with respect to spatial mesh of underground structure development may also provide a basis for CFD methods analysis. The insights presented will constitute future work by the authors in subsequent articles based on the methodology defined in this work.

Results acquired with the introduced method at the testbed in the closed mine in Złoty Stok prove the suitability of this solution for automatic evaluation of the corridor's geometry. Moreover, it was demonstrated that for similar use cases, a point cloud obtained with SLAM manifests sufficient accuracy for subsequent processing. This shows possibilities for further automation of the proposed method utilizing an autonomous mobile robotic system. Other important directions of research include automatization of inspection of more complex scenes, (e.g., a grid of underground corridors) and testing various measurement techniques for obtaining the point cloud for different use case scenarios.

## 7. Conclusions

In this paper, the authors have attempted to solve the problem of geometry assessment of large underground structures. As shown in the state of the art provided, this subject is a particular matter in the field of underground mining, but also tunneling. To meet the specific requirements defined by the complex nature of this problem, several issues needed to be considered during the solution development. It covers primarily a quick and reliable measurement method allowing geometry-related spatial data acquisition with relatively low effort and capital outlay. Secondly, processing obtained data should create a field for deviation detection and variation tracking and analysis employing informative features. Such an approach in the final stage would lead to a classification of certain tunnel segments, which in terms of geometry demands taking actions aimed at maintaining mine operation according to defined specifications.

The authors proposed a simple yet effective method of assessing the quality of excavated corridors. Input data constitute the point cloud measured by a LiDAR-based scanning system. This kind of data source is not very expensive and thus may be easily accessible. During the experiment conducted by the authors, the LiDAR sensor was attached to a mobile platform. Although this solution may be beneficial from different perspectives, the process of collecting data with such a device itself can also be carried out manually. The input data type does not influence the data processing methodology. However, depending on the desired purpose, it may result in different accuracy. In practice, a digital model of underground structures may be constructed based on data captured utilizing any method, such as TLS implementation or photogrammetry.

The initial part of the proposed methodology is related to 3D data preprocessing, which is required to apply statistical parameterization and analysis. The core of the method is based on creating a multivariate parameterization of a corridor section concerning its length. The calculation of selected statistics allowed the determination of more general features. Segmentation and classification of the created segments based on those features is performed concerning the quality defined by them. The set of cross sections arranged in groups provides a global and quantitative measure of the quality of tunnel geometry.

The overriding goal for the authors of this paper was to use simple statistics to determine some informative features. Obtaining information from them particularly was not of interest; however, even by themselves, they may be useful for some specific use cases. In terms of qualitative evaluation, one could go a step further and correlate the segmentation results with one of the statistics, i.e., deviation from the reference model, or a roughness coefficient. Based on this, one could conclude that one of the classes corresponds to a better quality of excavation, while the other class to a worse one. However, this consideration is dependent on the particular use case and is not of interest within this paper.

In the long-term perspective, the progressing excavation in a given location will produce new sections of a corridor that can be assessed as a continuation of a previously parameterized dataset. Such an approach can allow for quasi-real-time evaluation of an excavated corridor, paying attention to any deviations and eventually making corrections and maybe even adjustments to the excavation practices. Paying attention to the optimization of corridor geometry can have significant importance for the efficiency of key technological processes undertaken in underground mines.

The key advantage of the proposed approach is the ability to massively reduce the data volume at several levels of abstraction. Longitudinal and circumferential resampling allows us not only to standardize the operational structure for 3D data (which makes the analysis relatively easy and comfortable), but also to adjust the resolution to the needs of the particular case and inference, which allows making the dataset significantly smaller and the analysis significantly faster. Moreover, the parameterization step allows us to further reduce the operational dataset to only several variables in the longitudinal domain. It also makes the post-processing step (as the segmentation proposed in this paper) very efficient.

**Author Contributions:** Conceptualization, R.Z., J.W. and A.W.; methodology, J.W. and R.Z.; software, J.W.; validation, R.Z., J.W., A.W. and P.T.; investigation, A.W.; resources, P.T. and A.W.; data curation, P.T, A.W.; writing—original draft preparation, A.W., J.W., R.Z. and P.T.; writing—review and editing, A.W., J.W. and R.Z.; visualization, J.W.; supervision, R.Z.; funding acquisition, J.W. and A.W. All authors have read and agreed to the published version of the manuscript.

**Funding:** This activity has received funding from the European Institute of Innovation and Technology (EIT), a body of the European Union, under the Horizon 2020, the EU Framework Programme for Research and Innovation. This work is supported by EIT RawMaterials GmbH under Framework Partnership Agreement No. 21119 (V0T3D: Ventilation Optimizing Technology based on 3D-scanning) and No. 19018 (AMICOS. Autonomous Monitoring and Control System for Mining Plants). Scientific work published within the framework of an international project co-financed from the funds of the program of the Minister of Science and Higher Education titled "PMW" 2020-2021; contract no. 5163/KAVA/2020/2021/2.

**Institutional Review Board Statement:** Not applicable.

**Informed Consent Statement:** Not applicable.

**Data Availability Statement:** Not applicable.

**Acknowledgments:** Supported by the Foundation for Polish Science (FNP)—Jacek Wodecki.

**Conflicts of Interest:** The authors declare no conflict of interest.

## References

1. Person, P.A.; Holmberg, R.; Lee, J. *Rock Blasting and Explosives Engineering*; CRC Press: Boca Raton, FL, USA, 1993; p. 560.
2. AFENI, T.B. Optimization of drilling and blasting operations in an open pit mine—The SOMAIR experience. *Min. Sci. Technol. (China)* **2009**, *19*, 736–739. [CrossRef]
3. Yilmaz, O.; Unlu, T. An application of the modified Holmberg–Persson approach for tunnel blasting design. *Tunn. Undergr. Space Technol.* **2014**, *43*, 113–122. [CrossRef]
4. Navarro, J.; Sanchidrián, J.; Segarra, P.; Castedo, R.; Costamagna, E.; López, L. Detection of potential overbreak zones in tunnel blasting from MWD data. *Tunn. Undergr. Space Technol.* **2018**, *82*, 504–516. [CrossRef]
5. Bieniawski, Z. Principles and methodology of design for excavations in geologic media. *Res. Eng. Des.* **1993**, *5*, 49–58. [CrossRef]
6. Brady, B.H.; Brown, E.T. *Rock Mechanics: For Underground Mining*; Springer Science & Business Media: Berlin, Germany, 1993.
7. Jang, H.; Kawamura, Y.; Shinji, U. An empirical approach of overbreak resistance factor for tunnel blasting. *Tunn. Undergr. Space Technol.* **2019**, *92*, 103060. [CrossRef]
8. Islam, M.R.; Hayashi, D.; Kamruzzaman, A. Finite element modeling of stress distributions and problems for multi-slice longwall mining in Bangladesh, with special reference to the Barapukuria coal mine. *Int. J. Coal Geol.* **2009**, *78*, 91–109. [CrossRef]
9. Di Bartolo, S.; Salvini, R. Multitemporal Terrestrial Laser Scanning for Marble Extraction Assessment in an Underground Quarry of the Apuan Alps (Italy). *Sensors* **2019**, *19*, 450. [CrossRef]
10. Benton, D.J.; Iverson, S.R.; Martin, L.A.; Johnson, J.C.; Raffaldi, M.J. Volumetric measurement of rock movement using photogrammetry. *Int. J. Min. Sci. Technol.* **2016**, *26*, 123–130. [CrossRef] [PubMed]
11. Kajzar, V.; Kukutsch, R.; Heroldova, N. Verifying the possibilities of using a 3D laser scanner in the mining underground. *Acta Geodyn. Geomater.* **2015**, *12*, 51–58. [CrossRef]

12. Kukutsch, R.; Kajzar, V.; Konicek, P.; Waclawik, P.; Ptacek, J. Possibility of convergence measurement of gates in coal mining using terrestrial 3D laser scanner. *J. Sustain. Min.* **2015**, *14*, 30–37. [CrossRef]
13. Riveiro, B.; DeJong, M.; Conde, B. Automated processing of large point clouds for structural health monitoring of masonry arch bridges. *Autom. Constr.* **2016**, *72*, 258–268. [CrossRef]
14. Teri, S.S.; Musliman, I.A. Machine learning in big lidar data: A review. *Int. Arch. Photogramm. Remote Sens. Spat. Inf. Sci.* **2019**, *42*, 641–644. [CrossRef]
15. Ibarra, J.; Maerz, N.H.; Franklin, J.A. Overbreak and underbreak in underground openings part 2: Causes and implications. *Geotech. Geol. Eng.* **1996**, *14*, 325–340. [CrossRef]
16. Mei, J.; Zhang, W.; Xu, B.; Zhu, Y.; Wang, B. Optimization Methods of Blasting Parameters of Large Cross-Section Tunnel in Horizontal Layered Rock Mass. *Geotech. Geol. Eng.* **2021**, *39*, 5309–5323. [CrossRef]
17. Singh, S.P.; Xavier, P. Causes, impact and control of overbreak in underground excavations. *Tunn. Undergr. Space Technol.* **2005**, *20*, 63–71. [CrossRef]
18. Chen, J.; Qiu, W.; Zhao, X.; Rai, P.; Ai, X.; Wang, H. Experimental and numerical investigation on overbreak control considering the influence of initial support in tunnels. *Tunn. Undergr. Space Technol.* **2021**, *115*, 104017. [CrossRef]
19. Trisugiwo, M.; Zabidi, H.; Ahmad, F. Joint analysis to evaluate geological overbreak in excavation of surge chamber cavern. *Procedia Chem.* **2016**, *19*, 751–756. [CrossRef]
20. Jang, H.; Topal, E. Optimizing overbreak prediction based on geological parameters comparing multiple regression analysis and artificial neural network. *Tunn. Undergr. Space Technol.* **2013**, *38*, 161–169. [CrossRef]
21. Liu, K.; Liu, B. Optimization of smooth blasting parameters for mountain tunnel construction with specified control indices based on a GA and ISVR coupling algorithm. *Tunn. Undergr. Space Technol.* **2017**, *70*, 363–374. [CrossRef]
22. Mohammadi, H.; Barati, B.; Chamzini, A.Y. Prediction of blast-induced overbreak based on geo-mechanical parameters, blasting factors and the area of tunnel face. *Geotech. Geol. Eng.* **2018**, *36*, 425–437. [CrossRef]
23. Foderà, G.; Voza, A.; Barovero, G.; Tinti, F.; Boldini, D. Factors influencing overbreak volumes in drill-and-blast tunnel excavation. A statistical analysis applied to the case study of the Brenner Base Tunnel–BBT. *Tunn. Undergr. Space Technol.* **2020**, *105*, 103475. [CrossRef]
24. Maerz, N.H.; Ibarra, J.; Franklin, J.A. Overbreak and underbreak in underground openings Part 1: Measurement using the light sectioning method and digital image processing. *Geotech. Geol. Eng.* **1996**, *14*, 307–323. [CrossRef]
25. Franklin, J.A.; Ibarra, J.T.; Maerz, N.H. Blast overbreak measurement by light sectioning. *Int. J. Min. Geol. Eng.* **1989**, *7*, 323–331. [CrossRef]
26. Heritage, G.; Large, A. *Laser Scanning for the Environmental Sciences*; John Wiley & Sons: Hoboken, NJ, USA, 2009.
27. Trybała, P.; Blachowski, J.; Błażej, R.; Zimroz, R. Damage Detection Based on 3D Point Cloud Data Processing from Laser Scanning of Conveyor Belt Surface. *Remote Sens.* **2021**, *13*, 55. [CrossRef]
28. Lindenbergh, R.; Uchanski, L.; Bucksch, A.; Van Gosliga, R. Structural monitoring of tunnels using terrestrial laser scanning. *Rep. Geod.* **2009**, 231–238. Available online: https://yadda.icm.edu.pl/yadda/element/bwmeta1.element.baztech-article-PWAB-0005-0007/c/httpwww_rog_gik_pw_edu_plphocadownloadnr8727.pdf (accessed on 7 December 2021).
29. Loupos, K.; Doulamis, A.D.; Stentoumis, C.; Protopapadakis, E.; Makantasis, K., Doulamis, N.D.; Amditis, A.; Chrobocinski, P.; Victores, J.; Montero, R.; et al. Autonomous robotic system for tunnel structural inspection and assessment. *Int. J. Intell. Robot. Appl.* **2018**, *2*, 43–66. [CrossRef]
30. Amedjoe, C.G.; Agyeman, J. Assessment of effective factors in performance of an open stope using cavity monitoring system data: A case study. *J. Geol. Min. Res.* **2015**, *7*, 19–30.
31. Warneke, J.; Dwyer, J.; Orr, T. Use of a 3-D scanning laser to quantify drift geometry and overbreak due to blast damage in underground manned entries. In Proceedings of the 1st Canada-US Rock Mechanics Symposium, Vancouver, BC, Canada, 27–31 May 2007.
32. Zou, B.; Luo, Z.; Wang, J.; Hu, L. Development and Application of an Intelligent Evaluation and Control Platform for Tunnel Smooth Blasting. *Geofluids* **2021**, *2021*, 6669794. [CrossRef]
33. Che, E.; Jung, J.; Olsen, M. Object recognition, segmentation, and classification of mobile laser scanning point clouds: A state of the art review. *Sensors* **2019**, *19*, 810. [CrossRef]
34. Gikas, V. Three-dimensional laser scanning for geometry documentation and construction management of highway tunnels during excavation. *Sensors* **2012**, *12*, 11249–11270. [CrossRef]
35. Luo, Y.; Chen, J.; Xi, W.; Zhao, P.; Qiao, X.; Deng, X.; Liu, Q. Analysis of tunnel displacement accuracy with total station. *Meas. J. Int. Meas. Confed.* **2016**, *83*, 29–37. [CrossRef]
36. Chang, X.; Wang, H.; Zhang, Y.; Wang, F.; Li, Z. Bayesian prediction of tunnel convergence combining empirical model and relevance vector machine. *Meas. J. Int. Meas. Confed.* **2022**, *188*, 110621. [CrossRef]
37. Wang, D.; Luo, J.; Shen, K.; Gao, L.; Li, F.; Wang, L. Analysis of the causes of the collapse of a deep-buried large cross section of loess tunnel and evaluation of treatment measures. *Appl. Sci.* **2022**, *12*, 161. [CrossRef]
38. Tan, Z.; Li, S.; Yang, Y.; Wang, J. Large deformation characteristics and controlling measures of steeply inclined and layered soft rock of tunnels in plate suture zones. *Eng. Fail. Anal.* **2022**, *131*, 105831. [CrossRef]
39. Hou, G.Y.; Li, Z.X.; Hu, Z.Y.; Feng, D.X.; Zhou, H.; Cheng, C. Method for tunnel cross-section deformation monitoring based on distributed fiber optic sensing and neural network. *Opt. Fiber Technol.* **2021**, *67*, 102704. [CrossRef]

40. Puente, I.; González-Jorge, H.; Martínez-Sánchez, J.; Arias, P. Automatic detection of road tunnel luminaires using a mobile LiDAR system. *Meas. J. Int. Meas. Confed.* **2014**, *47*, 569–575. [CrossRef]
41. Vanneschi, C.; Mastrorocco, G.; Salvini, R. Assessment of a rock pillar failure by using change detection analysis and FEM modelling. *ISPRS Int. J. Geo-Inf.* **2021**, *10*, 774. [CrossRef]
42. Xu, T.; Xu, L.; Li, X.; Yao, J. Detection of Water Leakage in Underground Tunnels Using Corrected Intensity Data and 3D Point Cloud of Terrestrial Laser Scanning. *IEEE Access* **2018**, *6*, 32471–32480. [CrossRef]
43. Webber-Youngman, R.; Grobler, H.; Gazi, T.; Stroh, F.; van der Vyver, A. The impact of forensic laser scanning technology on incident investigations in the mining industry. *J. South. Afr. Inst. Min. Metall.* **2019**, *119*, 817–824. [CrossRef]
44. Voit, K.; Amvrazis, S.; Cordes, T.; Bergmeister, K. Drill and blast excavation forecasting using 3D laser scanning [Ausbruchprognose beim Sprengvortrieb mittels 3D-Laser scanning]. *Geomech. Tunnelbau* **2017**, *10*, 298–316.
45. Otero, R.; Lagüela, S.; Garrido, I.; Arias, P. Mobile indoor mapping technologies: A review. *Autom. Constr.* **2020**, *120*, 103399. [CrossRef]
46. Kang, Z.; Zhang, L.; Tuo, L.; Wang, B.; Chen, J. Continuous extraction of subway tunnel cross sections based on terrestrial point clouds. *Remote Sens.* **2013**, *6*, 857–879. [CrossRef]
47. Han, S.; Cho, H.; Kim, S.; Jung, J.; Heo, J. Automated and efficient method for extraction of tunnel cross sections using terrestrial laser scanned data. *J. Comput. Civ. Eng.* **2013**, *27*, 274–281. [CrossRef]
48. Yoon, J.S.; Sagong, M.; Lee, J.; Lee, K.s. Feature extraction of a concrete tunnel liner from 3D laser scanning data. *NDT E Int.* **2009**, *42*, 97–105. [CrossRef]
49. Adoko, A.C.; Jiao, Y.Y.; Wu, L.; Wang, H.; Wang, Z.H. Predicting tunnel convergence using Multivariate Adaptive Regression Spline and Artificial Neural Network. *Tunn. Undergr. Space Technol.* **2013**, *38*, 368–376. [CrossRef]
50. Fei, J.; Wu, Z.; Sun, X.; Su, D.; Bao, X. Research on tunnel engineering monitoring technology based on BPNN neural network and MARS machine learning regression algorithm. *Neural Comput. Appl.* **2021**, *33*, 239–255. [CrossRef]
51. Grilli, E.; Menna, F.; Remondino, F. A review of point clouds segmentation and classification algorithms. *Int. Arch. Photogramm. Remote Sens. Spat. Inf. Sci.* **2017**, *42*, 339. [CrossRef]
52. Remondino, F.; El-Hakim, S.; Gruen, A.; Zhang, L. Turning images into 3-D models: Developments and performance analysis of image matching for detailed surface reconstruction of heritage objects. *IEEE Signal Process. Mag.* **2008**, *25*, 55–64. [CrossRef]
53. Remondino, F.; Nocerino, E.; Toschi, I.; Menna, F. A critical review of automated photogrammetric processing of large datasets. *Int. Arch. Photogramm. Remote Sens. Spat. Inf. Sci.* **2017**, *42*, 591–599. [CrossRef]
54. Przylibski, T.A. Radon and its daughter products behaviour in the air of an underground tourist route in the former arsenic and gold mine in Złoty Stok (Sudety Mountains, SW Poland). *J. Environ. Radioact.* **2001**, *57*, 87–103. [CrossRef]
55. Muszer, A. Gold at Złoty Stok–history, exploitation, characteristic and perspectives. *Arch. Mineral. Monogr.* **2011**, *2*, 45–62.
56. Szrek, J.; Zimroz, R.; Wodecki, J.; Michalak, A.; Góralczyk, M.; Worsa-Kozak, M. Application of the infrared thermography and unmanned ground vehicle for rescue action support in underground mine—The amicos project. *Remote Sens.* **2021**, *13*, 69. [CrossRef]
57. Zimroz, P.; Trybała, P.; Wróblewski, A.; Góralczyk, M.; Szrek, J.; Wójcik, A.; Zimroz, R. Application of UAV in search and rescue actions in underground mine—A specific sound detection in noisy acoustic signal. *Energies* **2021**, *14*, 3725. [CrossRef]
58. Koide, K.; Miura, J.; Menegatti, E. A portable 3d lidar-based system for long-term and wide-area people behavior measurement. *IEEE Trans. Hum. Mach. Syst.* **2018**, *16*, 1729881419841532.
59. Trybała, P. LiDAR-based Simultaneous Localization and Mapping in an underground mine in Złoty Stok, Poland. *IOP Conf. Ser. Earth Environ. Sci.* **2021**, *942*, 012035. [CrossRef]
60. Velodyne Lidar Puck Data Sheet. Available online: https://velodynelidar.com/products/puck/#downloads (accessed on 7 December 2021).
61. Riegl VZ-400i Terrestrial Laser Scanner Data Sheet. Available online: http://www.riegl.com/uploads/tx_pxpriegldownloads/RIEGL_VZ-400i_Datasheet_2020-10-06.pdf (accessed on 7 December 2021).
62. Akima, H. A new method of interpolation and smooth curve fitting based on local procedures. *J. ACM* **1970**, *17*, 589–602. [CrossRef]
63. Silverman, B.W. *Density Estimation for Statistics and Data Analysis*; Routledge: London, UK, 2018.
64. De Loera, J.; Rambau, J.; Santos, F. *Triangulations: Structures for Algorithms and Applications*; Springer Science & Business Media: Berlin, Germany, 2010; Volume 25.
65. Guibas, L.J.; Knuth, D.E.; Sharir, M. Randomized incremental construction of Delaunay and Voronoi diagrams. *Algorithmica* **1992**, *7*, 381–413. [CrossRef]
66. Moore, B. Principal component analysis in linear systems: Controllability, observability, and model reduction. *IEEE Trans. Autom. Control* **1981**, *26*, 17–32. [CrossRef]
67. Peter, D.H. Kernel estimation of a distribution function. *Commun. Stat. Theory Methods* **1985**, *14*, 605–620. [CrossRef]
68. Silverman, B.W. Density estimation for statistics and data analysis. In *Monographs on Statistics and Applied Probability*; CRC Press: Boca Raton, FL, USA, 1986; Volume 26.

energies

Article

# Underground Mine Tunnel Modelling Using Laser Scan Data in Relation to Manual Geometry Measurements

Jakub Janus * and Piotr Ostrogórski

Strata Mechanics Research Institute of the Polish Sciences Academy, Reymonta 27, 30-059 Krakow, Poland; ostrogorski@imgpan.pl
* Correspondence: janus@imgpan.pl

**Abstract:** Underground mine tunnels, drifts, and mine headings are susceptible to the impact of convergence. The convergence has a big influence on further measurements such as airflow and the volume concentration of methane and other gases. In most cases, deformation of arch supports lead to getting a smaller cross-section area. A comparison is made between five methods of measuring the cross-sectional area of a mine tunnel. The reference size of the six cross-section mine drift areas were obtained by Terrestrial Laser Scanning, which were then compared with the cross-section areas obtained by four other methods. The following methods were considered: area calculation using CAD software, an empirical method, approximation by a semi-ellipse and approximation by a semi-ellipse with attached straight sections. This article presents the quantitative and qualitative differences of the obtained results. Differences in the calculated cross-sectional areas of the mine drift are discussed, and reasons for the differences are determined. In addition, the advantages and disadvantages of each method are indicated.

**Keywords:** arch support; laser scanning; mine drift area; mine tunnel; cross-section

check for updates

**Citation:** Janus, J.; Ostrogórski, P. Underground Mine Tunnel Modelling Using Laser Scan Data in Relation to Manual Geometry Measurements. *Energies* **2022**, *15*, 2537. https://doi.org/10.3390/en15072537

Academic Editor: Sergey Zhironkin

Received: 4 March 2022
Accepted: 28 March 2022
Published: 30 March 2022



## 1. Introduction

The modelling of underground mine tunnels, drifts and mine headings is of great importance in many tasks connected with maintenance, inventory, underground transport, CFD modelling and accuracy of measurements [1–3]. In these tasks, the 3D models are necessary to reconstruct the measurement or inventory object. In many scientific publications, we see very simplified 3D or only 2D models which were made using technical design documentation or manual measurements [4,5]. These models can be improved by laser scanning data and a reconstruction model from a data point cloud [6]. It takes a lot of time and requires a lot of effort to make it accurate and to a satisfactory degree of simplification. If we think about monitoring of geometry, we need to keep updates in a loop after any shape change. It can be done using machine learning and automatic recognition objects from a point cloud [7]. Underground mine spaces are susceptible to the impact of convergence. The convergence has a great influence on further measurements such as airflow and volume concentration of methane and other gases. In most cases, deformations of arch supports lead to getting a smaller cross-section area, which is an important factor in any flow measurements [8].

The convergence is a result of strata mechanics caused by stress and rock deformation. Stress and deformation can be derived from mining works [9–11].

In mine ventilation measurements, a necessary value that characterizes the flow is the flow rate. There are several methods to determine the flow rate, including the manual traverse method, which is the most commonly used method in underground mining [12,13]. This is an indirect method, where average air velocity and cross-sectional area are measured. The product of these two values gives the flow rate. In addition to the traverse method, and mainly for accurate measurements or scientific research, the multi-spot flow velocity field measurement system (SWPPP) is used [14]. This uses a system of spaced vane anemometers in the measurement cross-sections, which helps to measure air velocity at numerous points of the cross-sections at the same time. To calculate the air flow rate with the use of the SWPPP system, it is necessary to know the exact shape of the mine tunnel where the system is placed. Usually, the outline of the mine tunnel is obtained by measuring the tunnel's height and width and delineating the cross-section using CAD tools. Using equipment of this kind, such outlines can be drawn in several ways depending on the type of arch yielding support. In the simplest case, the cross-section outline can be approximated by a semi-ellipse, and in a slightly more complex case, by a semi-ellipse with attached straight sections. In the latter case, it is necessary to know the length of the straight sections or to take nominal lengths. The most complex case of delineation of the cross-section involves using the manufacturer's standards and catalog data and taking into account changes in the shape of the mine tunnel due to clamping [15]. The Creo Parametric software used by the authors allows one to define nodes between individual elements of the arch yielding support and to define moving dimensions, which, based on the many dependencies between individual elements of the arch, can be determined by the software. As a reference method, the use of laser scanning and point cloud processing tools is adopted. In many cases, the typical methods of measuring cross-sections in mines fail because of the large uncertainties. The source of these errors can be found in the strata forces which act on the arch yielding support [16]. The forces depend on the particular place in the mine: closer to the excavation process, the forces and strata tension are normally much greater. In the present case, we consider a ventilation tunnel which is sufficiently far from the main excavation area.

Information about the real shape and proper cross-sectional area of the mine tunnel are also very important in Computer Fluid Dynamics of air flow in the underground mine environment [17–20]. Proper cross-sectional area determination is necessary to obtain correct results of the numerical simulations. Very often, CFD simulations of air flow are carried out on simplified models. These simplifications consist of assuming the same cross-sectional area in the entire numerical model [21].

Constant development of new technologies has resulted in the proliferation of methods that were previously restricted to a small group of specialists. One example of this type of technology is Terrestrial Laser Scanning (in short TLS). The very fast determination of X, Y and Z point coordinates, called point cloud, is used to get the image of an object. From the spatially oriented point cloud obtained during the scanning process, it is possible to generate a scanned object 3D model. Development of the post-processing software has made the processing of such a large amount of data more efficient.

The capabilities of this technology, in conjunction with the CAD systems, create new possibilities. It is commonly used for land surveying to visualize changes taking place in the field [22,23], deep tunnels' evolution [24–26] and deformation of large-size devices such as wind turbine towers [27]. It can be in combination with a 3D printing technology to reproduce the actual shape of the object at a reduced scale [28,29]. This combination was used in [30] to reproduce the historical building ornamental components. It is also used in underground mining to observe temporary evaluation and changes of mine tunnels' geometry over time [31]. Another application of this technology is its use in CFD modelling. Accurate information about the shape of mine drift, in the form of a point cloud, was used in [32] to design the actual shape of the tested object. This work also presents differences in simulation results on simplified model geometries, the design of which was possible using the laser scanning technology. The combination of laser scanning technology with Virtual Reality gives even greater possibilities. Authors of work [33] show how effectively those technologies can improve mining safety.

In this article, the authors decided to use laser scanning technology to determine the exact cross-sectional area of a mine drift. These data were used to compare with the results of determining the cross-sectional area using other methods. This article presents the quantitative and qualitative differences of the obtained results.

## 2. Measurements of Mine Tunnel Geometry Using Terrestrial Laser Scanning

The functioning of a laser scanner is based on the phase method of measuring distances, based, in turn, on the properties of the wave emitted by the laser. The instrument emits a laser beam of known frequency, which—after meeting the object—comes back to the instrument. The phase of the returning light is compared with the phase of the known frequency. The difference between the two peak values is called the phase shift.

This is a polar measurement method, in which the location of a given point is determined by means of the horizontal and the vertical angle, as well as the distance from the measuring point. The angles are determined from the location of the mirrors scattering the laser beam. To establish the location of the point P, one has to know the length of the ray ρ, and the values of the angles α and ß (as illustrated in Figure 1). The scanners used to measure the phase shift are among the most accurate instruments for laser scanning intended for commercial purposes, as they make it possible to obtain the desired data very quickly, and generate scans with a very high resolution.

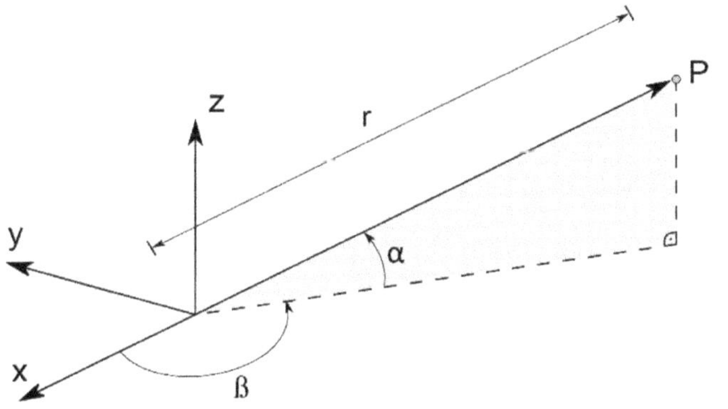


**Figure 1.** Schematic diagram of the functioning of TLS.

### *2.1. Measuring Device*

For measurements of a mine drift geometry, the FARO Focus 3D laser scanner was used [34]. The range of the instrument is 0.6–120 m; the laser beam falls at an angle of 90° on a surface whose reflectance is 90%. The scanner is capable of a high speed of measurement, from 120,000 to 980,000 points/s, depending on the scanning resolution. Uncertainty of measurement is ±0.002 m [35]. The scanning visual range of the device is 360° in the horizontal axial and 305° in the vertical axial with a wavelength of over 900 nm and beam divergence 0.16 mrd. During the measurements, data of the scanned points' coordinates are saved on a portable disk and can therefore be transferred to a computer easily.

### *2.2. The Measurement Site*

Measurements using the 3D scanner were performed in the ZG Sobieski mine, in the Grodzisko cross-cut gallery, level 300. The measurement site was chosen in the neighborhood of a turn, which gave the researchers wide-ranging measurement possibilities.

The mine tunnel was secured with 8-KS/KO-21 type arch supports. Due to the complex structure of the mine tunnel, authors had to choose appropriate places for the 3D laser measurements [36–38]. For the best mapping of the mine tunnel geometry, a double scanning site in the object's cross-section was established. The scanning sites were placed at the left and the right side of the mine tunnel, which minimizes the areas where no scanning was performed. Next, authors had to place the appropriate number of markers along the mine tunnel, thanks to which, in the preprocessing process, it was possible to combine subsequent scans placing them at a chosen set of coordinates. To obtain a full spatial model of the mine tunnel, the whole section of the gallery was divided into 11 scanning cross-sections, which resulted in 22 measurement sites. The first scanning cross-section was 23 m before the bend, and the last was 56 m after the bend (Figure 2).

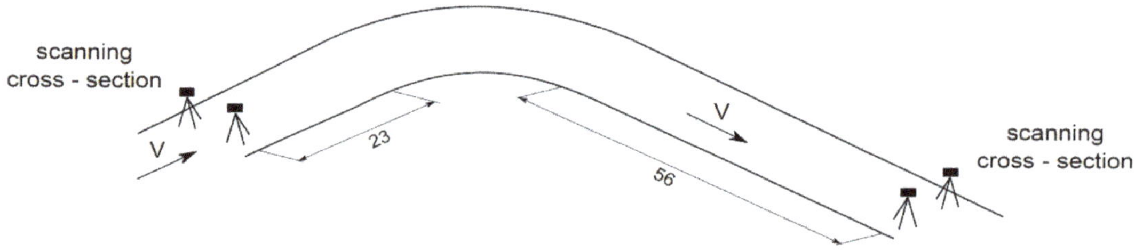


**Figure 2.** Explored section of the mining tunnel.

In situations where the floor is flat, 3D scanner can be placed on a special autonomous guided vehicle [39].

*2.3. Processing of Laser Scanning Data*

First stage of point cloud processing was preprocessing. As a part of the measurement data processing, the data had to be prepared for further use. The most important process of this stage was the point cloud's connection into one set of data. Obtained point cloud was filtered, which involved cleaning and removal of all measurement errors.

As a result of mine tunnel geometry laser scanning and point cloud preprocessing, a very detailed, digital mapping of the whole space of the scanned mine tunnel section was obtained, comprising over 150,000,000 points (Figure 3).

**Figure 3.** Point cloud view of the mine tunnel straight section near the bend.

During the measurements of the mine tunnel geometry, it was decided to divide the relevant part of the mine tunnel into three sections:

- Section 1, before the turn;
- Section 2, the turn;
- Section 3, after the turn.

In each section, the authors chose two cross-sections for further analysis, the first cross-section at the beginning of the section and second at the end of the section (Figure 4). The reason for choosing measurement cross-sections was the fact that the distance between sections reached tens of meters. The cross-sections from Section 1 were located 29 m and

6 m before the turn. The cross-sections from Section 2 were at the entrance to and exit from the turn. The cross-sections from Section 3 were located 6 m and 54 m after the turn. Standard measurements of the mine tunnel's height and width were taken.

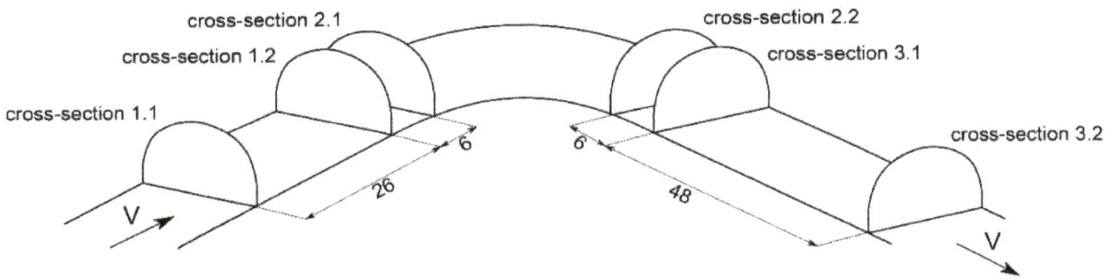


**Figure 4.** Measurement cross-sections.

## 3. Methods of Determining the Cross-Section Area of the Mine Tunnel

### *3.1. Reconstruction with Use of Point Cloud*

To obtain accurate data on the mine tunnel's cross-sectional area in selected measurement cross-sections, it was necessary to use reverse engineering [40,41], widely described in [17,42].

As a result of irregular deformation of the arch yielding supports, it was necessary to make cross-sections for each arch. This operation involves defining the vertical cross-section through the center of each arch. Next, the cross-section curve, consisting of several lines connecting the triangles obtained in the process of triangulation, was obtained (Figure 5). To obtain an accurate shape of the arch, based on the cross-section curve, the cross-section line was determined (Figure 6). Analogously, the shape line of the arch was determined using the horizontal cross-section. Using the "swept surface" function in the CAD software, after defining the arch shape line as a sketch profile and the cross-section line as a circular profile, it was possible to create the geometry of the arch support (Figure 7).

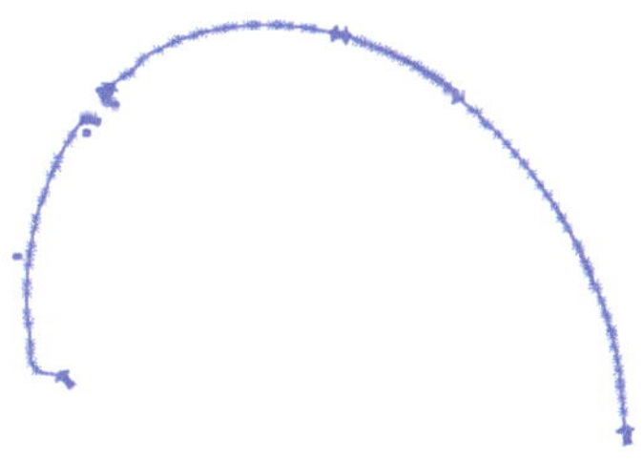

**Figure 5.** Cross-section curve.

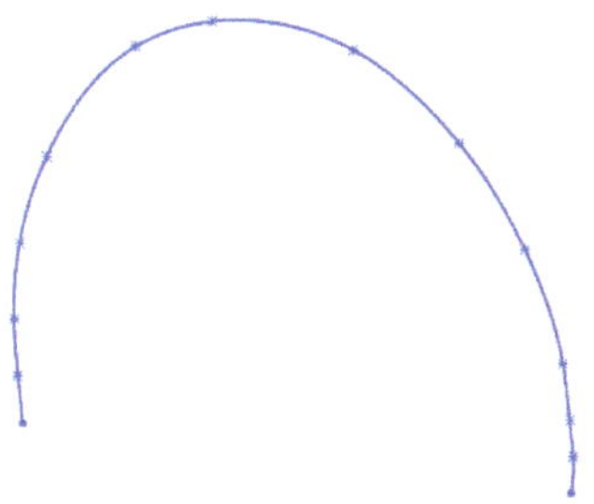

**Figure 6.** Cross-section line.

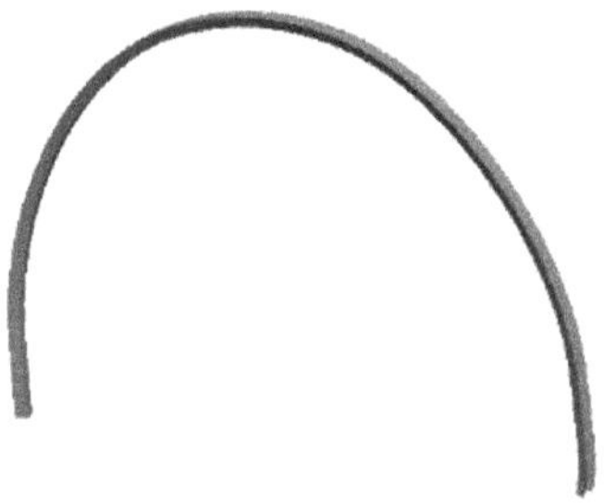

**Figure 7.** Arch yielding support geometry.

The floor and rails were mapped analogously to the arches. After obtaining the cross-section curves, the cross-sectional lines and shape lines of the floor and rails were determined, respectively. In the case of pipelines, a leading line was determined to map the arrangement of elements in the mine tunnel. Then, using the "swept surface" CAD function, the geometries of the floor, rails and pipelines were created.

As a result, the geometry of the 3D model was obtained with the exact representation of the shape of the real object (Figure 8), and with the exact values of the cross-sectional areas in the three measured cross-sections.

**Figure 8.** Obtained geometry model, view of the mine tunnel straight section near the bend.

### 3.2. *The Area as a Result of Parametric Sketch and Integration Method Using CAD*

Usually, to calculate the air flow rate, it is necessary to reconstruct the shape of the mine tunnel at the place where the measurements were carried out with the SWPPP system. Then, catalog data, such as the length of the arch yielding support, are used and adapted in accordance with the susceptibilities of the supports.

For approximation of the shape of the cross-section, catalog data from the manufacturer of ŁP arch yielding supports with KS/KO sections were used (Figure 9). ŁP arch yielding supports were made according to the PN-G-15001:1973 standard [43]. The table below contains data corresponding to the analyzed case.

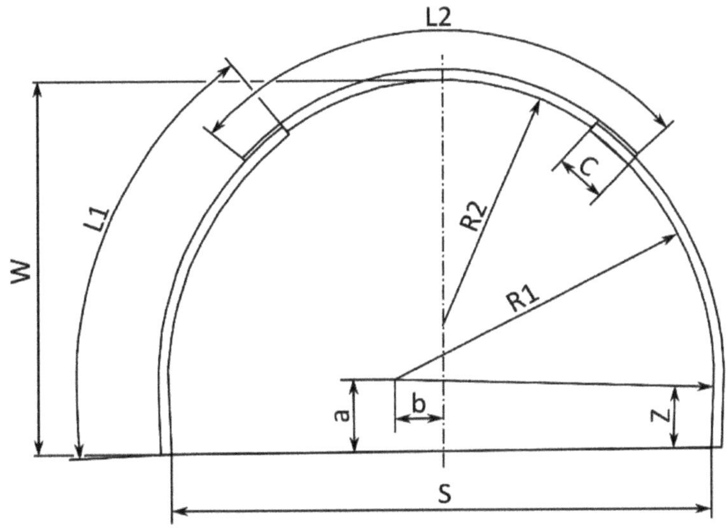


**Figure 9.** Dimensions of arch yielding support 8-KS/KO-21.

According to the catalog data (Table 1) the nominal width and height of this arch are 4.7 m × 3.3 m. Based on catalog data and the ability to define nodes and moving dimensions, the outlines of the cross-section area were delineated. Moving dimensions were limited values. The lengths of curves C and Z were taken as variable values. The input values were the height and width of the mine tunnel measured on the reference outline. The width was measured at a height 0.8 m under the floor. The floor was delineated as a straight line, due to the lack of sufficient information about its shape. The height was measured from the floor to the highest point on the arch.

**Table 1.** Data for arch yielding support 8-KS/KO-21.

| S [m] | W [m] | R1 [m] | L1 [m] | Z [m] | R2 [m] | L2 [m] | C [m] |
|---|---|---|---|---|---|---|---|
| 4.7 | 3.3 | 2.8 | 3.2 | 0.5 | 2.2 | 3.8 | 0.45 |

where:

S—width of the arch
W—height of the arch
R1—radius of the side part of the arch
L1—length of the side part of the arch
Z—length of the straight section of the side part of the arch
R2—radius of the roof part of the arch
L2—length of the roof part of the arch
C—length of the arch's tab
a, b—center coordinates of the arch's side part. These values depend on the height and width during fitting of the arch to the actual shape of the mine tunnel

### *3.3. Analytical Methods*

Methods of approximation of cross-sectional areas were analyzed in terms of accuracy relative to the reference method, namely the reconstruction of the real shape of the arch from many points obtained by the laser scanning method. Of the methods considered below, only the empirical method is used in daily ventilation measurements. The remaining methods are used for more complex air flow rate measurements, as in the SWPPP system.

All of these methods, other than the empirical method, can be used to reconstruct the shape of a mine tunnel for the purpose of CFD numerical calculations.

### 3.3.1. Empirical Method

The most frequently used method for calculating cross-sectional areas in underground mining involves measuring the mine tunnel's height and width, and multiplying their product by 0.8 [44]. The value 0.8 results from the shape of the arch yielding support, and is therefore called the cross-sectional shape coefficient. The height of the mine tunnel is measured in the middle of the tunnel's width, from the highest point on the arch to the corresponding point on the floor. The width is measured by placing measuring points on the straight section of the side part of the arch. In some cases, just a part of the straight section is above the floor, and it sometimes happens that this straight section is much shorter or completely inaccessible. This makes it difficult to perform the measurement properly, and increases the uncertainty of the result.

$$A_g = 0.8 \cdot H \cdot W \tag{1}$$

where:

*H*—height of mine tunnel
*W*—width of mine tunnel

### 3.3.2. Approximation by a Semi-Ellipse

Comparing the equations for the empirical method and the method based on approximation by a semi-ellipse, it is readily seen that they differ only in the cross-sectional shape coefficient: for the empirical method the coefficient is 0.8, and for approximation by a semi-ellipse the coefficient is 0.785.

At many measuring sites, it is found that a coefficient lower than 0.8 is more accurate in calculating the cross-sectional area. This applies to places with high rock stress, for example in walls. The relationship between the floor height and the cross-sectional shape coefficient is shown in the graph below. The floor height is understood as the depth of the arch's side part in the floor.

Figure 10 shows the direction of change in the cross-sectional shape coefficient. It shows that approximation by a semi-ellipse should give better results in places with a higher floor. This applies to places where the side part of the arch is completely embedded in the floor. The equation for approximation by a semi-ellipse is given below.

$$A_e = \frac{1}{4}\pi \cdot H \cdot W \tag{2}$$

where:

*H*—height of mine tunnel
*W*—width of mine tunnel

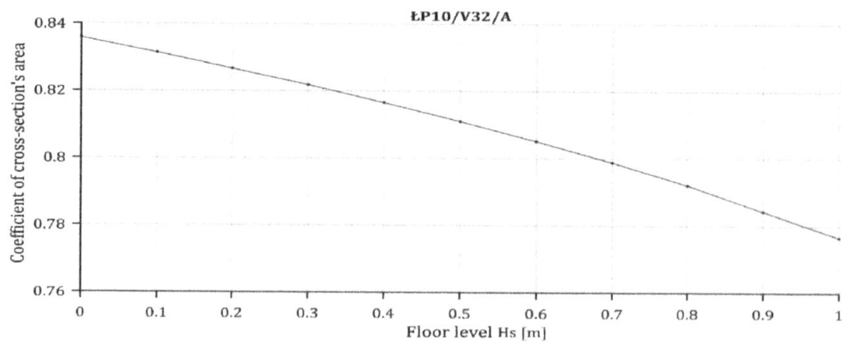


**Figure 10.** Change of cross-sectional shape coefficient as a function of floor height [13].

#### 3.3.3. Approximation by a Semi-Ellipse with Attached Straight Sections

This method improves the approximation based on a semi-ellipse by moving the minor axis of the ellipse up to the level where the straight sections end. The lower part of the cross-section is approximated by a rectangle. Here, the straight sections are assumed to be parallel to each other, which is not the case with the ŁP arch yielding supports. Therefore, the approximation of this section by a rectangle gives rise to certain errors in this method. Greater accuracy would be obtained with approximation by a trapezoid.

The cross-section's outline is approximated by a curve consisting of a semi-ellipse with straight sections attached at the ends.

$$A_{ezop} = \frac{1}{4}\pi \cdot W \cdot (H - Z) + W \cdot Z \tag{3}$$

where:

*H*—height of mine tunnel
*W*—width of mine tunnel
*Z*—length of straight section of side part of arch

The value of Z was taken to be the nominal value of the straight section for the arch type 8-KS/KO-21, which, according to Table 1, is 0.5 m.

## 4. Results

Six outlines of cross-sections were obtained using the laser scanning method. Figure 11 shows the shape of each cross-section. On each of them, a nominal outline was marked according to the considered arch type (Table 1).

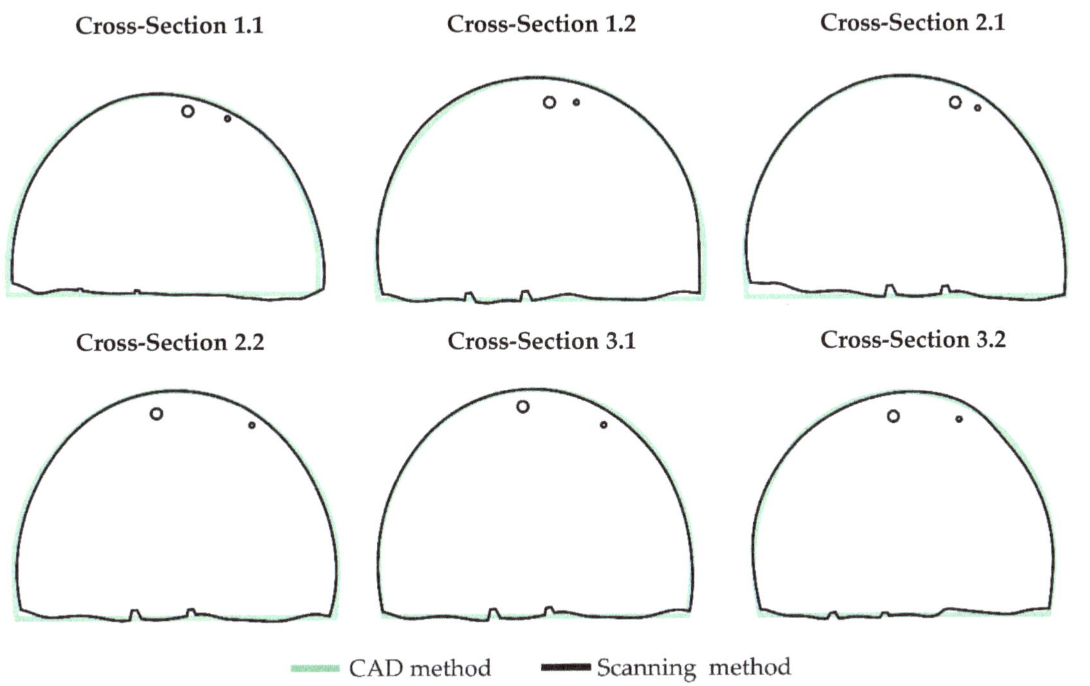


**Figure 11.** Reference cross-section outlines obtained by the laser scanning method presented on the cross-section outlines obtained by CAD method.

Tables 2 and 3 give the results of cross-sectional area measurements calculated by four different methods. Each case was compared with the reference cross-sectional area obtained by laser scanning. The results include the cross-sections of the pipelines under the roof. The total cross-sectional area of the pipelines is 0.048 $m^2$. This does not cause a significant difference in the cross-section of the mine tunnel; however, comparison of the results with the cross-sectional area obtained by laser scanning required their inclusion.

**Table 2.** Results of cross-sectional area measurements.

| | Cross-Section 1.1 [$m^2$] | Difference [$m^2$] | Cross-Section 1.2 [$m^2$] | Difference [$m^2$] | Cross-Section 2.1 [$m^2$] | Difference [$m^2$] |
|---|---|---|---|---|---|---|
| Scanning method | 11.26 | 0 | 13.20 | 0 | 12.66 | 0 |
| CAD method | 11.38 | 0.12 | 13.30 | 0.101 | 13.00 | 0.34 |
| Empirical method | 11.08 | −0.18 | 12.65 | −0.547 | 12.52 | −0.14 |
| Semi-ellipse | 10.88 | −0.38 | 12.42 | −0.779 | 12.26 | −0.40 |
| Semi-ellipse with straight sections | 11.39 | 0.13 | 12.93 | −0.266 | 12.77 | 0.11 |

**Table 3.** Results of cross-sectional area measurements.

| | Cross-Section 2.2 [$m^2$] | Difference [$m^2$] | Cross-Section 3.1 [$m^2$] | Difference [$m^2$] | Cross-Section 3.2 [$m^2$] | Difference [$m^2$] |
|---|---|---|---|---|---|---|
| Scanning method | 13.18 | 0 | 12.99 | 0 | 13.18 | 0 |
| CAD method | 13.49 | 0.31 | 13.03 | 0.046 | 13.49 | 0.31 |
| Empirical method | 12.87 | −0.31 | 12.53 | −0.462 | 12.87 | −0.31 |
| Semi-ellipse | 12.61 | −0.57 | 12.30 | −0.692 | 12.61 | −0.57 |
| Semi-ellipse with straight sections | 13.12 | −0.06 | 12.80 | −0.193 | 13.12 | −0.06 |

Figures 12–14 show differences in the shape of cross-section 1.2. The differences are given for three methods: the CAD method (using catalog data), approximation by a semi-ellipse, and approximation by a semi-ellipse with attached straight sections. Using the empirical method, it was not possible to obtain outlines of cross-sections because there is no defined shape. In each of the figures, fragments that increase and decrease the result obtained for the cross-sectional area are marked. Regions that increase the cross-sectional area are marked in a green color, and those that decrease the cross-sectional area in a purple color.

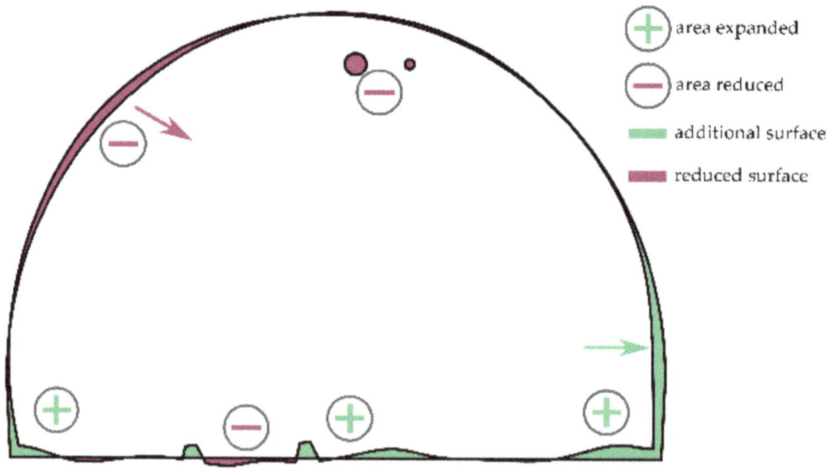


**Figure 12.** Differences in cross-sections between reference method and CAD method; cross-section 1.2.

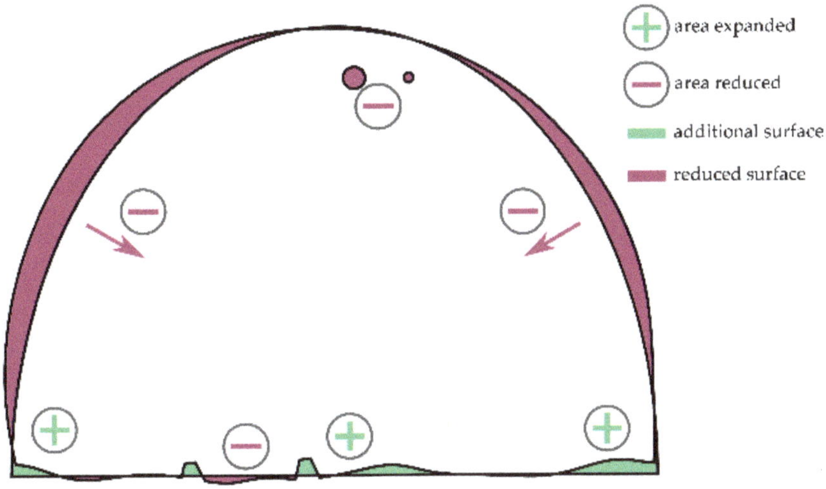


**Figure 13.** Differences in cross-sections between reference method and approximation by a semi-ellipse; cross-section 1.2.

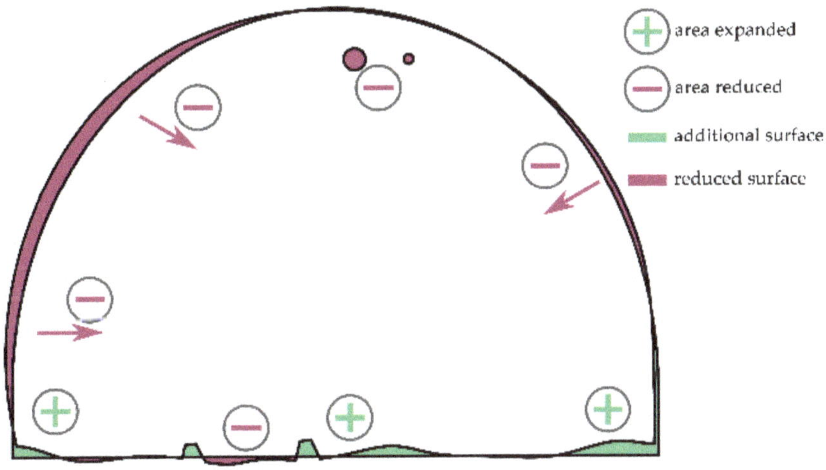


**Figure 14.** Differences in cross-sections between reference method and approximation by a semi-ellipse with attached straight sections; cross-section 1.2.

Figures 12–14 show a common feature: an identically mapped roof. In application of the surface, the symmetry line of cross-sections was used. This line was imposed on the symmetrical section connecting the top of the right and left part of the walls. Vertically, points were placed at the maximum height. This positioning of the two surfaces causes the bottom line to be determined at the same level in the case of each of the cross-sections.

Table 4 shows that out of the four methods analyzed, the smallest differences compared with the scanning method were obtained for the CAD method. It should be noted that the result obtained by this method was larger than the reference result, while in the other cases the results were smaller.

**Table 4.** Percentage differences for each cross section, calculated by four methods compared with the scanning method.

| | CAD Method [%] | Empirical Method [%] | Approximation by a Semi-Ellipse [%] | Approximation by a Semi-Ellipse with Attached Straight Sections [%] |
|---|---|---|---|---|
| Cross-Section 1.1 | 1.07 | −1.60 | −3.37 | 1.15 |
| Cross-Section 2.1 | 2.35 | −2.35 | −4.32 | −0.46 |
| Cross-Section 2.1 | 0.77 | −4.15 | −5.90 | −2.02 |
| Cross-Section 2.2 | 2.69 | −1.11 | −3.16 | 0.87 |
| Cross-Section 3.1 | 0.35 | −3.56 | −5.33 | −1.49 |
| Cross-Section 3.2 | 1.00 | −3.78 | −5.55 | −1.63 |

To present a quantitative assessment of used determining the cross-section area methods a decision was made to use a mean relative error measure, which is a sum of relative errors at each cross-section (Table 4) divided by the number of cross-sections:

$$\delta_{sr} = \frac{1}{n}\sum_{i=1}^{n} \frac{|x - x_0|}{x} \cdot 100\% \quad (4)$$

where:

$n$—cross-section number
$x$—calculated cross-section area using CAD and analytical methods
$x_0$—measured cross-section area using scanning method

The results of mean relative errors, shown in Table 4, shows that the biggest differences of calculated cross-section area in competition with cross-section area measured by scanning methods is for approximation by a semi-ellipse method. The reason for such a result reaching almost 5% is bad representation of arch supports (Figure 13). The smallest differences of calculated cross-section area in competition with cross-section area measured by scanning methods is for approximation by a semi-ellipse method with attached straight sections and it is exactly 1.27% (Table 5).

**Table 5.** Mean relative errors for four methods compared with the scanning method.

| | CAD Method [%] | Empirical Method [%] | Approximation by a Semi-Ellipse [%] | Approximation by a Semi-Ellipse with Attached Straight Sections [%] |
|---|---|---|---|---|
| Mean relative error | 1.37 | 2.76 | 4.61 | 1.27 |

Each method has some degree of difficulty in applying them. The easiest is empirical in which we can easily calculate the result during the measurements. The simple ellipse method is also easy to calculate, but it gives less accurate results. The method of ellipse plus straight sections is a reasonable compromise between more demanding methods and accuracy. The CAD method is only post-factum to use; it demands CAD software, and we need to know the arch type and choose the proper standard to draw the sketch of it. The most demanding and complicated method is the laser scanning method. It obviously demands a special device (laser scanner) and sometimes special permission for its use. Therefore, it is not possible to use it in every place of the mine and at all times. The laser scanning method gives the most accurate results and it can be applied to measure complicated cross-sections, especially in places where the convergence is significant. It works well as a reference method for comparing different measurement methods.

Comparing these results with the shape of chosen cross-sections, we noticed that all methods, except the scanning method, have problems with reflecting the correct shape of the arch support where any deformations appear (for example cross-section 3.2). The floor in these methods is always a straight line, so any unbalancing deformations increase measurement uncertainty.

## 5. Conclusions

Each of the presented methods is susceptible to inaccuracies caused by deformation of the arch supports and floor. The outline of the mine tunnel can be approximated using the semi-ellipse method, which gives a better approximation than the empirical method. In the analyzed cross-sections, the difference was reduced by more than half. The best approximation, however, is given by the CAD method. It is expected that the cross-sectional area measured by this method will be higher than the true value, because of the rock mass pressure on the arches. It is helpful to use nodes to determine the relationship, for example, the condition of contact at the connection of the arch's side part and roof part, and the symmetry condition of the arch. This makes it possible to adjust the nominal height and width of the arch and automatically adjust the remaining dimensions. The results show differences between methods at maximum level 5.9% by the semi-ellipse approximation method according to the laser scanning method. This level of differences in determining of the mine tunnel cross-section area can gradually extend depending on the deformation degree of the object geometry.

The smallest differences between the scanning method and four other methods is for the CAD method and semi-ellipse approximation method with attached straight sections.

It is common in every method to reconstruct the floor shape as a straight line. In many cases this is not sufficient. To improve the accuracy of determination of the cross-sectional area in case of a deformed floor, the floor height can be included. This leads to adjustment of the length of the straight parts of the arch and allows better shape reconstruction.

Considering the time of carrying out the measurement and calculations determining the mine drift cross-sectional area, it is justified to use CAD methods. However, a decision should be made each time to choose the appropriate method. This decision should be dictated by the degree of mine drift deformation.

**Author Contributions:** Conceptualization, J.J. and P.O.; Methodology, J.J. and P.O.; Software, J.J. and P.O.; Validation, J.J. and P.O.; Formal Analysis, J.J. and P.O.; Investigation, J.J. and P.O.; Resources, J.J. and P.O.; Data Curation, J.J. and P.O.; Writing—Original Draft Preparation, J.J.; Writing—Review and Editing, J.J. and P.O.; Visualization, J.J. and P.O.; Supervision, J.J.; Project Administration, J.J.; Funding Acquisition, J.J. All authors have read and agreed to the published version of the manuscript.

**Funding:** This paper is financed from the statutory funds of the Strata Mechanics Research Institute of the Polish Sciences Academy.

**Institutional Review Board Statement:** Not applicable.

**Informed Consent Statement:** Not applicable.

**Data Availability Statement:** Not applicable.

**Acknowledgments:** This paper presents results of the statutory research of the Strata Mechanics Research Institute of the Polish Sciences Academy.

**Conflicts of Interest:** The authors declare no conflict of interest.

## References

1. Hasheminasab, F.; Bagherpour, R.; Aminossadati, S.M. Numerical simulation of methane distribution in development zones of underground coal mines equipped with auxiliary ventilation. *Tunn. Undergr. Space Technol.* **2019**, *89*, 68–77. [CrossRef]
2. Wang, X.; Kulatilake, P.; Song, W.-D. Stability investigations around a mine tunnel through three-dimensional discontinuum and continuum stress analyses. *Tunn. Undergr. Space Technol.* **2012**, *32*, 98–112. [CrossRef]
3. Wang, X.; Cai, M. Numerical modeling of seismic wave propagation and ground motion in underground mines. *Tunn. Undergr. Space Technol.* **2017**, *68*, 211–230. [CrossRef]
4. Mo, S.; Sheffield, P.; Corbett, P.; Ramandi, H.L.; Oh, J.; Canbulat, I.; Saydam, S. A numerical investigation into floor buckling mechanisms in underground coal mine roadways. *Tunn. Undergr. Space Technol.* **2020**, *103*, 103497. [CrossRef]
5. Ozturk, C. Support design of underground openings in an asphaltite mine. *Tunn. Undergr. Space Technol.* **2013**, *38*, 288–305. [CrossRef]
6. Xie, H.; Zhao, J.; Zhou, H.; Ren, S.; Zhang, R. Secondary utilizations and perspectives of mined underground space. *Tunn. Undergr. Space Technol.* **2020**, *96*, 103129. [CrossRef]

7. Gallwey, J.; Eyre, M.; Coggan, J. A machine learning approach for the detection of supporting rock bolts from laser scan data in an underground mine. *Tunn. Undergr. Space Technol.* **2021**, *107*, 103656. [CrossRef]
8. Jiang, Q.; Zhong, S.; Pan, P.-Z.; Shi, Y.; Guo, H.; Kou, Y. Observe the temporal evolution of deep tunnel's 3D deformation by 3D laser scanning in the Jinchuan No. 2 Mine. *Tunn. Undergr. Space Technol.* **2020**, *97*, 103237. [CrossRef]
9. Xue, Y.; Liu, J.; Ranjith, P.G.; Zhang, Z.; Gao, F.; Wang, S. Experimental investigation on the nonlinear characteristics of energy evolution and failure characteristics of coal under different gas pressures. *Bull. Eng. Geol. Environ.* **2022**, *81*, 38. [CrossRef]
10. Kukutsch, R.; Kajzar, V.; Konicek, P.; Waclawik, P.; Ptacek, J. Possibility of convergence measurement of gates in coal mining using terrestrial 3D laser scanner. *J. Sustain. Min.* **2015**, *14*, 30–37. [CrossRef]
11. Khalymendyk, I.; Baryshnikov, A. The mechanism of roadway deformation in conditions of laminated rocks. *J. Sustain. Min.* **2018**, *17*, 41–47. [CrossRef]
12. Dziurzyński, W.; Skotniczny, P.; Krawczyk, J.; Gawor, M.; Pałka, T.; Ostrogórski, P.; Kruczkowski, J.; Janus, J. Wytyczne Rozmieszczenia Anemometrów Stacjonarnych Wzdłuż Długości Wyrobiska Kopalni Jak I W Samym Polu Przekroju Poprzecznego Wyrobiska. In *Zasady Pomiarów Przepływów Powietrza W Wyrobiskach Kopalnianych. Wybrane Sposoby Kontroli I Kalibracji Przyrządów Pomiarowych*; Strata Mechanics Research Institute of the Polish Academy of Sciences: Kraków, Poland, 2017.
13. Ostrogórski, P. Consideration of the shape and height of the floor in result of measuring the cross-sectional area for rings type ŁP. *Trans. Strata Mech. Res. Inst.* **2015**, *17*, 1–2.
14. Krach, A.; Krawczyk, J.; Kruczkowski, J.; Pałka, T. Variability of the velocity field and volumetric flow rate in air ways of underground mines. *Arch. Min. Sci.* **2006**, 1.
15. Walentek, A.; Janoszek, T.; Prusek, S.; Wrana, A. Influence of longwall gateroad convergence on the process of mine ventilation network-model tests. *Int. J. Min. Sci. Technol.* **2019**, *29*, 585–590. [CrossRef]
16. Pytlik, A. Comparative bench testing of steel arch support systems with and without rock bolt reinforcements. *Arch. Min. Sci.* **2019**, *64*, 4.
17. Janus, J. The Application of Laser Scanning in The Process of Constructing A Mine Drift Numerical Model. In Proceedings of the 24th World Mining Congress-Underground Mining, Brazilian Mining Association, Rio de Janeiro, Brazil, 18–21 October 2016.
18. Janus, J. Modelling of Flow Phenomena in Mine Drifts Using the Results of Laser Scanning. Ph.D. Thesis, Strata Mechanics Research Institute of Polish Academy of Sciences, Kraków, Poland, 2018.
19. Wierzbiński, K. Wpływ geometrii chodnika wentylacyjnego i sposobu jego likwidacji na rozkład stężenia metanu w rejonie wylotu ze ściany przewietrzane sposobem U w świetle obliczeń numerycznych CFD. *Zeszyty Naukowe Instytutu Gospodarki Surowcami Mineralnymi i Energią Polskiej Akademii Nauk.* **2016**, *94*, 217–228.
20. Janus, J.; Krawczyk, J. Numerical modeling of flow phenomena in mine drift using laser scan results. In *Proceedings of the 11th International Mine Ventilation Congress*; Science Press: Beijing, China, 2018. [CrossRef]
21. Skotniczny, P. Three-dimensional numerical simulation of the mass exchange between longwall headings and goafs, in the presence of methane drainage in A U-type ventilated longwall. *Arch. Min. Sci.* **2013**, *58*, 705–718. [CrossRef]
22. Klapa, P.; Mitka, B. Application of terrestrial laser scanning to the development and updating of the base map. *Geodesy Cartogr.* **2017**, *66*, 59–71. [CrossRef]
23. Gargoum, S.A.; Karsten, L. Virtual assessment of sight distance limitations using LiDAR technology: Automated obstruction detection and classification. *Autom. Constr.* **2021**, *125*, 103579. [CrossRef]
24. Cao, Z.; Chen, D.; Shi, Y.; Zhang, Z.; Jin, F.; Yun, T.; Xu, S.; Kang, Z.; Zhang, L. A flexible architecture for extracting metro tunnel cross sections from terrestrial laser scanning point clouds. *Remote Sens.* **2019**, *11*, 297. [CrossRef]
25. Xu, L.; Gong, J.; Na, J.; Yang, Y.; Tan, Z.; Pfeifer, N.; Zheng, S. Shield tunnel convergence diameter detection based on self-driven mobile laser scanning. *Remote Sens.* **2022**, *14*, 767. [CrossRef]
26. Jia, D.; Zhang, W.; Liu, Y. Systematic approach for tunnel deformation monitoring with terrestrial laser scanning. *Remote Sens.* **2021**, *13*, 3519. [CrossRef]
27. Helming, P.; von Freyberg, A.; Sorg, M.; Fischer, A. Wind Turbine tower deformation measurement using terrestrial laser scanning on a 3.4 MW wind turbine. *Energies* **2021**, *14*, 3255. [CrossRef]
28. Farahani, B.V.; Barros, F.; Sousa, P.; Cacciari, P.P.; Tavares, P.; Futai, M.M.; Moreira, P. A coupled 3D laser scanning and digital image correlation system for geometry acquisition and deformation monitoring of a railway tunnel. *Tunn. Undergr. Space Technol.* **2019**, *91*, 102995. [CrossRef]
29. Michalak, D.; Gomez Herrero, J.A. Innovative solutions need an innovative approach–3D printing technology, example of use and conclusion from implementation in an organization. *Min. Mach.* **2020**, *162*, 2.
30. Xu, J.; Ding, L.; Love, P. Digital reproduction of historical building ornamental components: From 3D scanning to 3D printing. *Autom. Constr.* **2017**, *76*, 85–96. [CrossRef]
31. Pejić, M. Design and optimisation of laser scanning for tunnels geometry inspection. *Tunn. Undergr. Space Technol.* **2013**, *37*, 199–206. [CrossRef]
32. Janus, J.; Krawczyk, J. Measurement and simulation of flow in a section of a mine gallery. *Energies* **2021**, *14*, 4894. [CrossRef]
33. Rozmus, M.; Tokarczyk, J.; Michalak, D.; Dudek, M.; Szewerda, K.; Rotkegel, M.; Lamot, A.; Rošer, J. Application of 3D scanning, computer simulations and virtual reality in the redesigning process of selected areas of underground transportation routes in coal mining industry. *Energies* **2021**, *14*, 2589. [CrossRef]
34. *Faro Laser Scanner FOCUS 3D—Manual*; FARO Technologies Inc.: Lake Mary, FL, USA, 2010.

35. Roca-Pardiñas, J.; Argüelles-Fraga, R.; de Asís López, F.; Ordóñez, C. Analysis of the influence of range and angle of incidence of terrestrial laser scanning measurements on tunnel inspection. *Tunn. Undergr. Space Technol.* **2014**, *43*, 133–139. [CrossRef]
36. Chen, B.; Deng, K.; Fan, H.; Hao, M. Large-scale deformation monitoring in mining area by D-InSAR and 3D laser scanning technology integration. *Int. J. Min. Sci. Technol.* **2013**, *23*, 555–561. [CrossRef]
37. Monsalve, J.J.; Baggett, J.; Bishop, R.; Ripepi, N. Application of laser scanning for rock mass characterization and discrete fracture network generation in an underground limestone mine. *Int. J. Min. Sci. Technol.* **2018**, *29*, 131–137. [CrossRef]
38. Olszyna, G.; Sioma, A.; Tytko, A. Laser measurement system for the diagnostics of mine hoist components. *Arch. Min. Sci.* **2014**, *59*, 337–346. [CrossRef]
39. Chi, H.; Zhan, K.; Shi, B. Automatic guidance of underground mining vehicles using laser sensors. *Tunn. Undergr. Space Technol.* **2012**, *27*, 142–148. [CrossRef]
40. Heyduk, A. Laser triangulation in 3-dimensional granulometric analysis. *Arch. Min. Sci.* **2016**, *61*, 15–27. [CrossRef]
41. Lenda, G.; Lewińska, P.; Siwiec, J. Accuracy of merging point clouds at the maximum range of a scanner with limited pos-sibilities of target placement. *Arch. Civ. Eng.* **2019**, *65*, 4.
42. Janus, J. *Construction the Numerical Models Geometry by Using Terrestrial Laser Scanning*; Selected Issues Related to Mining and Clean Coal Technology; AGH University of Science and Technology: Krakow, Poland, 2016.
43. Horyl, P.; Šňupárek, R.; Maršálek, P.; Pacześniowski, K. Simulation of laboratory tests of steel arch support. *Arch. Min. Sci.* **2017**, *62*, 163–176. [CrossRef]
44. Roszczynialski, W.; Trutwin, W.; Wacławik, J. *Kopalniane Pomiary Wentylacyjne*; Śląsk: Katowice, Poland, 1992.

MDPI
St. Alban-Anlage 66
4052 Basel
Switzerland
www.mdpi.com

*Energies* Editorial Office
E-mail: energies@mdpi.com
www.mdpi.com/journal/energies

www.ingramcontent.com/pod-product-compliance
Lightning Source LLC
Chambersburg PA
CBHW061441210926
51850CB00008B/1577
* 9 7 8 3 7 2 5 8 1 0 2 6 0 *